D0431607

HOUSING POLICY IN EUROPE

This book provides a comprehensive text on housing policy in Europe, comparing government intervention in the housing markets of fifteen different countries with diverse geographical, historical, political and social attributes.

The countries under consideration are grouped into those where the private rental sector is dominant, or where the social rental stock is disproportionately large; where owner-occupation is the principal tenure; or where housing policy is in transition from being an instrument of a centrally planned economy to a component of the free market.

Within each grouping, housing policy is assessed against a background of post-1945 problems and political change. Policy is examined with specific reference to tenurial shifts, housebuilding and housing rehabilitation, housing investment, finance and subsidies, and recent developments in housing policy. The book combines a rigorous empirical examination of past and present policy with a theoretical explanation of tenurial development, focusing on contemporary theories of welfare provision.

Housing Policy in Europe charts the recent shift of emphasis from housebuilding to selective rehabilitation, from 'bricks and mortar' subsidies to demand subsidies (particularly housing allowances), and from policies of intervention to policies in support of the free market.

Paul Balchin is Reader in Urban Economics at the University of Greenwich.

HOUSING POLICY IN EUROPE

edited by Paul Balchin

London and New York

11246866-3

Learning Resources
Centre

First published 1996
by Routledge
11 New Fetter Lane, London E4P 4EE

Simultaneously published in the USA and Canada
by Routledge
29 West 35th Street, New York, NY 10001

© 1996 edited by Paul Balchin

Typeset in Garamond by
Pure Tech India Limited, Pondicherry
Printed and bound in Great Britain by Biddles Ltd, Guildford and King's Lynn

All rights reserved. No part of this book may be reprinted or
reproduced or utilized in any form or by any electronic,
mechanical, or other means, now known or hereafter
invented, including photocopying and recording, or in any
information storage or retrieval system, without permission in
writing from the publishers.

British Library Cataloguing in Publication Data
A catalogue record for this book is available from the British Library

Library of Congress Cataloguing in Publication Data
A catalogue record for this book has been requested

ISBN 0–415–13512–5(hbk)
ISBN 0–415–13513–3(pbk)

CONTENTS

CONTENTS

Part III The dominance of owner-occupation

Part IV Housing in transition

FIGURES

TABLES

NOTES ON CONTRIBUTORS

Baralides Alberdí Head of Research Department, Banco Hipotecario de España – Argentaria, Madrid. Member of the European Network for Housing Research.

Paul Balchin Reader in Urban Economics, School of Land and Construction Management, Faculty of the Environment, University of Greenwich, London. Member of the European Network for Housing Research.

Laurence Bertrand Senior Research Officer, Nancy urban planning agency (ADUAN), France.

Maurice Blanc Professor of Sociology, Head of LASTES research group, University of Nancy 2, France. Member of the European Network for Housing Research.

Peter Boelhouwer Researcher, OTB Research Institute for Policy Sciences and Technology, Delft University of Technology, Netherlands, Member of the European Network for Housing Research.

Wolfgang Förster Housing Researcher, Deputy Director of WBSF (Vienna Land Procurement and Urban Renewal Fund), Vienna. Member of the European Network for Housing Research.

Henryk Hajduk Professor of Investment Analysis, Warsaw School of Economics, and Director of Housing Research Institute, Warsaw, Member of the European Network for Housing Research.

József Hegedüs Managing Director, Metropolitan Research Institute, Budapest. Member of the European Network for Housing Research.

Ivo Lavrač Professor of Regional Economics and National Accounting, University of Ljubljana, Slovenia.

Roderick J. Lawrence Lecturer and Researcher, Centre for Human Ecology and Environmental Services, University of Geneva. Member of the European Network for Housing Research.

Gustavo Levenfeld Professor in the Faculty of Applied and Theoretical Economics, Universidad Autónoma de Madrid, Spain. Adviser, Banco de España – Argentaria.

Patrick McAllister Senior Lecturer in Finance and Valuation, School of Land and Construction Management, Faculty of the Environment, University of Greenwich, London.

Liliana Padovani Vice-Chairman, European Network for Housing Research, and Associate Professor in Urban Politics, Department of Urban, Regional and Environmental Planning, Istituto Universitario di Architettura di Venezia.

Zorislav Perković City Planning Officer (1956–94), Zagreb, Croatia.

Hugo Priemus Vice-Chairman, European Network for Housing Research, and Managing Director of OTB Research Institute for Policy Sciences and Technology, Delft University of Technology, Netherlands.

Dusica Seferagić Senior Research Associate, Institute for Social Research, University of Zagreb, Croatia.

Luděk Sýkora Assistant Professor in Urban Geography, Department of Social Geography and Regional Development, Faculty of Science, Charles University, Prague. Member of the European Network for Housing Research.

Horst Tomann Professor of Economics, Institute of Economic Policy and Economic History, Free University of Berlin, Berlin. Member of the European Network for Housing Research.

Iván Tosics Managing Director, Metropolitan Research Institute, Budapest. Member of the European Network for Housing Research.

Bengt Turner Chairman, European Network for Housing Research, Institute for Housing Research, Uppsala University, Gavle, Sweden.

Harry van der Heijden Researcher, OTB Research Institute for Policy Sciences and Technology, Delft University of Technology, Netherlands. Member of the European Network for Housing Research.

Barbara Verlič Christensen Urban researcher and Lecturer, Institute for Social Sciences, University of Ljubljana, Slovenia. Member of the European Network for Housing Research.

PREFACE

Despite or because of the traumas of two world wars and the rise to power of fascism and communism within the first half of the twentieth century, economic and social policies in much of Europe during the last fifty years have been formulated and applied within a more stable political environment than hitherto, and conditioned to a greater or lesser extent by the parameters of western liberal democracy. Within most of western Europe, specific policies were often the outcome of either Keynesian or social-market economic policy wedded to a belief in a welfare state – reinforced in the early post-war years by Marshall Aid in those countries worst afflicted by the destruction of 1939–45, and subsequently by the Monnet–Schuman plan for an eventual European political union. Housebuilding on a massive scale (often heavily subsidised) began to erode serious shortages by the 1960s, the development of the social-rental sector in many countries was considered to be a crucial means of alleviating housing need, and wider home-ownership as a long-term goal was increasingly assisted by tax relief and exemptions. There was, moreover, a general 'consensus' between 'left of centre' and 'right of centre' parties in respect of housing policy and its future direction.

By the mid-1970s, however, the consensus began to break down. The oil crisis of 1973–74 heralded a period of 'stagflation' throughout much of the world when both the level of unemployment and the rate of inflation began to soar towards unprecedented heights. Monetarism was increasingly applied as a means of combating inflation, and public expenditure cuts as a perceived way of reducing the money supply and resurrecting a 'free market' was soon adopted as the central strategy of 'right of centre' governments – most notably in the United Kingdom under Thatcherism in the 1980s. Housing investment and housebuilding consequently diminished, the size of the social-rented sector was reduced in a number of countries – in large part through processes of privatisation, subsidies were diverted from supply to targeted demand, whereas, for reasons of ideology, the owner-occupied sector continued to receive substantial fiscal assistance. Only in a few countries, and particularly in the Netherlands and Sweden, was this reaction to unfavourable macroeconomic indicators largely absent in the arena of housing policy – a reflection of social-democratic values rather than the ideology of the New Right.

In central Europe after the Second World War, state direction supplanted market mechanisms in the allocation of resources, in the production of goods and services, and in pricing. Throughout most of the period of communist government, social housing was the principal recipient of state housing expenditure – largely in the form of supply subsidies and the demand-equivalent, low sub-market rents. With the onset of economic difficulties in the 1970s and 1980s, notably unacceptable levels of inflation, low rates of economic growth, and unfavourable trade balances, the state began to implement plans to reduce public expenditure (not least within the field of housing) and to introduce programmes of privatisation. These policies gathered momentum as an outcome of political change and the adoption of the 'free market' in the late 1980s and early 1990s – the latter process being already under way in Hungary during the final years of communist government.

This book, a companion volume to *Housing Policy: An Introduction* (Paul Balchin, 1995, London, Routledge), consists of chapters written by housing specialists from a number of western and central European countries, and is essentially both an analysis of the factors which determine the pattern of tenure across Europe and a critical review of the direction of contemporary housing policy. The countries examined are grouped together into four clusters: first, those in which the proportion of private rented housing is above the average for the European Union (EU); second, those in which the size of the social-rental housing sector is proportionately larger than the EU average for the tenure, or at least has been expanding rapidly towards the average in recent years; third, those in which owner-occupation is by far the dominant tenure and well above the EU average; and fourth, those in which housing policy has generally ceased to be an instrument of state planning and had become increasingly a component of a market economy.

An overarching consideration of the book is household choice. In the absence of policies designed to favour one tenure or another, it is probable that, in most developed market-dominated mixed economies, there would be a balance between the proportion of owner-occupied and rented housing, with the financial advantages and disadvantages of each tenure approximately equal. There would also be a balance between private-sector and social-sector rental housing – each competing with the other on broadly equal terms. In a number of countries in western Europe, however (with the United Kingdom being the prime example), household choice is very limited by subsidy and tax policies positively discriminating in favour of owner-occupation at the expense of the rental sectors, and normally within these countries the social-rental sector is marginalised, or is in the process of being marginalised. In other countries in Europe – such as Germany, the Netherlands and Sweden, household choice is very real since there is more of a balance between tenures, with tenure-neutral policies resulting in an adequate supply of both social and private rental housing in comparatively effective competition with owner-occupation. It is perhaps ironic that most of the former communist-controlled countries of central Europe, in a process of economic transition, have – to date – embarked on housing privatisation, with the aim of expanding the owner-occupied sector and depleting the stock of social rental housing – with predictably adverse medium- and long-term effects on household choice. Much of the book, therefore, examines policy with reference to discriminatory or non-discriminatory subsidies and with regard to 'integrated' and 'dualist' rental systems – adopting Jim Kemeny's hypothesis that the pattern and extent of renting determines the degree of attraction and the size of the owner-occupied sector (*From Public Housing to Social Market*, J. Kemeny, 1995, Routledge, London).

Within this context, and after a brief historical introduction, each chapter therefore considers the pattern of housing tenure; examines how both new housebuilding and rehabilitation have an impact on tenure distribution; explores the intricacies of investment, finance and subsidies as determinants of tenure and tenure-shift; critically reviews developments in policy; and concludes by emphasising similarities and differences in policy between countries and by suggesting some future directions of policy which might be necessary to ameliorate current problems of housing need. While consistency is broadly maintained by each author adhering to this framework – an essential prerequisite for a comparative study, within each country often very different housing issues and policies have been selected for particular consideration.

In producing this book, all the contributors were continually aware of the dangers of being overtaken by economic and political events. Housing policy, perhaps more than any other policy, can be subject to sudden and sometimes radical change, though the degree of change varies from one part of Europe to another – being less marked in those countries where policy aims and objectives have been based on consensus (as in Sweden and the Netherlands), but being very evident where 'left of centre' and 'right of centre' political parties have divergent agendas (as in the United Kingdom historically, or in the states of central Europe before and after the political and economic changes of 1989–91). However, every effort has been made to ensure that statistical and legislative detail is accurate at the time of going to press.

All the authors of this volume are indebted a wide range of people in the fields of research, teaching or government who have co-operated in the preparation of this book, and I am particularly grateful to Sue Brimacombe and Cherie Apps of the Faculty of the Environment, University of Greenwich, for converting much of the material on disk into print and for typing and retyping the introductory and concluding chapters of the manuscript. I would also like to thank Sue Lee and Peter Stevens for processing much of the artwork within the text. Last, but not least, I very much indebted to my wife Alicia for assisting me with the production of the book and most of all for her cheerful encouragement and forbearance.

Paul Balchin, London
Autumn 1995

1

INTRODUCTION

Paul Balchin

Throughout western Europe, as elsewhere in the world, housing policy reflects the political ideology of the government in power. Despite considerable variations in the aims and objectives of housing policy from one country to another, governments 'right of centre' generally tend to favour less state intervention, give only limited support to the social-rented sector, and promote owner-occupation and private landlordism. Governments to the 'left of centre' normally accept the need to intervene in the market, give responsibilities and funds to local authorities and non-profit housing organisations to enable them to provide affordable housing, and attempt to ensure that housing resources are distributed fairly equitably across and within tenures. Nevertheless, prior to the formulation and application of housing policy, broad demographic and macroeconomic trends need to be fully taken into account by governments, of whatever political predilection, to ensure that workable and politically relevant solutions are devised for the many different problems of housing market dysfunction.

MARKET DETERMINANTS OF HOUSING POLICY

The underlying factors influencing the level of housing demand or need within a country (aside from state intervention) are essentially confined to its population size and growth, its standard of living as indicated by gross domestic product (GDP) per capita, and expenditure on housing as a proportion of total private consumption. The principal underlying determinant of supply is the overall level of investment in the domestic economy – as measured by gross fixed-capital formation (GFCF) – and, derived from this amount, the level of housing investment. The quantitative and qualitative outcome of this investment includes, for example, the number of dwellings built, the size of the housing stock in relation to the number of households, the number of dwellings per thousand of the population, the area of habitable floorspace and number of rooms per dwelling, and the age and condition of dwellings. Although house prices and rents, the number of housing transactions, and the volume of resources allocated to the production of housing within specific periods of time are determined by the interaction of demand and supply, so too is the pattern of housing tenure – the ultimate outcome of market forces.

THE DEMAND FOR HOUSING

In western Europe the population of countries varies greatly, ranging in 1991 from 384,000 in Luxembourg to 3.5 million in Ireland to 57.8 million in the United Kingdom and nearly 80 million in Germany (Table 1.1), but whereas population size has an influence on the total amount of resources allocated to housing, the rate of population growth is often a more important determinant of housing policy.

Population growth in western Europe from 1945 to 1991 showed a marked spatial variation – being greatest in the Netherlands, Spain, Belgium, France and the former West Germany (with growth rates ranging from 44 to 61 per cent), and being least in the United Kingdom and Austria (with growth rates of only 16 and 15 per cent respectively). In the former East Germany, the population declined by nearly 8 per cent over that period (Table 1.1). The growth in population throughout western Europe in recent decades was attributable less and less to natural increase (both birth and death rates were diminishing, particularly in the 1980s), but in some countries, for example, the Netherlands and West Germany, net immigration from time to time has been the principal cause of growth. In East Germany, however, population decline was a result of net emigration – mainly to West Germany (McCrone and Stephens, 1995).

Table 1.1   Population growth, western Europe, 1945–91

	Population 1991 (000s)	Population 1991 (as % 1945)
Netherlands	15,060	161.3
Spain	38,872	145.5
Belgium	10,022	144.5
France	56,634	144.3
Former West Germany	79,984[a]	143.9
Luxembourg	384	135.7
Finland	5,029	130.0
Sweden	8,644	129.0
Denmark	5,129	126.8
Greece	10,200	126.7
Italy	57,746	122.6
Ireland	3,526	119.4
Portugal	10,400	118.4
United Kingdom	57,800	116.4
Austria	7,796	114.7
Former East Germany	–	92.2

Source: CEC, Demographic statistics
Note: [a] Whole of Germany

2

The standard of living of countries in western Europe, as indicated by gross domestic product (GDP) per capita, also varied substantially from one country to another – being highest in Switzerland and Denmark with estimated GDPs per capita of over $36,000 and $29,000 per capita respectively in 1995, and lowest in Greece and Portugal with per capita GDPs of only $8,400 and $6,900 (Table 1.2). Relative levels of standards of living, of course, change with time: for example, in the 1970s and 1980s Sweden had the highest GDP per capita in Europe but fell to seventh place by 1995, whereas in the 1950s the United Kingdom had a higher per capita income than any of the other 11 members of the pre-1995 European Union (EU) (Gilbert and Associates, 1958; McCrone and Stephens, 1995).

Table 1.2 Estimated gross domestic product per capita, and expenditure on housing, western Europe

	GDP per capita 1995[a] $	Expenditure on housing as % total private consumption 1989
Switzerland	36,430	–
Denmark	29,190	27.4
Norway	26,590	–
Germany	26,000	18.4[b]
Austria	25,010	17.9
France	23,550	18.9
Sweden	23,270	20.0
Belgium	22,260	16.7
Netherlands	21,300	18.6
United Kingdom	18,950	19.5
Italy	18,400	12.5
Ireland	15,100	10.9
Spain	12,500	13.1
Greece	8,400	11.1
Portugal	6,900	–

Source: Economist Publications (1994) *The World in 1995*; CEC, Statistics on housing in the European Community
Notes: [a] Forecast
       [b] West Germany only

Derived from GDP, the level of private consumption is, in part, attributable to expenditure on housing. There is, moreover, a broad positive correlation between GDP per capita and expenditure on housing as a proportion of total private consumption. As Table 1.2 reveals (albeit in respect of different years), housing expenditure as a proportion of personal consumption exceeded 19 per cent in all countries with GDPs in excess of about $19,000 per capita, whereas in countries with GDPs per capita of about $18,000 or less the proportion of housing expenditure fell to as low as 10.9 per cent (McCrone and Stephens, 1995). It is sometimes suggested that housing expenditure is a reflection of climate; for example, it is argued that Scandinavians spend proportionately more on housing than Mediterranean households because of the greater need for heating in areas of low winter temperatures, but on this basis one would expect Germany and Austria to spend more than the United Kingdom, but the reverse is apparent, while the proportionately lowest spending country is not Italy, Greece or Spain, but Ireland. Personal expenditure on housing is thus undoubtedly determined by complex scales of preference – cultural as much as economic.

THE SUPPLY OF HOUSING

As with GDP per capita and housing consumption expenditure, there were marked variations throughout western Europe in the proportion of the GDP which was invested in recent years – as measured by gross fixed-capital formation (GFCF) (Oxley and Smith, 1993). Tables 1.2 and 1.3 suggest that (with the possible exceptions of Italy, Portugal and the United Kingdom) there is very little positive correlation between the standard of living in 1994 and the annual average level of GDCF over the period 1970–90.

Table 1.3   Estimated gross fixed capital formation,
western Europe, 1970–90

	GFCF as a % gross domestic product: av. per annum 1970–90
Luxembourg	26.34
Ireland	23.26
Portugal	23.21
Greece	22.70
West Germany	21.73
France	21.65
Spain	21.47
Netherlands	21.11
Denmark	20.58
Italy	20.14
Belgium	19.40
United Kingdom	18.14

Source: *UN Annual Bulletin of Housing and Building Statistics*

Spatial variations in housing investment (as a proportion of both total investment and GDP) are also very apparent (Tables 1.4 and 1.5), but again (with the exceptions of the United Kingdom, Portugal and Italy) there is very little similarity between levels of housing investment and variations in GDP per capita (Oxley and Smith, 1993).

Table 1.4 Housing investment, western Europe

	% total investment av. 1970–89	% total investment 1992
France	29	27
Sweden	–	27
Germany[a]	28	26
Italy	27	26
Greece	27	23
Netherlands	27	23
Spain	26	20
Denmark	26	20
Luxembourg	–	20
Belgium	25	21
Ireland	23	22
Austria	–	19
United Kingdom	20	18
Portugal	19	17

Source: UN Annual Bulletin of Housing and Building Statistics
Note: [a] West Germany until 1989

Table 1.5 Housing investment: average per annum, 1970–89

	% gross domestic product
Greece	6.34
France	6.20
West Germany	5.98
Spain	5.82
Ireland	5.67
Netherlands	5.52
Italy	5.51
Denmark	5.22
Belgium	4.52
Portugal	4.22
United Kingdom	3.59

Source: UN Annual Bulletin of Housing and Building Statistics

Undoubtedly, housing investment has been reflected in the volume of housebuilding – with the number of dwellings constructed being proportionately the greatest in Greece, the Netherlands and France and the least in the United Kingdom, Portugal and Italy throughout the period 1972–89 (Table 1.6). Although, by 1991, there was some change in the order of housebuilding at the top (Ireland and Denmark becoming proportionately the largest housebuilders), the volume of housebuilding remained proportionately low in the United Kingdom and Italy (Oxley and Smith, 1993).

Table 1.6 Dwellings constructed, western Europe

	Dwellings constructed per 1000 population (average per annum 1972–88)	Completions as % stock	
		1980	1991
Greece	14.19	3.4	1.4
Netherlands	8.45	2.4	1.4
France	8.06	1.2 (1985)	1.2 (1990)
Spain	7.77	1.8	1.6
Ireland	7.33	3.1	1.9
Germany	6.49	1.5	1.2
Denmark	6.41	1.4	1.9
Luxembourg	6.18	1.5	–
Belgium	5.12	1.3	1.1 (1990)
United Kingdom	4.71	1.1	0.8
Portugal	4.46	1.2	–
Italy	3.60	1.3	0.8

Sources: UN Annual Bulletin of Housing and Building Statistics; the Netherlands' Ministry of Housing, Physical Planning and Environment (1992), Statistics on Housing in the European Community

Taking into account dwellings falling into serious disrepair or being demolished, Oxley and Smith (1993) calculated that net additions to the housing stock also suggested that a low level of housebuilding in the United Kingdom is a reflection of a low level of investment. Net additions to the housing stock (per 1000 population) were greatest in the Netherlands (at 6.8) and lowest in the United Kingdom (at 3.8).

Comparative levels of housebuilding and the resulting size of housing stocks, however, do not by themselves indicate the extent to which household needs are adequately satisfied. The number of dwellings must be

Table 1.7 Housing surpluses and deficits, European Community: 1991

	Number of dwellings (000s)	Number of households (000s)	Housing surplus/ deficit (000s)
Spain	17,173	12,040	5,133
France (1990)	26,237	21,535	4,702
Italy	23,232	20,646	2,586
Greece	4,690	3,344	1,346
Portugal	4,181	3,176	1,005
United Kingdom	23,622	22,800	822
West Germany (1990)	7,017	6,652	365
Belgium (1990)	3,805	3,610	295
Denmark	2,375	2,251	124
Ireland	1,039	1,029	10
Luxembourg	135	145	−10
Netherlands	5,965	6,135	−170

Source: the Netherlands' Ministry of Housing, Physical Planning and Environment (1992), Statistics on Housing in the European Community

compared with the number of households to establish whether there is a crude housing surplus or deficit. Within the European Community (EC) in the early 1990s, surpluses ranged from over 5 million dwellings in Spain and more than 4 million dwellings in France (many of which were second or holiday homes) to only 10,000 dwellings in Ireland, whereas in Luxembourg and the Netherlands there were deficits of 10,000 and 170,000 respectively (Table 1.7).

Crude housing surpluses and deficits, however, are not in general clearly reflected in the number of dwellings per 1000 population or in the number of persons per household; nevertheless according to Tables 1.7 and 1.8 four countries in particular – Belgium, Ireland, Luxembourg and the Netherlands – appear to suffer from inadequacies of supply.

Table 1.8   Dwellings per population and persons per household: 1991

	Dwellings per 1000 population	Persons per household
Sweden	475	2.1
Denmark	465	2.2
France	463	2.6
Greece	457	3.1
Former East Germany	445	–
Spain	441	3.3
Austria	435	2.5
Portugal	424	3.1
Former West Germany	421	2.3[a]
United Kingdom	411	2.5
Italy	404	2.8
Luxembourg	404	2.6
Belgium	395	2.5
Netherlands	393	2.4
Ireland	286	3.3

Source: CEC, Statistics on housing in the European Community
Note: [a] All Germany

Qualitatively, as well as quantitatively, the housing stock of Europe varies substantially. Whereas over 70 per cent of dwellings in the Netherlands, Greece and Spain have been built since 1945, in France, Ireland and the United Kingdom nearly 30 per cent of the housing stock dates back to before 1919 (Table 1.9). Variations in habitable floorspace (Table 1.10) are also very marked – dwellings in Luxembourg, Denmark and the Netherlands had over 98 m^2 of space in 1992, compared to floorspaces of less than 80 m^2 in the United Kingdom, Greece and former East Germany. The number of rooms per dwelling (Table 1.10), however, ranged from 5 in the United Kingdom to as few as 3.7 in Denmark (McCrone and Stephens, 1995). It is notable that although dwellings in the United Kingdom were

Table 1.9 Building period of the dwelling stock, western Europe: 1981–82

	% built			
	Before 1919	1919–45	1945–70	After 1970
Netherlands	13	15	35	37
Greece	9	13	49	29
Spain	15	12	46	27
France	29	14	31	26
Ireland	29	17	28	26
Portugal	26	19	30	25
Luxembourg	23	19	36	22
Italy	18	12	49	21
West Germany	20	13	49	18
Denmark	23	23	36	18
Belgium	27	23	33	17
United Kingdom	29	21	35	15

Source: the Netherlands' Ministry of Housing, Physical Planning and Environment (1992), *Statistics on Housing in the European Community*

Table 1.10 Housing floorspace and number of rooms per dwelling, western Europe

	Habitable floorspace m^2	Average number of rooms/ dwellings	
	1992	1980	1990
Luxembourg	107.0	5.4	–
Denmark	106.9	3.8	3.7 (1991)
Netherlands	98.6	4.2	4.1
Ireland	95.0	5.0	–
Sweden	92.0	–	–
Italy	92.0	4.0	4.2
Former West Germany	86.6	4.2	4.4
Belgium	86.3	5.0	4.9
France	85.4	3.7	3.8
Austria	85.0	–	–
Spain	83.6	4.7	4.8 (1991)
United Kingdom	79.7	4.9	5.0
Greece	79.6 (1990)	3.8	3.8
Former East Germany	64.6	3.8	3.8

Source: the Netherlands' Ministry of Housing, Physical Planning and Environment (1992), *Statistics on Housing in the European Community*

disproportionately old and had comparatively small habitable floorspaces, they contained a relatively large number of rooms.

Further qualitative indicators such as the proportion of dwellings which are unfit, lacking amenities or in serious disrepair, and the proportion of dwellings without a bath also show distinct spatial variations in the condition of housing in western Europe (Table 1.11). Dwellings in France, former West Germany (probably), Denmark, Ireland and the United Kingdom

Table 1.11 Housing condition, western Europe, 1991

	% dwellings unfit, lacking amenities or in serious disrepair	% dwellings without bath
France	9	5
Former West Germany	–	4
Denmark	13	10
Ireland	13	10
United Kingdom	14	1
Austria	16	16
Netherlands	20	1
Portugal	23	16
Belgium	32	12
Greece	34	16
Luxembourg	49	14
Italy	49	14
Spain	51	6
Former East Germany	–	18

Source: CEC, Statistics on housing in the European Community

were in the best condition in 1991, whereas those in Luxembourg, Italy, Spain and (probably) former East Germany were in the worst condition; the proportion of dwellings with a bath was highest in the United Kingdom and the Netherlands but lowest in Austria, Portugal, Greece and former East Germany (Table 1.11).

GOVERNMENT INTERVENTION IN THE HOUSING MARKET

Both the demand for and supply of housing are, in varying ways, much influenced by government policy in all European countries. Demand can be increased if macroeconomic stimulants to growth such as lower rates of interest on borrowing, easier credit, an increase in public expenditure and a decrease in taxation result (in part) in an increase in housing consumption. More specifically, and at a microeconomic level, a range of 'subject subsidies' increase the level of demand in targeted areas of the housing market: for example, housing allowances are available often to tenants of rented and sometimes to owner-occupied housing to facilitate demand among relatively low-income households; while mortgage-interest tax relief, exemption from capital gains tax (and possibly from tax on imputed rent income), and discounts on the purchase of social housing boost the demand for owner-occupation, often regardless of the income of the recipient. Clearly the reduction or withdrawal of these macroeconomic stimulants and/or micro-subsidies would decrease demand.

Supply can also be increased if the macroeconomic stimulants to growth generate an increase in housing investment. At a microeconomic level, 'object subsidies' (sometimes referred to as 'bricks and mortar subsidies') increase the level of supply in specific areas of housing provision; for example, recurrent subsidies might be available to cover in whole, or in part, loan charges on capital borrowed for the construction or renovation of social housing. In addition, social landlords might be able to qualify for one-off low-interest loans or capital grants; while private developers, landlords and owner-occupiers are sometimes eligible for tax allowances, grants and low-interest loans for investment in new housing or renovation, and landlords and owner-occupiers might both be able to claim depreciation allowances on past investment (although this subsidy could also reinforce demand). Supply would clearly be reduced if the macroeconomic stimulants and/or micro-subsidies were reduced or withdrawn.

Although in some European countries (notably Sweden and Germany) there is an attempt to be 'tenure-neutral' in the distribution of subsidies, in others (for example, the United Kingdom and Spain) support might be concentrated on one or two tenures or targeted at specific groups of households or geographical areas. Nevertheless, 'in all countries, regardless of the average standard of living, there is a large section of the population that cannot afford the full economic cost of what would generally be regarded as an adequate or tolerable standard of housing' (McCrone and Stephens, 1995). In each west European country, it is probable that up to a third of the total population is 'unable to pay the full economic cost of the housing it occupies' (McCrone and Stephens, 1995), and it would therefore be unrealistic to assume that housing could be left completely to the vagaries of the free market without any form of moderating government intervention. Thus the cost of government support for housing ranges from 1 to 4 per cent of the gross national product.

TENURE

Market forces and government intervention combine to produce a relationship in western Europe between levels of GDP per capita and different housing sectors (Table 1.12). Where, in 1995, GDPs per capita were highest (from $22,260 to $36,430), the proportion of rented housing ranged from 33 to 70 per cent. Switzerland had the highest proportion of private rented housing, 60 per cent, compared to the EU average of 21 per cent, while the Netherlands had proportionately the most social rented housing, 36 per cent, compared to the EU average of 18 per cent. Where GDPs per capita were generally at their lowest (from $20,410 down to $6,900), the proportion of owner-occupied housing was highest – upwards from 65 per cent. Ireland had proportionately the largest owner-occupied sector, 80 per cent compared to the EU average of 56 per cent. Only in respect of Norway was there both a comparatively high GDP per capita ($26,590) and an above average level of owner-occupation (60 per cent) – a result of the recent

Table 1.12 Estimated gross domestic product per capita and housing tenure, western Europe, 1995

	GDP $ per capita	Owner occupation %	Private rental %	Social rental %	Total rental %	Other tenure %
Private rented sector above the EU average:						
Switzerland[a]	36,430	31	60	3	63	6
Germany	26,000	38	36	26	62	–
Luxembourg	–	67	31	2	33	–
Belgium	22,260	62	30	7	37	–
Social-rented sector above or broadly at the EU average:						
Netherlands	21,300	47	17	36	53	–
Austria	25,010	41	22	23	45	14
Sweden	23,270	43	16	22	38	19
Denmark	29,010	50	24	18	42	8
France	23,550	54	21	17	38	8
Owner-occupation above the EU average:						
Ireland	15,100	80	9	11	20	–
Spain	12,500	76	16	2	18	–
Finland	20,410	72	11	14	25	3
Greece	8,400	70	26	0	26	4
Italy	18,400	67	8	6	14	19
United Kingdom	18,950	66	10	24	34	–
Portugal	6,900	65	28	4	32	3
Norway[b]	26,590	60	18	4	22	18
EU average:		56	21	18	39	5

Sources: CECODHAS (European Liaison Committee for Social Housing) (1995); Economist Publications (1994), *The World in 1995*
Notes: [a] 1990
   [b] 1988

North Sea oil boom, and, until recently, an essentially agricultural economy with a tradition of home-ownership.

There is thus little statistical evidence to support the view that, in terms of comparative patterns of tenure in western Europe, owner-occupation is necessarily a sign of affluence or that renting in the private or social sectors is an indication of relative poverty. Indeed, as Figures 1.1 and 1.2 show, there is an inverse relationship between owner-occupation and high per capita GDPs, and a positive relationship between renting and high GDPs per capita.

PATTERNS OF TENURE: A POLITICAL EXPLANATION

Although market forces and government intervention are instrumental in determining the specific size of each of the housing tenures in any

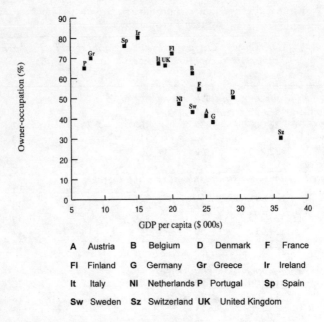

Figure 1.1   Owner-occupation and gross domestic product per capita, western Europe, 1994

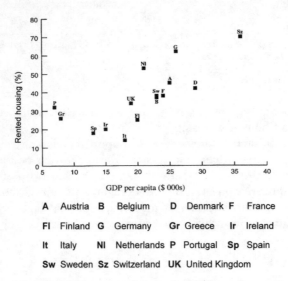

Figure 1.2   Rented housing and gross domestic product per capita, western Europe, 1994

economically advanced country, the socio-political system in operation in that country provides the arena in which the relationships between the market and policy develop.

Esping-Andersen (1990), in examining social policy in 18 countries, concluded that there were three distinct regimes: social-democratic welfare states, corporatist welfare states and liberal welfare states. The first group, the social-democratic welfare states, were the 'dominant political force behind reforms based on universalism and decommodification which were extended to all classes' (Barlow and Duncan, 1994). Countries within this group aim to provide a 'one nation' system of welfare based on equality of high standards of welfare for all, not, as elsewhere, an equality of minimum needs. Within this group, Sweden is the archetypal case, with the other Scandinavian countries being other members of the cluster. The second group, the corporatist welfare states, having generally eschewed a recent obsession with free markets and commodification (privatisation), attempt to reinforce the rights attached to different classes and professions, and to this end are willing to replace the market as a provider of welfare. This group includes Germany (the archetypal case), Austria, the Netherlands and France (although in this last case there are strong social-democratic tendencies). The third group, the liberal welfare states, provide little more than a means-tested 'safety net' of limited benefits for low-income, working-class state dependants, and include such countries as the United States (the archetypal case within this group), Canada, Australia and New Zealand. In Europe, the list includes the United Kingdom (although by the mid-1990s it was still in transition from its pre-1979 social-democratic base) and Ireland.

Barlow and Duncan (1994) suggested that there is a fourth group of countries – rudimentary welfare states – that should be added to Esping-Andersen's categorisation. To an extent, these countries are similar to those in the liberal welfare grouping, although benefits are at best even more residual and at worst non-existent. There is generally a strong agricultural bias to the economy with much evidence of household or family subsistence, a situation reinforced in the twentieth century by the emergence of authoritarian and militaristic governments unwilling to develop a welfare state of equal citizens.

According to Barlow and Duncan's analysis, Greece and Portugal are archetypal cases, whereas Spain is in a state of transition from this group to the liberal welfare cluster, while Italy 'can be seen to be straddling the rudimentary and corporatist regimes' (Barlow and Duncan, 1994).

Although Esping-Andersen (1990) did not include housing tenure in his analysis, it is comparatively straightforward to match the social-democratic group with the promotion of various forms of rented and co-operative housing as alternative sectors available to all and on a long-term basis. The corporatist welfare group can be identified as those countries in which both the social and private rental sectors are overtly promoted by the state and where either one or the other consequently becomes the dominant sector. This, however, unlike the social-democratic regime, does not disturb social differentiation, nor is state promotion of the rented sectors regarded as anything other than a temporary measure to remedy market imperfection (although in practice promotion might extend over many decades). The liberal welfare countries tend to be those where owner-occupation is, by far, the dominant sector, and where state intervention in housing is 'limited to a stigmatised provision for a residual population who are unable to adequately participate in markets' (Barlow and Duncan, 1994).

In terms of the tenure patterns set out in Table 1.12, it is difficult from data alone to distinguish between social-democratic and corporatist countries. By EU standards, all have disproportionately large private- and/or social-rented sectors. The distinction is also blurred since countries can move from one regime to another over a comparatively short period of time. The United Kingdom, for example, moved more centrally into the liberal welfare regime in the 1980s, possibly being followed by France in the 1990s, while the Netherlands, with its high level of social rental provision and low rates of residualisation may already have moved out of the corporatist regime and into the social-democratic arena (Barlow and Duncan, 1994).

Esping-Andersen's categorisation of welfare regimes, as applied to housing tenure, tends to be descriptive rather than theoretical. Kemeny (1995), however, attempts to offer a more analytical though broadly compatible explanation of the distribution of tenure, particularly *vis-a-vis* the relative importance of social housing (Table 1.13). With regard to the maturation of

Table 1.13 Social welfare regimes and rental markets, western Europe

Social welfare regimes	Rental markets	Countries
Social democratic	Unitary	Sweden
	Unitary	Denmark
	Unitary	Norway
	Unitary	Finland
Transitional	Unitary	Netherlands
Corporatist	Unitary	Germany
	Unitary	Austria
	Unitary	Switzerland
Transitional	Unitary	France
Liberal welfare	Dualist	Ireland
	Dualist	United Kingdom
Rudimentary welfare	Dualist	Italy
	Dualist	Spain
	Dualist	Portugal
	Dualist	Greece

Sources: After Esping-Andersen (1990) and Kemeny (1995)

14

social (non-profit) rented housing (as measured by an inflation-induced decline in the outstanding debt on the existing stock compared to the outstanding debt on newly built, acquired or renovated dwellings), Kemeny distinguishes between two rental systems: *'unitary'* rental systems, in which social and private renting are integrated into a single rental market – with the social-democratic and corporatist states of Sweden, the Netherlands, Germany, Switzerland and Austria (at least Vienna) being principal examples; and the *'dualist'* system in which the state controls and residualises the social-rented sector to protect private (profit) renting from competition – as, for example, in the liberal welfare states of the United Kingdom, Ireland, Finland and Spain (Kemeny, 1995).

Unitary rental systems are clearly attributes of 'social market' economies, where the state encourages social rented housing to compete directly with the private-rental sector in order to dampen rents and to provide good-quality housing on secure tenancy terms. Clearly, if within a social market the private or profit rental sector is to compete with social or non-profit housing (the maturation process enabling the latter sector to set low levels of rent), the private rental sector will need to be a recipient of equivalent subsidies to those allocated to the social sector in order to ensure an adequate return on investment, but a flexible form of rent control might be necessary across both sectors if the market shows signs of imperfection (Kemeny, 1995). If subsidies are also comparable with those received by owner-occupiers, subsidisation will be tenure-neutral, and each of the tenures will be equally attractive to a large proportion of households. In a unitary market, therefore, cost-rental social housing (under deregulated conditions) competes with the private sector supported by the state on equal or near-equal terms to cost renting.

Kemeny (1995) claims that in the dualist system – for reasons of social and political expedience – the state has introduced social housing or encouraged its development to provide a safety net for the relatively poor. The non-profit rental sector is protected from the profit sector by being segregated from the private market and organised as a residualised, stigmatised and often means-tested sector – a process which is particularly well advanced in the United Kingdom and generally supported across the political spectrum (Kemeny, 1995). There is often no attempt to ameliorate the undesirable effects of the market by creating a balance between the profit-motive and social priorities. Within the dualist system, policy intentionally or unintentionally steers all but the lowest-income households towards owner-occupation. For most households there is only a choice between profit renting (at high rent and with little long-term security of tenure) and owner-occupation with its many perceived advantages. Kemeny (1995) suggests that in this situation, far from preferences for owner-occupation determining policy, as Saunders (1990) would argue, policies create preferences for home-ownership. In a dualist system, therefore, renting at cost is

15

provided by a state-controlled social-housing sector, while private profit renting is left largely to sink or swim.

Although dualist rental markets in west European liberal welfare regimes are broadly similar, unitary systems develop differently from one corporatist or social-democratic country to another. In the Netherlands, cost-rental social housing *dominates* the rental market. Since the private-rented sector consisted of only 17 per cent of the total stock of housing in 1995 (Table 1.12), market rents were largely determined by the cost structures of a very mature social-rented sector (Kemeny, 1995). In Sweden, cost-rental housing *leads* the market since municipal housing companies are increasingly charging demand-sensitive rents and are consequently becoming market leaders – dampening rents in the relatively small profit-oriented private sector (Kemeny, 1995). In Germany and Switzerland, cost-rental housing only *influences* private rental markets by marginally dampening profit rents. In Germany this is more evident than in Switzerland since the cost-rental sector constitutes a much greater proportion of the total rented stock.

THE DEVELOPMENT OF HOUSING POLICY IN WESTERN EUROPE SINCE 1945

Clearly, regardless of the social-welfare regime existing in a country and irrespective of whether a unitary or dualist system of renting was in operation, housing policies of west European countries have, to an extent, shared a common history. Similarly, 'left of centre' governments across western Europe attempted to employ broadly common policies – normally compatible with those expected within a social-democratic welfare regime, whereas 'right of centre' administrations (some of them having previously adhered to either social-democratic or corporatist parameters) increasingly appeared to adopt policies associated with liberal-welfare regimes.

In western Europe since the Second World War, there have been four distinct stages in the development of housing policy (Boelhouwer, 1991). The *first stage* of policy development was presided over by both 'left' and 'right of centre' governments broadly united in their aim to eradicate large-scale housing shortages during the post-war period (Figure 1.3). The *second stage* mainly coincided with 'left of centre' government in several west European countries. Social democratic parties were in power in Sweden until 1976, in Denmark until 1982, in West Germany from 1970 to 1982, in Belgium from 1970 to 1974 and from 1977 to 1980, and in the Netherlands from 1973 to 1977; while the Labour Party was in office in the United Kingdom from 1974 to 1979. At least during the early years of this stage, all 'left of centre' governments continued to employ object subsidies to promote large-scale housebuilding particularly in the social sectors, for example the Swedish 'one-million dwellings programme' of 1964–75, before switching over to housing renewal and subject subsidies after the 1973 oil

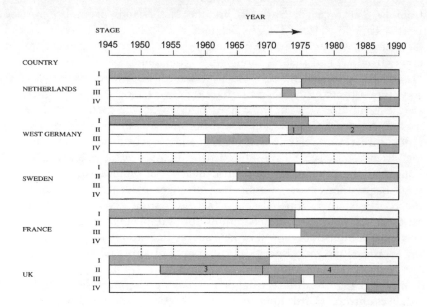

I High degree of government involvement, particularly in order to alleviate housing shortages.

II Greater emphasis on housing quality.

III Greater emphasis on problems of housing distribution and targeting specific groups, and the withdrawal of the state in favour of the private sector.

IV Reappearance of quantitative and/ or qualitative housing shortages; state involvement increases.

1 Improvement in the quality of new housing construction.

2 Improvement in the quality of the housing stock.

3 Improvement in the quality of the stock by slum clearance programmes and substitute new construction.

4 Emphasis on maintenance and improvement instead of slum clearance

Figure 1.3 The general development of housing policy during the period 1945–90 (after Boelhouwer, 1991)

17

crisis heralded cuts in state expenditure. The *third stage* was mainly associated with 'right of centre' government (or right-dominated coalitions) – from 1976 in Sweden, from 1977 in the Netherlands, from 1979 in the United Kingdom, from 1981 in Belgium, and from 1982 in Denmark and Germany; however, in France a wholly 'left of centre' government was in power from 1981 to 1986. Although in Sweden policies of the former Social Democratic government were largely retained, elsewhere the free market and other aspects of the liberal welfare state were increasingly embraced as state expenditure was cut, new housebuilding in the social-rented sector was superseded by renovation, social housing was increasingly privatised, rent controls were relaxed or abolished, and object subsidies were increasingly replaced by subject subsidies benefiting those already reasonably well housed at the expense of those inadequately accommodated. The *fourth stage* has witnessed the continuation of 'right of centre' government in much of western Europe, although there was some reversion in the early 1990s to 'left of centre' administrations, for example in Sweden. Whether or not these administrations will be able successfully to reduce absolute housing shortages and more specifically shortages of affordable housing for the less well-off remains to be seen.

It is clear that the different stages of policy development did not coincide in all countries, whilst some countries experienced more than one stage concurrently (Figure 1.3). It is also evident that the degree to which 'left of centre' and 'right of centre' policy diverged was variable. In West Germany, France and particularly in the United Kingdom, as 'right of centre' governments took over from 'left of centre' administrations there was a general reduction in state intervention in housing markets and a greater emphasis placed on deregulation and market forces; but in Sweden and the Netherlands there was far more of a political consensus with little change of policy when 'right of centre' governments replaced their 'left of centre' counterparts in the 1980s (Boelhouwer, 1991). Whether or not these differences will continue after future changes of administration will depend substantially on whether the ideologies of the governing parties continue to diverge from or converge with those of their political opponents.

CENTRAL EUROPE

From the aftermath of the Second World War until 1989 or the early 1990s, communist governments were in power throughout the countries of central Europe. The production, pricing and allocation of goods and services, the resources needed to produce them, and the distribution of revenues subsequently gained were all determined by varying forms of state management within a command economy. Hungary (particularly in the 1980s) and Yugoslavia were partial exceptions to centralised decision-making and rapidly embraced a market economy – a process which cautiously got under way elsewhere within the Eastern bloc in reaction to a deteriorating economic situation, prior to the introduction of multi-party political systems in the 1990s. In contrast to western Europe, however, housing provision (like the provision of any other product) was not normally determined by the free interaction of market forces and government policy, but by government policy alone.

As in most of the former Eastern bloc, adequate housing provision in the Czech Republic, Hungary and Poland was, and is, severely constrained by relatively low gross domestic products per capita and low proportions of total consumer expenditure allocated to housing (Tables 1.2 and 1.14). In recent years, housebuilding has in consequence been at a particularly low level and has been decreasing, comparing unfavourably with countries of the EC (Tables 1.6 and 1.15). There are therefore still substantial housing deficits, in contrast to the situation in the EU (Tables 1.7 and 1.16).

Although the average size of households was not significantly different in the Czech Republic, Hungary and Poland from that in the EC, there were far fewer dwellings per thousand of the population in the former countries

Table 1.14 Gross domestic product per capita and expenditure on housing in Hungary, the Czech Republic and Poland, 1995

	GDP per capita $ 1995[a]	Expenditure on housing as % of total private consumption
Hungary	2,970	8.7 (1987)
Czech Republic	2,450	–
Poland	1,910	4.4 (1986)
EU range	6,900–36,430	10.9–27.4

Sources: Economist Publications (1994), The World in 1995; CEC, Statistics on Housing in the European Community; Hungarian Central Statistical Office, Housing Statistics Yearbook; Polish Central Statistical Office, Statistical Yearbook; SNTL, Statistical Yearbook of the Czech Republic
Note: [a] Estimated

Table 1.15 Dwellings constructed in Hungary, the Czech Republic and Poland

	Dwellings constructed per 1000 population (average per annum) 1985–91	Completions (000s) 1985	1991	Change % 1985–91
Hungary	5.1	72.5	33.2	−54.2
Czech Republic	4.9	66.7	41.7	−37.5
Poland	4.5	189.6	136.8	−27.8
EC/EU range (1972–89)	3.6–14.2			

Sources: Hungarian Central Statistical Office, Housing Statistics Yearbook; UN Annual Bulletin of Housing and Building Statistics; Polish Central Statistical Office, Statistical Yearbook; SNTL, Statistical Yearbook of Czechoslovakia

Table 1.16 Housing deficits in Poland, the Czech Republic and Hungary

	Number of dwellings (000s)	Number of dwellings (000s)	Housing deficits (000s)
Poland (1988)	10,716	11,970	−1,254
Czech Republic (1991)	3,706	4,052	−346
Hungary (1990)	3,688	3,890	−202

Sources: Hungarian Central Statistical Office, *Housing Statistics Yearbook*; Polish Central Statistical Office, *Statistical Yearbook*; SNTL, *Statistical Yearbook of Czechoslovakia*

than in western Europe (Tables 1.8 and 1.17). Housing floorspace and the average number of rooms per dwelling were also less (Tables 1.10 and 1.18), and the proportion of dwellings unfit, lacking amenities or in serious disrepair (or specifically without a bath) were higher than in western Europe (Table 1.11).

Because of a faster transition to a market economy in Hungary and parts of the former Yugoslavia (such as Slovenia and Croatia), than in the Czech Republic and Poland (at least in respect of housing), it is not altogether

Table 1.17 Dwellings per population and persons per household in Poland, Hungary and the Czech Republic

	Dwellings per 1000 population	Persons per household
Poland (1988)	350	3.2
Hungary (1990)	280	2.7
Czech Republic (1991)	280	2.5
EC range (1991)	286–475	2.1–3.3

Sources: Hungarian Central Statistical Office, *Housing Statistics Yearbook*; the Netherlands' Ministry of Housing, Physical Planning and Environment (1992), *Statistics on Housing in the European Community*; Polish Central Statistical Office, *Statistical Yearbook*; SNTL, *Statistical Yearbook of Czechoslovakia*

Table 1.18 Housing floorspace and number of rooms per dwelling in the Czech Republic, Hungary and Poland

	Habitable floorspace m²		Average number of rooms/dwelling 1980	1991
Czech Republic	70.5	(1991)	2.4	2.7
Hungary	69.0	(1990)	–	–
Poland	59.6	(1988)	–	2.4 (1988)
EC range (1992)	64.4–107.0		3.7–5.4	3.8–5.0

Sources: Hungarian Central Statistical Office, *Housing Statistics Yearbook*; the Netherlands' Ministry of Housing, Physical Planning and Environment (1992), *Statistics on Housing in the European Community*; Polish Central Statistical Office, *Statistical Yearbook*; SNTL, *Statistical Yearbook of Czechoslovakia*

surprising that owner-occupation (as a proportion of the total housing stock) is noticeably greater in the pacemaking countries than in the latter; indeed the proportion of home-ownership is greater in Hungary, Slovenia and Croatia than in many west European countries (Tables 1.12 and 1.19). Also, because of the post-1945 history of the central European countries under consideration, it is not altogether remarkable that the social-rented sector is likewise larger than the EU average. The obvious explanation of this situation is, of course, that the private-rented sector is still very nearly non-existent throughout most of central and eastern Europe despite attempts to rehabilitate the private landlord. It clearly cannot be argued that a unitary rental system is applicable to former communist countries.

Table 1.19 Tenure in central European countries

	Owner-occupied %	Social-rented %	Other tenure (including private rented) %
Poland (1988)	41.1	56.3	2.6
Czech Republic (1991)	40.5	59.4	0.1
Hungary (1990)	77.3	22.6	0.1
Croatia (1991)	64.1	24.7	11.2
Slovenia (1991)	67.8	31.0	1.2
EU average (1995)	56.0	18.0	26.0

Sources: CECODHAS (European Committee for Social Housing); Hungarian Central Statistical Office, *Housing Statistics Yearbook*; Polish Central Statistical Office, *Statistical Yearbook*; SNTL, *Statistical Yearbook of Czechoslovakia*

In all former communist countries there are varying degrees of privatisation – not least within the sphere of housing. For reasons both political and economic (rents often fail to cover maintenance costs), tenants of public-sector stock have been encouraged to buy their homes, very often at a substantial discount. Owner-occupation is thus expanding rapidly in central European countries, while the social-rented sector is destined to become a relatively minor and increasingly marginalised sector. In many former communist countries a liberal welfare state is being rapidly constituted (as in the United Kingdom in the 1980s and early 1990s), in which a dualist rental system will increasingly operate, rather than a corporatist or social-democratic regime being developed (as in Germany, the Netherlands or Sweden) with a unitary system of renting. It is clear that transition is occurring 'without any regard to the existence of possible alternatives, such as converting state-owned housing into smaller and competing autonomous non-profit housing organisations...there is barely any awareness that an alternative to neo-liberalism even exists, let alone...any debate...over the strengths and weaknesses of the social market alternative' (Kemeny, 1995).

Clearly the distinction between housing policy in western Europe and in the former communist states has been greater than the differences in policy within the various west European countries, but during the period of transition to a market economy in central Europe, the distinction between the direction of policy chosen by post-communist governments and policies existing in the liberal welfare regimes of western Europe (and in the United States) are getting less and less.

Part I

THE PRIMACY OF PRIVATE RENTED HOUSING

2

INTRODUCTION TO PRIVATE RENTED HOUSING

Paul Balchin

As a proportion of the total housing stock, the private rented sector in both Switzerland and Germany is considerably larger than the average for Europe as a whole. Although in terms of space, demography and economic, social and political attributes, Switzerland and Germany are markedly different, the two countries – in terms of housing – are very similar. Both Switzerland and Germany have, by European levels of tenure, a very high proportion of rented housing (and particularly private rented accommodation), and in both countries (within the rented sector) the private and social components of the system of renting are very largely integrated.

Despite both Switzerland and Germany having high gross domestic products per capita, the contexts in which the Swiss and German housing markets function are generally different. In Switzerland magnificent natural landscape has resulted in policy measures aimed at preserving the physical environment: for example, strict building codes and zoning restrictions. These, together with elements of monopoly in the construction and building materials industries have resulted in high construction costs and land prices and have led to housing shortages and constraints on 'greenfield' development particularly in the owner-occupied sector.

In contrast to Switzerland, the former West Germany faced considerable housing shortages after the Second World War, and had problems of housing need greater than any other country in western Europe. The first and second Housing Construction Acts of 1950 and 1956 imposed a duty on federal and *Land* governments and on the communes (*Gemeinden*) to promote a substantial level of housebuilding for a sizeable proportion of the population. Subsequently West German housing policy worked closely with the market, probably to a greater extent than in any other country in Europe. Thus, until reunification in 1990, the state gradually withdrew from direct intervention, but subsequently it has adopted a more active role (McCrone and Stephens, 1995). Nevertheless, although the policies of the major parties, when in office, were markedly different (the Social Democrat Party favouring rent regulation, and the Christian Democrats/ Free Democrats demonstrating a preference for the free market) in contrast

PAUL BALCHIN

with the United Kingdom, policies showed a high degree of continuity, as in Switzerland.

TENURE

Switzerland has the highest proportion of rented housing in western Europe – 63 per cent in 1990 – compared to an average of 39 per cent in the European Union (EU). Of the Swiss total stock, 60 per cent is in the private rented sector, while the social sector contains only 3 per cent (Table 2.1). Because of a considerable degree of confidentiality regarding the ownership of property in Switzerland, it is difficult to be very precise about the distribution of tenure, but it is probable that, in addition to private rental apartments, 3 per cent of the stock was owned by co-operatives (Gurtner, 1988). The amount of the private stock owned by cost-rental organisations is unknown but it is likely to be small. The small social sector is targeted at the poorest households – access being means-tested.

Table 2.1 Housing tenure in Switzerland and Germany (percentage)

	Private rented	Social rented	Total rented	Owner- occupation	Other tenure
Switzerland (1990)	60	3	63	31	6
Germany (1994)	36	26	62	38	–

Sources: CECODHAS (1995); Federal Statistical Office, *Statistical Yearbook of Switzerland, 1994*

It may seem that 'a dualist rental system' characterises rented housing in Switzerland. The small social sector appears to be very marginal and has many of the attributes of public renting in the liberal welfare economies of Europe, but this is an illusion since both the private and social rented sectors are subject to the same housebuilding standards (there are no ghettos in social housing) and both social and private sectors are subsidised and subject to rent regulation (Kemeny, 1995). As such, rented housing in Switzerland conforms with the 'unitary rented system' found elsewhere in western Europe.

The owner-occupied sector is the lowest in western Europe – 31 per cent in 1990 – compared to an average of 56 per cent in the EU. Proportionately the sector has been in decline in recent years, diminishing from 37 per cent in 1950 to its present level largely because of the need for rented housing to accommodate immigrant workers and their families. The development of owner-occupation has been restricted not only by planning constraints but by the high cost of house purchase, the cost of mortgage repayments often being higher than rents and (until 1965) restrictions on the conversion of large houses into multiple dwellings for sale.

26

Like Switzerland, Germany has a very high proportion of rented housing, 62 per cent of its total stock compared to an EU average of 39 per cent in 1994. The private rented sector is also high by EU standards – 36 per cent in Germany compared to 21 per cent in the EU, and the social housing sector is also proportionately higher than the average for the EU – 26 per cent compared with l8 per cent (see Table 1.1). Whereas the private rented sector, however, has been very stable since the 1970s, the social rented stock has diminished in size over the past two decades. Although undoubtedly a 'unitary' rental system still prevails, if the social rental sector continues to decline to residual proportions (within a liberal welfare regime rather than as a constituent part of a corporatist state), a 'dualist' system might emerge.

In contrast to all other countries of the EU and Switzerland the distinction between tenures, however, is very blurred. 'Social housing' depends not on ownership but on whether or not the owner receives subsidies and provides dwellings at social rents (Figure 2.1). Recently, the proportion of social housing has been in decline. Since 1990 housing associations (except for co-operatives) have lost their tax-exempt status and are now private-sector organisations. Also, when social landlords have repaid their subsidised loans after 15 years their housing transfers to the private rented sector. There is also a transfer of some private rented housing to the owner-occupied sector (McCrone and Stephens, 1995).

In Germany the owner-occupied sector is much smaller than in all other EU countries – comprising only 38 per cent of its housing stock compared to an EU average of 56 per cent. Owner-occupied houses are expensive in Germany compared to the United Kingdom or France, in large part because of high building standards being imposed by the state. Financial institutions give encouragement for house purchase to be delayed until later in life, while investment in a home is not considered more attractive than investment elsewhere (McCrone and Stephens, 1995).

Overall, in former West Germany, the pattern of tenure was remarkably stable over the years in contrast with very marked tenure shifts in other west European countries. However, with the reunification of Germany, 3.5 million social rented houses (amounting to twice the West German stock of this tenure) were added to the total supply of housing, altering the traditional pattern of ownership.

HOUSEBUILDING

As in many European countries, housebuilding activity in Switzerland has been cyclical, with peaks in 1962, 1973 and 1984, and troughs in 1968, 1976 and 1992. Although 81,000 dwellings were completed in 1973 there was a general decline subsequently, with only 31,000 completions being reported in 1992.

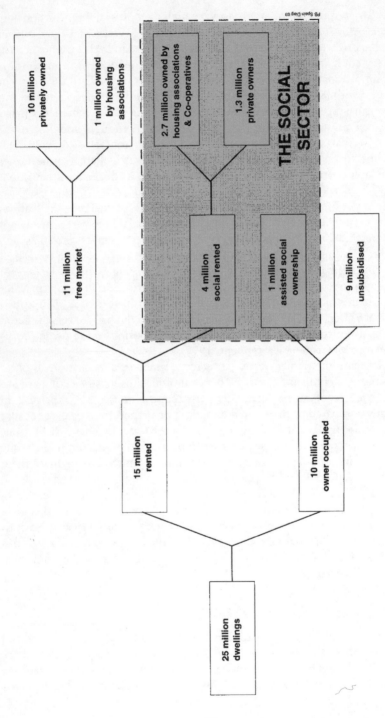

Figure 2.1 The West German housing stock, 1994 (after Jaedicke and Wollmann, 1990 and McCrone and Stephens, 1995)

According to the Organisation of Economic Co-operation and Development (OECD), the decline in housebuilding is attributable first to high construction costs (costs have been one third higher than the OECD average in recent years), and second to restrictive planning control and consequential escalation of land prices. The resulting housing shortage reflects the low level of housebuilding, low vacancy rates, an inefficient system of housing finance, and rent regulation which deters tenants from seeking new housing (OECD, 1994).

In Germany, unlike Switzerland, there was a massive loss of dwellings in the Second World War. Of a total stock of 10.5 million dwellings in 1939, 2.3 million were severely damaged or destroyed by 1945 (Duvigneau and Schönefeldt, 1989; Jaedicke and Wollman, 1990). In the post-war years the shortage of housing (of up to 6 million dwellings) was exacerbated by large-scale immigration of German peoples from eastern and central Europe (Leutner and Jensen, 1988). There was further immigration into the western part of Germany following unification. The large population of foreign workers (*Gastarbeiter*) amounting to 4.4 million in the late 1980s also increased housing demand (Duvigneau and Schönefeldt, 1989).

Large-scale housebuilding was thus essential. From 325,000 housing completions in 1950, West German housing output increased to a peak of 714,000 in 1973, converting a housing deficit into a surplus by the early 1970s. There was subsequently a decrease in housebuilding with only 140,000 dwellings being completed in 1987. Following reunification, output in Germany as a whole reached 374,000 in 1992. In recent years there has been a shift of emphasis from new housebuilding to maintenance and rehabilitation; for example, by the early 1980s over 200,000 units per annum within the social stock were being modernised. Of the investment in social rented housing about 94 per cent was allocated to new housebuilding in 1965 compared to 6 per cent allocated on maintenance and modernisation, but by 1985 only 53 per cent was spent on new housebuilding and 47 per cent on maintenance and modernisation. House renovations clearly necessitated higher rents and therefore tenants would either have expected renovation expenditure to be constrained or to have received 'good value for money' (Emms, 1990). The total value of output, however, diminished from DM 13.5 billion in 1965 to DM 6.3 billion in 1985 (at 1976 prices).

HOUSING INVESTMENT, FINANCE AND SUBSIDIES

Housing investment in Switzerland is facilitated by object subsidies in both of the rented sectors and in owner-occupation. Regardless of whether rented housing is developed for social or private landlords there are two levels of subsidy. The upper level of subsidy is conditional upon the landlord abiding by rent regulations and minimum construction and equipment standards; the lower level of subsidy is targeted at rental housing for the poor, particularly in the social sector, which in recent years assisted up to 20 per cent of new housebuilding (Kemeny, 1995).

Since 1974 the extension of owner-occupation has been encouraged by a federal programme of loan guarantees and interest subsidies (often targeted at low-income housing) for the improvement and development of sites, although in most years this accounts for less than 10 per cent of total building output. In total 140,000 dwellings have been constructed or renovated with object subsidies over 1974–94, subsidies amounting to SFr 140 million in 1992 (OECD, 1994). Subject subsidies also facilitate the development of the owner-occupied sector. Federal guarantees are extended to mortgages and other loans (amounting to SFr 4.3 billion in 1992), and the cantons and municipalities also provide a variety of assistance. Together with federal subsidies these payments amounted to SFr 500 million per annum in the early 1990s but are modest sums when compared to the SFr 17 billion incurred each year in housebuilding (OECD, 1994).

Schulz *et al.* (1993) claimed that the 1974 programme achieved its objectives, but suggested that up to 50 per cent of households benefiting from assistance would probably have bought their houses without help. It was also clear that the programme's aim of increasing the size of the owner-occupied sector conflicted with rent regulation which maintained the attractiveness of renting.

While owner-occupiers are liable to tax on imputed rent incomes (the percentage of the tax varying from one canton to another) and while they are eligible neither for any tax privileges or premiums on savings for house purchase, nor for depreciation or housing allowances, they do enjoy full mortgage-interest tax relief – in effect a subject subsidy.

As in Switzerland, housing supply in Germany is, in part, facilitated by the provision of object subsidies for new housebuilding and renovation, but they are not confined to the rented sectors; being tenure-neutral they are also available for owner-occupation. Originating from the Housing Acts of 1950 and 1956, there are at present three incentive schemes: the First Incentive Scheme, although tenure-neutral, largely benefits rented accommodation subject to its being let at agreed social rents. Either interest-free loans are provided, repayable over 35 years, or interest-free subsidies are available decreasing over 15 years. The Second Incentive Scheme is essentially an interest subsidy on loans to assist the development of owner-occupied housing, particularly for higher-income households. The Third Incentive Scheme (introduced only in 1989) is targeted mainly at the private rented sector to benefit tenants on intermediate incomes. They are allocated at the discretion of the *Länder* and are of only 7 to 10 years' duration. Rents return to market levels when the loan is repaid (McCrone and Stephens, 1995).

Because of cuts in public expenditure the number of social units assisted by the relevant subsidies diminished from 326,000 in 1960 to 39,000 in 1988 rising (after reunification) to 82,000 in 1992. For the same reason, subsidised housebuilding as a proportion of total completions diminished from 57 per cent in 1960 to 22 per cent in 1992. The federal government also relinquished complete responsibility for paying object subsidies in 1986 – delegating this role to the *Länder* until reunification took place, after which it re-assumed joint responsibility for expenditure in this area (McCrone and Stephens, 1995). In total 43 per cent of the 18 million dwellings built in West Germany from 1950 to the late 1980s were supplied with the assistance of object subsidies (Duvigneau and Schönefeldt, 1989).

Although house purchase is financed largely by loans from private institutions (notably first mortgage loans from a variety of banks, and loans stemming from contract savings schemes from the *Bausparkassen*), these sources cannot provide more than 80 per cent of the total cost of house purchase. Owner-occupiers and private landlords are thus eligible for interest-free loans (subject subsidies) from the state to top up most of the difference. The larger private landlords may, however, raise funds from equity finance and fixed-interest long-term loans.

Subsidies are also provided through the medium of tax allowances and exemptions. Depreciation allowances provide the principal incentive for investment in both the private-rented and owner-occupied sectors and as such are object subsidies. Owner-occupiers are also able to set loan interest against taxable income while their houses are being built or their homes being renovated before occupation. Both private landlords and owner-occupiers are generally exempt from capital gains tax, and owner-occupiers are exempt from tax on imputed rent income. House buyers also benefit from limited mortgage-interest tax relief (available only for new housing since 1991, having been phased out completely in 1984) (McCrone and Stephens, 1995).

Rent control

Whereas housing subsidies normally involve transfer payments from tax payers to the producers or consumers of housing, rent control – in effect – results in a transfer of payment from landlords to tenants representing the difference between market rents and controlled rents, and is thus also a subsidy.

Strict rent controls were first instituted in Switzerland in 1936 and were liberalised in 1954. The monitoring of rents by the federal government was introduced in 1962 but subsequently phased out. The subsequent escalation of rent resulted in legislation being introduced in 1972, the provisions of which were amended and incorporated into ordinary civil law in 1989 (OECD, 1994). Based on this legislation, rent increases were permitted if they compensated for rising costs (for example, higher mortgage-interest rates or the cost of renovation), if they maintained the purchasing power of the investor and if they were in line with rent levels locally.

Rents are controlled for ten years where dwellings were built or renovated with federal loans at subsidised rates of interest, but after this time tenants have the right to have their rents reviewed by the courts. In general, controlled rents are determined in relation to use-value rents, that is, cost rents modified by local comparative rents (similar to German *Mietspiegel* – 'mirror rents') (Kemeny, 1995). Rent controls have significantly regulated the market. Since older dwellings are subject to heavy regulation, newer dwellings are let at 'dampened' current market prices. The overwhelming number of older dwellings with lower costs therefore influence the market, rather than the market being led by newer property as in the case of an uncontrolled market. The system of rent control in Switzerland therefore has a significant amount of cost-based use-value rent setting, enabling households to spend on average only 20 per cent of their incomes on rent (Kemeny, 1995). Since all rented housing qualified for public subsidy and was consequently liable to rent control the Swiss rental system was clearly unitary.

This system, however, differs from that of Germany since cost-rental (social) housing accounts for only a very small part of the Swiss rented stock. Nor does it perform a market leadership role or have a dampening effect on general rent levels. It must be borne in mind, however, that there is a lack of reliable information on property ownership and therefore the precise size of the cost-rental sector is unknown.

Critics of rent control in Switzerland argue that it leads to a sub-optimal use of dwellings and reduces the mobility of labour. It is also claimed that since older properties are not let at market rents, allocation is often based on social or personal criteria rather than the willingness to pay, and control lowers the return on rented dwellings over time and is therefore a disincentive to invest.

As in many other countries of western Europe, rent control in West Germany was a cause of decline in the size of the private rented sector, although the severity of regulation was less than in France or the United Kingdom. Control was gradually relaxed after 1960 since it was considered incompatible with the social market philosophy of the Christian Democrat Party (although it remained in force in West Berlin, Hamburg and Munich because of shortages in rented housing). As a result of rapidly rising rents in the 1960s, the Social Democrat government introduced the present system of regulation by the Tenancy Protection Act of 1971 (McCrone and Stephens, 1995).

Both the private and social rented sectors are controlled by a two-tier system of rent regulation. With the upper tier, all landlords of newly built houses and in receipt of 25-year subsidies, are obliged to keep rents at cost-covering levels throughout that period, but thereafter the subsidy is withdrawn and rents become more flexibly regulated. With the lower tier, rent increases are permitted if they do not exceed local rent levels for comparable housing; vacant dwellings cannot be let at more than 20 per cent above local rents; rent increases of up to 30 per cent (over three years) are permitted in respect of existing tenancies, and all *Gemeinden* with populations in excess of 15,000 must establish *Mietspiegel* for housing of different sizes and standards. Landlords can therefore only raise rents if generally they conform with local *Mietspiegel* (Kemeny, 1995). After 25 years most rented housing transfers from the upper to the lower tier.

Housing allowances

Housing allowances (a subject subsidy) are not available in Switzerland but perform an essential role in the German system of housing finance. First introduced in 1965 at a time of decontrol and soaring rents, allowances (*Wohngeld*) have increased substantially and have more than matched the reduction in object subsidies. However, although the value of *Wohngeld* has risen from DM 1.8 billion to DM 3.8 billion over 1980–91, the total number of recipients has only increased from 1.62 million to 1.76 million over the same period, while the proportion of tenants in receipt of *Wohngeld* has diminished from 95 to 93 per cent (McCrone and Stephens, 1995). Although *Wohngeld* can be claimed by both tenants and owner-occupiers, tenants (with incomes of less than one-third of average incomes) are the main recipients. *Wohngeld* is intended to cover only two-thirds of any rent increase – sufficient to ensure that housing costs do not exceed 15–25 per cent of tenant income. Eligible owner-occupiers receive *Wohngeld* to compensate for rising interest rates and higher maintenance costs. Since there are substantial local variations in rents, *Wohngeld* (with ceilings based on space standards and family size) often fails to ensure that claimant housing costs do not exceed 25 per cent. Some *Länder* have therefore introduced supplementary schemes to assist *Wohngeld* recipients. Because of a generous system of social security in Germany, housing allowances are low by EU standards (McCrone and Stephens, 1995).

Overall, public expenditure on housing in Germany is a little lower than in France and less than half the level of expenditure in the United Kingdom. In 1991, it amounted to DM 36.8 billion of which DM 10.2 billion was spent on social rented housing, DM 10 billion on tax relief for owner-occupiers, DM 10 billion on depreciation allowances for landlords, and DM 3.8 billion on *Wohngeld*. These sums will inevitably rise when major housebuilding programmes get undertaken in the eastern *Länder*.

DEVELOPMENTS IN HOUSING POLICY

In Switzerland, concern about rising rents led to the strengthening of rent control in 1990. However, to partly ensure that landlords were able to receive a fair return on their investment, rents were permitted to increase by 40 per cent of the rate of inflation, and rent increases were permitted to compensate for increases in the rate of interest on mortgages.

By 1994 the adverse affects of rent control were beginning to be recognised. A federal commission proposed measures that would need to accompany deregulation, were it to be introduced. To minimise any adverse affect on tenants, decontrol (in the view of the commission) should take place over ten years and compensation be paid to the economically disadvantaged over this period from the proceeds of a temporary tax on landlords (Schips and Müller, 1993).

An aim of the Federal government was to increase the level of owner-occupation and housebuilding in this sector. From 1995 people covered by occupational pension schemes were permitted to use part of their accumulated equity to acquire an owner-occupied dwelling, to reduce existing mortgage debt, or to purchase shares in a housing co-operative. If employees were under the age of 50 they would be able to use all of their equity, but if they were over 50 they could obtain either the value of their equity as it was at the age of 50, or half of their accumulated equity at the time of request (OECD, 1994). It was estimated that in the short term this would increase demand for owner-occupation by up to 20 per cent, falling to only 3 per cent in the long term.

Policy in Germany in recent years has focused on the problems of re-unification. The housing stock of the former German Democratic Republic is relatively old (60 per cent being built before 1939), and since 1945 low rents have been insufficient to ensure an adequate level of maintenance and have resulted in the accumulation of large housing debts. In contrast to West Germany, the eastern *Länder* contained a much higher proportion of social rented housing and smaller proportions of private housing, although the private rented stock (at 17 per cent of the total) was notably larger than that of the United Kingdom in 1992 (9 per cent) (Table 2.2). The average size of dwellings in the eastern *Länder*, however, is smaller than in the West, 64 m² compared to 84 m² (*Statistiche Jahrbuch, 1992*).

Although the number of dwellings per thousand inhabitants is higher in the East than in the West (426 as opposed to 415), since reunification there has been a large migration of population from the eastern *Länder* to western Germany. The consequential increase in housing needs in west Germany has led to the return of interventionist policies (McCrone and Stephens,

Table 2.2 Housing tenure in West and East Germany before reunification

	Private rented %	Social rented %	Total rented %	Owner-occupation %	Other %
West Germany (1987)	43	15	58	38	4
Eastern Länder (1988)	17	59	76	24	–

Source: Tomann (1992)

1995). There has been a substantial increase in public expenditure and, because of escalating rents, pressure to tighten rent regulation. There may also be the need to introduce a major housebuilding programme for social rented housing because if more and more dwellings are transferred to the private-rented sector *Wohngeld* would rise continually and even be extended to higher-rent properties without necessarily stimulating any notable increase in supply.

CONCLUSIONS

In both Switzerland and Germany private rented housing in recent years has been maintained as the dominant tenure. Since it exerts a significant influence over social rented housing, particularly in terms of rents, the rental systems of both countries are clearly unitary.

In Switzerland, however, recent policy has concentrated on constraining the market for private rented housing by means of rent control while promoting the increase in the supply of, and the demand for, owner-occupied housing. This has resulted in the need to reduce the cost of housebuilding in this sector and to facilitate demand by permitting the release of equity in occupational pension schemes. There remains the need to increase competitive pressures in the housebuilding and related industries and to liberalise the Swiss system of land-use control to ensure an adequate supply of sites for housing development (OECD, 1994). If, however, the private rented sector is to be maintained as a dominant tenure, rent controls may also need to be liberalised.

In contrast with most other countries in western Europe, housing policy in Germany has worked largely with rather than against the grain of the market (McCrone and Stephens, 1995). Notwithstanding a substantial programme of housebuilding in the social rented sector especially in the early 1970s (Emms, 1990), the dominance of the private rented sector contrasts with all other EU countries despite pressures to strengthen rent controls in the 1990s. The owner-occupied sector still remains small compared to the EU average, in large part because of the availability of good private rented accommodation and the recognition that other forms of investment are more attractive in terms of performance and as a hedge against inflation (McCrone and Stephens, 1995).

3

SWITZERLAND

Roderick J. Lawrence

Switzerland is a relatively small country with a surface area of 41,293 square kilometres. The area is similar to that of Denmark or the Netherlands. However, owing to the Alpine environment of Switzerland, about two-thirds of the territory is unsuitable for building construction. Switzerland comprises 26 cantons, the Helvetic Confederation being initially formed by the cantons of Uri, Schwyz and Unterwalden (Oberwalden and Niedwalden) in 1291. The last canton was founded by dividing the canton of Bern and forming the canton of Jura in 1978.

The present Federal Constitution of the Confederation was adopted in 1848 when it was agreed to establish a three-tier system of government. A central authority still needed to respect the wishes of the formerly independent cantons, which wanted to retain the autonomy they had acquired during the course of history. The Federal Constitution states that the federal government can only acquire new responsibilities if and when a majority of the electorate and a majority of the cantons are agreed. Within each canton the hierarchical structure of government encompasses the communes, which are the smallest and most basic democratic units in Switzerland. Today, there are 3,072 communes. It should be borne in mind that the Federal Constitution is currently undergoing revision.

The power and responsibilities of the authorities at each of the three levels of government form a complex matrix. In principle, each of the three levels of government has the right to participate in the definition and the implementation of housing and land use-planning. In practice, however, the involvement of public authorities in housing has been marginal, except with respect to recent debates concerning the relationship between rents and housing mortgage interest rates. These rates have become a political issue because the balance between the supply and demand for credit from the mortgage market was not attained solely through increases in the interest rate. Consequently, attempts were made to remove credit from the housing sector.

This chapter will examine the economic role and functions of housing and building construction in Switzerland in terms of economic productivity and new construction output. These characteristics will be related to trends in the composition and structure of the housing stock according to size and tenure status. Then developments in the size and composition of households will be presented. These societal characteristics provide the background for an overview of government housing policy since the Second World War. Some of the outcomes of policy, including housing cost and the requirements of underprivileged groups, are discussed prior to a brief synthesis and conclusion.

TENURE

The ownership, provision and tenure of the housing stock can be examined in relation to a range of parameters, including the size of dwelling units, the number of inhabitants per habitable room (or per unit of habitable surface area), as well as tenure. Analysis of Federal Census statistics provides information on these parameters.

In contrast to many countries with market economies, the predominant form of housing tenure in Switzerland has been, and still is, the rental sector. This sector has varied from 64.1 per cent of occupied units in 1970 through 63.0 per cent in 1980 to 62.8 per cent in 1990. Concurrently, it is noteworthy that co-operative tenure in Switzerland has been relatively insignificant and invariant, comprising only 3.8 or 3.7 per cent in 1970 and 1990. The proportion of owner-occupied dwelling units has decreased marginally from 33.7 per cent of all occupied housing units in 1960, to 28.1 per cent in 1970, 30.1 per cent in 1980, and 31.3 per cent in 1990 (Table 3.1). This proportion of owner-occupation is the lowest of all countries, with either socialist or market economies, that are participants in the Economic Commission for Europe. These national averages do not indicate disparities between urban and rural regions, nor between large cities and small towns. In 1980, for example, the proportion of owner-occupied dwellings in the canton of Geneva was merely 11.2 per cent and increased to 13.8 per cent in 1990 and in the canton of Zurich it was 19.7 per cent in 1980 and 20.9 per cent in 1990, whereas in the canton of Valais, a region dominated by

Table 3.1 Tenure, Switzerland, 1960–90 (percentages)

	Rented			Owner-occupation	Co-operatives	Other tenure
	Private	Social[a]	Total			
1960	53.4	3.5	56.9	33.7	3.8	5.6
1970	60.5	3.6	64.1	28.1	3.8	4.0
1980	59.8	3.2	63.0	30.1	3.9	3.0
1990	60.1	2.7	62.8	31.3	3.7	2.2

Source: Swiss Federal Statistical Office
Note: [a] Owned by federal, cantonal and municipal authorities

agricultural and tourist activities, this proportion was as high as 59.5 per cent in 1980 and 59 per cent in 1990. In sum, these regional comparisons indicate that as the degree of urbanisation increases, the proportion of owner-occupation sharply declines. Moreover, the longitudinal perspective briefly outlined here shows that the composition and structure of the housing stock in terms of tenure have not changed significantly since the Second World War.

Given the large proportion of the housing stock attributed to the rental sector it is instructive to examine the types of ownership within that sector. Analysis of Federal Census statistics indicates that at least two-thirds of the rental housing stock has been owned by private individuals. However, there has been a steady decline in this type of ownership, from 79 per cent in 1950 to 68.7 per cent in 1990. Concurrently, limited property companies, associations and institutions have assumed an increasing role in the rental-housing sector, whereas the ownership of housing units by the Swiss Confederation, cantons and communes has rarely exceeded 4 per cent: in fact this proportion declined from 3.7 per cent in 1970 to 2.7 per cent in 1990. Again, it is noteworthy here that regional differences are important. Even in urbanised cantons like Geneva, in 1990, public authorities did not own more than 4.9 per cent of the housing stock.

Collectively, these figures indicate some structural characteristics of the housing market and the composition of the housing stock. First, social-sector housing in Switzerland has been and still is a very small proportion of the housing stock. Consequently, it is not representative. Therefore, it is difficult to make direct comparisons with other European countries that traditionally have had a significant proportion of the housing stock in the social-rental sector. Second, given the gradual increase in the ownership of the housing stock by property investors in financial institutions and limited property companies, there has been a steady reduction in the number of private landlords, and a growing number of anonymous building owners. Consequently, this trend has meant that direct landlord–tenant relations have been replaced by negotiations between stewards, caretakers and tenants (Lawrence, 1986). Both stewards and caretakers are commonly employed by limited property companies and financial institutions to administer and maintain residential buildings.

HOUSEBUILDING

The construction sector has an important impact on the Swiss national and regional economies. It is the largest category of gross investment and it provides a major input into other sectors. From 1990 to 1992, expenditure in the construction sector accounted for 16.8 per cent of gross domestic product (GDP). Expenditure in the construction sector can be classified according to residential, non-residential and public works. Residential construction accounts for about 35 per cent of total expenditure in the construction sector, whereas non-residential construction comprises about 46 per cent and public works about 19 per cent (OECD, 1994).

The cost of building construction in Switzerland is about one-third more than the average for all member countries of the OECD, but less than that for Sweden. A recent comparative study of new housing construction found that costs were 29 per cent higher in Switzerland than in Germany for comparable flats. Higher costs in Switzerland were attributed to local customs in the construction sector, building construction standards and regulations, different bathroom and kitchen equipment, and higher professional fees (Office fédéral du logement, 1993). According to the OECD, however, these factors need to be supplemented by others, including relatively high prices in all sectors of the Swiss economy as well as the labour market. This situation can be attributed to a lack of competition in national and regional markets, protectionism reflected in administrative and legal barriers and procedures, cartels in numerous ancillary industries and markets, and the fragmented organisation of the construction sector (OECD, 1994).

In Switzerland there is no informal construction sector like that in Italy or Portugal. All new housing and building construction in Switzerland is regulated by the issue of land-use planning permits and authorisations for the development of specific sites according to the principles of federal, cantonal and municipal legislation. These permits and authorisations prescribe plot coverage, maximum building heights, setbacks from street alignments and minimum distances between neighbouring buildings. In addition, the functional use of specific buildings and vehicular access to precise sites can be prescribed according to specific, localised characteristics and measures to ensure environmental protection. Although this administrative and legal framework has direct implications on land markets and the development of specific sites it is much less clear how it is implicated in the relatively high cost of housing construction. Clearly, other factors including housing finance are relevant.

At the beginning of the 1970s, housing production was relatively high. It reached a peak of 82,000 new units in 1973. From that year annual production declined steadily to 54,900 units in 1975 and 32,000 units in 1977. Annual production was between 40,000 and 45,000 units between 1980 and 1990, except during 1984 when it reached 45,300 units. During the period of high production in the 1970s not more than 10 per cent of all new residential buildings were constructed with public assistance, about 4 per cent were constructed by the public sector, and about 83 per cent were in multi-family residential buildings. Although reliance on public assistance declined with the slump in housing production after 1973, the share of new housing of the villa or semi-detached house type increased substantially from a low of about 16 per cent in 1973 to reach 41.5 per cent in 1980. These trends challenge simplistic interpretations of the relationships between housing costs and housing production, as well as 'supply and demand' in the housing sector.

Renovation

Fiscal measures for encouraging the construction of rental housing units at affordable rental costs are only applicable for newly constructed multi-family residential dwellings. However, public-welfare building authorities can purchase extant housing units using the basic rental reduction scheme if these flats are not in need of renovation or upgrading works.

The Federal Housing Bill in Support of Housing Production and the Promotion of Owner-Occupation of Dwellings was instigated in 1974 primarily to encourage the construction of new housing units. Subsequent demographic and economic trends, which are related to a growing number of relatively large, old residential buildings, led the Confederation to launch a special programme in July 1975 to encourage the renovation of extant housing and increased employment opportunities in the building construction industry. This federal assistance is a form of grant towards paying interest on capital. The subsidies are guaranteed for six years, bearing an annual interest of 2 per cent on the total cost of upgrading the housing units. This subsidy is intended to benefit tenants by having a direct impact on reduced rents.

Improvements in the condition of housing

During the twentieth century the size of rental housing units has increased whereas the number of persons per household has declined. Concurrently, the number of tenement buildings has increased, especially since the Second World War; for example, the number of storeys in tenement buildings, and the number of flatted dwelling units on each floor level have increased. The Federal Census returns for 1870 to 1990 tabulate these trends (Lawrence, 1989a, 1989b). When census data are examined with respect to cities and towns they show that as the degree of urbanisation increases so the number of tenement buildings increases. Consequently, the number of flats and the number of households in each tenement building increase in tandem with urbanisation.

Analysis of the composition of the housing stock in terms of the proportion of habitable rooms indicates the distribution of housing units with one to six rooms from the period 1970 to 1990. These figures indicate that three-roomed dwelling units have consistently been the most predominant size in recent decades, varying slightly between 29.9 per cent in 1970 and 27.4 per cent in 1990. During the same period, the proportion of housing units with four rooms increased gradually from 23.4 per cent in 1970 to 25.7 per cent in 1980 and 26.8 per cent in 1990. Concurrently the share of housing units with one or two rooms has remained virtually constant at about 20 per cent. Those trends leading to a slight yet steady increase in the proportion of housing units with three or four rooms can be contrasted with the steady decline in household size and especially with the growing proportion of households with one or two persons. In 1990, there was a national average of 39 m^2 of habitable floorspace per person, which is an increase of 5 m^2 for the same average in 1980.

These general findings do not reveal the differences in the size of housing units located in urban and rural regions. In principle, the analysis of census data shows that, as urbanisation increases, the average size of housing units declines.

From the late nineteenth century until today the size and composition of households in Switzerland have changed considerably. In 1870, for example, the average number of persons per household was 4.8 whereas the figure had declined to 4.5 in 1910. At the same time 9 per cent of households comprised only one person, and 26 per cent of households were childless. The most significant decreases in household size occurred from 1920 to 1950, with a reduction from 4.4 to 3.6 persons. From 1950 to 1990 the decline continued from 3.6 to 2.4 persons per household. These steady reductions in the average size of households are attributed to a range of factors, but especially a slow decline in the fecundity rate, which can be traced back to the late nineteenth century. Consequently, in 1960, 40 per cent of all households in Switzerland included one or two persons whereas this proportion had increased to 60 per cent in 1980 and 63.4 per cent in 1990. Concurrently, 20 per cent of households in 1960 included five or more persons, whereas this proportion had declined to 10 per cent in 1980 and 6.7 per cent in 1990.

The constant decline in the number of persons per household is valid for the whole of Switzerland, but this trend has been more pronounced in cities and towns than in rural areas. These trends can be related to the housing conditions of the population, such as the number of persons per housing unit and per habitable room. In 1960, for example, census returns indicate that there was an average of 3.27 inhabitants per housing unit, and this average declined to 2.60 persons in 1980 and to 2.40 persons in 1990. During the same period the average number of inhabitants per room declined from 0.86 in 1960 to 0.70 in 1980 and then to 0.63 in 1990. These averages do not indicate variations in the size of housing units. In 1980, 37 per cent of the population lived in housing units comprising one to three habitable rooms, 30.2 per cent lived in four rooms, 16.4 per cent lived in five rooms, and 16.3 per cent lived in six or more rooms. These statistics collectively show that during recent decades the amount of habitable space per person has increased, especially for those households of relatively small size. This trend is not only related to the steadily decreasing size of the number of persons per household but also to an increase in the average size of housing units from 88 m² in 1980 to 93 m² in 1990.

HOUSING INVESTMENT, FINANCE AND SUBSIDIES

The major share of new housing construction in Switzerland is financed privately by loans. In 1985, the extent of the loan was about 70 per cent of construction costs for private individuals, between 80 per cent and 95 per cent for co-operative owner-occupiers, and about 10 per cent for pension funds, whereas insurance companies were generally self-funded. Loans for housing are made in the form of mortgages. In 1984, Swiss banks lent around 85 per cent of all mortgages, while insurance companies accounted for 6.2 per cent and pension funds lent about 5.2 per cent.

Until the end of the 1970s savings deposits at banks and mortgages were relatively the same. After 1979, however, an imbalance grew because mortgages increased much more than the money deposited in banks. In 1980 and 1981, for example, only about 39 per cent and 45 per cent respectively of new mortgages were covered by public savings. This can be attributed to the custom that mortgage-interest rates are variable, that first mortgages are not expected to be amortised, and that repayment of second mortgages is made over a period of 15 to 20 years. This way of financing housing encourages debts and dependency. In fact, housing rents are largely determined by amortisation and bank interest charges which are expenditures that are not tied to either construction or maintenance costs in the housing sector. This is also shown by comparing the relative cost of housing construction to the costs of other goods and services. Since 1980, although the relative cost of new housing construction declined continually, except during 1986 and 1988 rents increased steadily. These trends have no direct relationship to housing production.

The Federal Office of Statistics recently presented an overview of the cost of rental housing in relation to the cost of living for the period from 1975 to 1988. The results of this study indicate that housing costs increased at a greater rate than the cost of living from 1980 until 1992. An in-depth sample survey of 471 households in 1988 found that expenditure for housing rent has been the second most important item in household budgets and, in that year, it totalled 13.8 per cent of household expenses (plus an additional 2.8 per cent for house heating and electricity). In 1992, the Federal Housing Office reported that the cost of housing in the rental sector had risen to a national average of 21 per cent of household income. Those who paid less than 15 per cent of their income decreased from 43 per cent in 1980 to 34 per cent in 1990. Concurrently, the share of households that paid more than 30 per cent of their income remained virtually unchanged at about 10 per cent (Office fédéral du logement, 1993). Another recent survey undertaken by the Office of Statistics in the canton of Vaud concluded that 22 per cent of household expenditure was allocated to housing by households in that canton. Both surveys confirm that as household income increases, the proportion of household expenditure on education and leisure increases, whereas expenditure on housing, electricity and heating declines.

The increase in housing costs from 1980 to 1992 can be explained by sharp increases in mortgage interest rates. In addition, there are two factors, not considered in the surveys, that are equally important in understanding market rents. In particular, the length of tenancy and the age of the dwelling unit are crucial factors that ought to be examined in relation to the condition of the dwelling unit (e.g. whether it has been partly or wholly renovated). Numerous newspaper reports in the local press have illustrated the pertinence of examining these factors. There can be as much as a threefold difference in the rental charges for apartments of the same size, in the same building, when that building is more than 20 years of age and when some apartments have been completely renovated whereas others have not and they are inhabited by the initial tenants. This practice is growing in importance and it has become allied to another custom known as 'buy the flat or move elsewhere'. An increasing number of building owners are presenting this option to tenants, especially elderly citizens who have been living for decades in the same dwelling unit. The building owner has already paid off his or her borrowed capital when the construction, or purchase, was made at least two decades previously and, given the spiralling costs during the 1980s, the owner has the opportunity to resell at a very favourable price. In essence, the tenant is in an unfavourable position.

The divergence between rentals charged for older and newer dwelling units, and the insecurity of tenants who are obliged to live in recently constructed residential buildings, have been a major preoccupation of the federal and cantonal governments and the Swiss Association of Tenants. In this respect, housing costs can be examined in relation to mortgage rates. One interpretation proposes a linear relationship between mortgage interest rates and housing rents. This simplistic model has commonly been used by private investors and politicians in Switzerland. Nonetheless, it does not account for the amount of borrowed capital required by the property owner to acquire (or renovate) a residential building, nor the contribution of income from rents to cover the cost of repaying this capital and the maintenance costs of the property. Notwithstanding these shortcomings, this model was endorsed by the federal government in 1980 and it has been increasingly used since then by property owners and estate agents to calculate and charge housing rents. Although this method is simple to apply, it leads to the increasing differentiation between the rents charged for housing in new and old buildings. Moreover, it encourages the practice of modifying rents in tandem with fluctuations in the cost of borrowed money. Given that such changes have been numerous since 1987, they have provoked a sense of economic insecurity for many tenants (Biélier *et al.*, 1993). Consequently, the Swiss Association of Tenants contested this state of affairs, which led the Federal Parliament to modify legislation concerning security of tenure in 1989. From then it was possible for tenants to examine whether preceding decreases in mortgage interest rates had led to a decrease in rents.

MEASURES TO SUBSIDISE RENTS

There are three interrelated sets of measures to reduce rent changes which the Federal Office of Housing (Office fédéral du logement, 1990) has summarised as follows:

1 *Basic rent reductions*
These include direct financial assistance, in the form of negotiating and guaranteeing loans up to 90 per cent of the investment costs. These negotiations by the Confederation on behalf of the applicant include arranging loans through banks, if an applicant is unable to secure such financial backing through banks in his or her own region.

The allocation of repayable loans to enable rents to be reduced during the initial phase of occupancy. The rent can be reduced during this period to approximately 23 per cent below the basic rental fee necessary to cover costs by fixing an annual rent increase of 3 per cent for a period of 25 years. The deficit thus initially incurred by the landlord is covered by interest-bearing loans from the Confederation or the banking institutions involved. After ten years, the rent attains a basic level necessary to cover the capital outlay, while surpluses which accumulate from increases in rent over the next 15 years serve to cover the repayment of the loans originally granted by the Confederation or the banks.

2 *Supplementary rent reductions (I)*
In order to assist lower-income households by lowering initial rents, the Confederation provides non-repayable subsidies which remain constant over a period of ten years. This supplementary rent reduction represents a further reduction of the initial rent by approximately 7 per cent. Together with the loans granted in connection with the basic rent reductions it accumulates to an overall initial reduction of 30 per cent – that is, the rent payable in the first year of occupancy amounts to 70 per cent of the basic fee required to cover costs.

3 *Supplementary rent reductions (II)*
These apply to apartments occupied by elderly people, invalids, people in need of care, and people undergoing educational training. Rental fees for such apartments are reduced with the aid of constant non-payable subsidies for a maximum period of 25 years. This supplementary reduction amounts to about 19 per cent. Together with the loans granted under the basic rent

reductions scheme, this subsidy provides a total initial rent reduction of approximately 40 per cent of the basis cost-covering rent – that is, the rent payable in the first year of occupancy amounts to approximately 60 per cent of the basic fee required to cover costs.

These supplementary reductions are only granted when the tenant's annual income does not exceed 40,000 francs and his or her assets do not exceed 100,000 francs. Allowances for each dependent child include an additional 3,700 francs per child in additional income plus 12,000 francs per child for assets. These allocations are adapted periodically.

The Confederation only grants these forms of assistance when there is a proven shortage of certain categories of apartment in the applicant's district. An appropriate set of regulations has been developed to investigate whether this precondition is fulfilled. In addition, the Housing Evaluation System is applied to assess whether a projected residential building meets established requirements.

DEVELOPMENTS IN HOUSING POLICY

In contrast to other European countries, the customary provision, management and tenure of housing in Switzerland has been attributed to the private sector. Housing finance is also commonly provided by the private financial sector. These customs indicate that national, cantonal and municipal governments have not been actively involved in the provision of housing. Although it is not unfair to claim that housing policy has not been a high priority, the control and regulation of housing rents has been explicit since the First World War, except between 1970 and 1972. In general, legislation has sought a compromise between encouraging investors in the housing and building sector while maintaining low increases in housing rents. In a housing sector dominated by private rental tenure, a partnership approach between property owners and tenants has been formulated. This constitutional and legal framework coincides with the financial interests of institutional and private investors who are attracted by a secure investment and a steady yield. The private rental housing stock has remained dominant because financial investments in this sector have continued to be profitable.

Federal housing policy in Switzerland since the Second World War can be divided into five periods. During the first period from 1942 to 1949 two goals, of increasing the labour force and providing housing for households with families, were meant to stimulate national economic growth. Legislation was enacted for rent controls and the protection of tenants against the unjustified resolution of leases. During this period subsidised housing construction exceeded that financed by private capital.

From 1950 to 1958 federal policies to reduce rent controls and abolish subsidies for low-cost housing construction were introduced primarily in order to stimulate new housing construction, reduce post-war housing shortages and overcome unemployment, which had been compounded by the effects of the war. The promotion of new low-cost housing by a deregulated housing market was meant to counteract the shortage of affordable housing in urban areas for both Swiss and foreign workers. While reduced rent controls did remain effective for those residential buildings constructed before 1946, the legally permitted maximum rent levels were raised several times during this period. These adjustments reflected the climate of a political and public debate about the role of government in the housing market. On the one hand, private investors and right-wing politicians blamed the remaining rent controls for insufficient returns on capital investment. They argued for a complete liberalisation of the housing market. On the other hand, tenant associations and left-wing politicians maintained that, given recurrent shortages of affordable housing, rent controls were necessary to protect tenants against unjustified increases. The advocates of government intervention in the housing market also argued that, alone, the private sector did not cover the needs of low-income households nor special groups, such as the elderly and the handicapped.

Federal housing policy between 1958 and 1974 tried to account for these opposing viewpoints. The Federal Parliament enacted a new Federal Housing Act in 1965. This Act stipulated that the promotion of new housing construction and land acquisition, plus general rent policy, were no longer temporary but rather permanent responsibilities of the Confederation (Bassand et al., 1984). Nonetheless, the overriding goal of policies during this period was to deregulate the housing market by 1970. In order to achieve this goal a compromise agreement was formulated: those in favour of progressively removing rent controls by 1970 accepted the promotion of low-income housing construction programmes. These were to be implemented by the usual actors operating in the housing market, using subsidised returns on invested capital, secured mortgages and the regulated supply of land for new building construction. In return, subsidised housing projects were subjected to four sets of restrictions:

1 Total construction costs should not exceed a legally fixed amount per housing unit.
2 Floors areas of housing units should conform to minimum requirements.
3 Tenants' income should not exceed six times the rent and a prescribed upper limit that is adjustable over time.
4 Rents are related to effective costs and cannot be raised simply to reflect market trends.

Studies of the implementation of federal housing policy from 1965 indicate that some policies were ineffective, that there were significant differences between subsidised new housing construction in the Swiss cantons, and that the implementation of policies did not reflect regional or local

conditions of the housing market. The failure to respond to spatially concentrated housing shortages and needs is an outcome of homogeneous
policy incentives which are indifferent to localised variability (Bassand *et
al.*, 1984).

From December 1970 until March 1972 there was no federal legislation
that enforced rent controls or regulated the contractual terms of tenancy
agreements. During this short period there was a rapid increase in rent
charges. Following a referendum on 5 March 1972, which was overwhelmingly accepted by the population, the Confederation was granted constitutional and legal authority to protect tenants against malpractice. The
subsequent legislation enacted on 30 June 1972 was ambiguous. Although
it was meant to eliminate sharp increases in rents, it did not reintroduce
statutory rent controls. Concurrently, it sought to promote private investments and owner-occupation by introducing fiscal incentives.

In 1974, the Federal Parliament enacted the Federal Housing Bill in
Support of Housing Production and the Promotion of Owner-Occupation
of Dwellings. This Bill came into force on 1 January 1975 and it is still
operational today. Its clauses explicitly state that measures will only be
taken insofar as they prove necessary within the framework of national
social and economic policies. The 'intervention' of the federal government
in the construction industry and the housing market is intended to encourage an increase in owner-occupation and promote the construction of
housing for the aged and the disabled. This law has two main goals. First,
it stipulates those direct and indirect fiscal measures that are meant to
encourage the production of housing units by the private sector while
regulating those conditions that can enable a reduction of production costs.
Second, the law is intended to encourage owner-occupation by applying
fiscal measures that aid citizens to purchase flatted dwellings or houses. In
both these respects there are decrees concerning aid for land development,
the acquisition of land for social housing construction, reductions in rent
for new flats, assistance for owner-occupiers, and measures to encourage the
renovation of extant housing units. Personalised aid is not prescribed in this
legislation because 'supply side' subsidies are considered more suitable for
promoting the construction of new housing. However, this does not mean
that housing subsidies in the form of personal allocations are not available
from cantonal and municipal authorities.

In order to be eligible for fiscal aid, a set of requirements related to the
design, construction and site layout of new residential buildings was elabor-
ated and applied by the Federal Office of Housing (in 1979 and 1986). In
fact, each housing unit and residential building must satisfy certain minimum requirements. In addition, each projected building is assessed using a
housing evaluation system.

The preceding paragraphs show that the ongoing tasks of the Confederation in the sphere of housing are defined in terms of encouraging housing

construction and renovation by the private sector, while ensuring that housing costs do not escalate. According to the Federal Constitution, the functions of the Confederation are:

1 To facilitate the purchasing and development of sites for housing construction.
2 To support efforts aimed at improving housing and environmental conditions for families, persons with limited earning capacity, the elderly, the disabled, and persons in need of care.
3 To research into the housing market and into building methods as well as to encourage rationalisation in building.
4 To ensure that capital is obtained for housing construction.

These responsibilities have a bearing on contemporary housing policy, financing and housing costs in both the rental and owner-occupied sectors.

CONCLUSIONS

Since 1980, the construction of housing in Switzerland has been characterised by three developments:

1 New federal legislation that has sought to regulate the quality of the external environment, including nuisances in residential areas stemming from excessive air pollution and noise emissions.
2 A growing disparity between the costs of new and old housing units which has commonly been tied to fluctuations in mortgage interest rates.
3 A professional and political debate reported by the mass media about those means and measures required to stimulate building construction in general, and the housing sector in particular.

The most significant characteristics of new housing built during the last decade do not concern their architectural, technical or functional aspects, but their cost. From 1989, the cost of housing rents increased at a much greater rate than the cost of building construction, the cost of living index, and wages. The increasing disparity between these units is shown in Figure 3.1. It was stated earlier that although numerous factors are involved, the sharp increase in the purchase price of land and borrowed capital during the 1980s is largely responsible for this trend. The consequence of successive increases in mortgage interest rates has meant that more money must be borrowed for the construction of new housing: in fact, between 1985 and 1991, accumulated debts in this sector increased by 78 per cent! (Biélier *et al.*, 1993).

The theoretical model of a linear relationship between mortgage interest rates and housing rents has been endorsed by government and applied by property owners during the last decade. Concurrently, there has been an increasing differentiation between the rents charged for housing units in

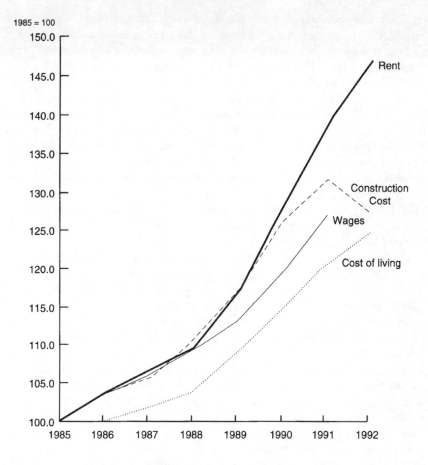

Figure 3.1 Relationship between rent, construction costs, wages and the cost of living, Switzerland, 1985–92

new and old residential buildings, as well as between similar flats in the same residential building. In order to overcome inequity of this kind, the following principles could be borne in mind. First, housing legislation should consider the lessor and the tenant as legally equitable. Both parties could have the right to argue for an increase or a reduction in rent. Second, the notion of a fair rent should not be solely related to the purchase price of housing units, because that approach relies on a 'market mechanism' that encourages financial speculation. Rather it should be based on a comparison of housing units, in the same or an adjacent residential building, and the right of the property owner to receive a return on the invested capital in that building. Any method for the calculation of a fair rent should also account for the date of construction of the building; its location, size, facilities and services; and its state of repair. Third, tenancy agreements should be enforced for a minimum of five years in order to eliminate successive fluctuations of rents which create uncertainty for tenants. Any fluctuation in rent should be prescribed by a maximum that is specific to a locality (Biélier *et al.*, 1993).

Bearing in mind recent demographic and economic developments (especially related to costs, decreasing household size and an ageing population) it is noteworthy that their implications on the construction of new housing have not been as significant as one might expect. Indeed, the increasing share of three- and four-roomed housing units does not appear to reflect these trends. Nonetheless, with respect to the architectural design of new housing units intended for owner-occupation, the detached house type (the villa) has been built relatively infrequently, owing to the construction of an increasing proportion of semi-detached and row houses on relatively small plots of land (e.g. 600 to 800 m^2).

This chapter has shown that although the main responsibility for the provision, management and tenure of housing has been attributed to 'the market place', it has not been an unregulated market. Since the 1970s, federal and cantonal policies have been enacted to apply means and measures that are not meant to penalise investors in the housing sector, that encourage owner-occupation, and that maintain a low inflation rate of housing costs, especially in the private-rental sector. Furthermore, the federal government has oscillated between applying strict rent controls, monitoring and supervising rents, and renouncing intervention in transactions between landlords and tenants. Nonetheless, in 1989 a new law concerning tenancy contracts was enacted. Collectively, these approaches have supported the financial interests of institutional and private investors (including the capital of foreigners) who are attracted by a secure investment in a housing sector of a relatively high quality that provides a profitable return.

During the same period, cantonal and municipal authorities have been given legal and administrative responsibilities and obligations for the formulation and the application of federal regulations related to land-use planning, housing construction standards, traffic circulation, nuisances stemming from air pollution and noise emissions, and energy-saving measures. These recent initiatives reflect the relatively advanced environmental legislation in Switzerland.

Compared with other European countries, these practices reflect a characteristic of the Swiss constitution which distributes these and other obligations and responsibilities between the federal, cantonal and municipal governments and institutions. This *modus operandi* ought to be contrasted with the centralised approach advocated and applied in countries such as Britain, France and Sweden.

4

GERMANY

Horst Tomann

The main objectives of post-war German housing policy have been to provide affordable housing for a broad spectrum of people and to offer special assistance to discriminated groups. Traditionally, social housing subsidisation and rent regulation have been the instruments to pursue these ends. In recent years, housing policies have shifted more and more to means-tested housing allowances made directly to private households.

In addition, housing policy has been directed towards distortions in the housing market. Two different kinds of distortions combine to create large and lasting market disequilibria. First, due to its long life-time, housing investment implies an extremely low turnover on capital. It is long-term expectations which command investment, hence investment does not visibly respond to demand shocks. Secondly, flexibility in adjusting the utilisation of existing housing stock to demand shocks may be restricted, which in turn aggravates the problem of market disequilibria. The instruments directed towards improving the performance of housing markets are urban land-use planning, supply subsidisation (by grants or tax expenditure), and indirect measures which increase the flexibility of markets. Housing allowances also have an indirect effect.

However, housing policies may themselves have detrimental effects on the market's performance. In particular, market regulations in the existing housing stock may destabilise long-term expectations of investors. Furthermore, household mobility is restricted by the system of rent regulation. With regulated rents, excess demand in the rental housing market drives market rents up, but the rents in existing contracts hardly respond. Tenants have a disincentive to move, with the consequence of lasting market disequilibria and substantial price effects. Hence, as a consequence of inappropriate housing policies, house prices and rents may overshoot their long-term trends.

In view of lasting market disequilibria and overshooting, governments tend to stabilise housing markets by direct intervention. In this respect, rent regulation – either with the existing stock or with social housing programmes – is a case in point. The problem is that the government may take as a dysfunction of the market what is actually a consequence of its intervention. That interrelationship of government intervention and market inefficiency is the fundamental problem of housing policy.

The German housing market is divided into two very different scenarios. West German households enjoy high-quality housing, with only minor differences by income or region. There are virtually no slums or abandoned residences. Vacancy rates are low. The old housing stock in inner cities has to a large extent been rehabilitated or upgraded. There is a shortage of low-standard housing, however, as a consequence of different factors including rent regulation, rehabilitation policies and, in particular, a tremendous increase of in-migration during the 1990s.

On the other hand, the East German housing stock still bears the mark of state socialism. Approximately half the multi-storey buildings in inner cities are severely damaged; many are no longer usable. There are large derelict areas, partly because of decay, partly as a consequence of socialist urban planning which gave priority to 'strategic' purposes. Vacancy rates in the pre-war housing stock are extremely high, in particular because of unresolved restitution claims. Financial and administrative restrictions impede the conduct of rehabilitation programmes.

The post-war suburbs, where one-third of the East German population resides, also need rehabilitation, partly as a result of low-quality standards and construction deficiencies, in particular poor insulation, and partly through their monotony. There is no urban variety, as services, jobs and urban infrastructure are lacking. Although these suburbs are still the preferred residential areas in East Germany, they might become problem areas as soon as incomes rise and inner cities are restructured.

TENURE

In West Germany, the private-rented sector grew gradually in the period 1978–87, whereas the non-profit and social sectors declined. By 1987, although the rented sectors were still larger than the owner-occupied sector, the latter tenure had grown markedly since 1987 (Table 4.1).

Table 4.1 Tenure in Germany, 1978–94

	Private-rented	Non-profit rented	Social housing	Owner-occupied	Total stock
	%	%	%	%	million
1978	40.8	4.0	17.7	37.5	22.6
1987	41.7	3.7	12.6	42.0	26.3
1994					
West	n/a	–	6.5[a]	n/a	28.3
East	n/a	–	0.9[a]	n/a	7.0

Sources: Federal Office of Statistics; Expert Commission on Housing Policy
Notes: 1978 and 1987 West Germany only
[a] Estimated

Private landlords

West Germany has a relatively large private-rented sector where rents respond to market forces (Table 4.2). It is true that the 1971 Rent Regulation Act regulates rent increases for those with existing tenancies and provides security of tenure. According to this regulation, a rent increase must not exceed an upper limit given by an index or by reference to existing comparable tenancies. However, there is in principle no restriction on the rent level at which a new letting is initially made. The rent structure remains flexible, therefore, but the general rent level responds to variations of housing demand only with some delay.

Table 4.2   Rents in Germany

| | West Germany | | | |
| | Private-rented sector | | Social housing | |
	Index 1985 = 100	Index deflated[a]	Index 1985 = 100	Index deflated
1980	80.8	94.6	79.0	n/a
1986	102.1	98.9	101.7	98.5
1987	103.4	98.4	103.3	98.3
1988	105.6	99.0	106.0	100.9
1989	108.5	99.3	110.1	100.7
1990	112.3	99.6	113.4	100.6
1991	117.4	100.2	117.3	100.1
1992	123.9	101.3	123.5	101.0
1993	130.9	103.6	132.5	104.7
1994	136.8	105.8	138.2	106.8

| | East Germany | |
| | Transitory rent regulation | |
	Index 1991 = 100	Index deflated
1990	56.5	n/a
1991	100.0	100.0
1992	226.5	193.6
1993	359.2	278.8
1994	398.8	301.4

Sources: Federal Office of Statistics; own calculations
Note: [a] By GNP deflator.

In addition to the smoothing effect of the rent regulation system, the social consequences of rent increases in the private sector are mitigated by a scheme of housing allowances for low-income households which was introduced in 1965. The importance of housing allowances has grown greatly since then.

Part of the private rented housing stock is subject to rent control. These are dwellings for low-income households constructed during the post-war period as part of the social housing programmes. The federal and state governments provided direct financial aid to investors in order to induce investments in the new low-rent housing sector. The market share of social rented housing is decreasing for two reasons. First, the investor's commitment to means-tested letting at low rents expires after a predetermined, although long period (up to 30 years). Second, social housing programmes have been steadily reduced since 1960 (from 300,000 per annum during the 1950s to less than 100,000 per annum during the last ten years), partly because of changed policy objectives and partly because of the boost in subsidies required to provide new social housing at affordable rents. As a consequence of the resulting imbalance in the provision of new private and social rental housing, the social rented stock is being diminished at a rate of – in the mid-1990s – approximately 10 per cent per annum.

Housing associations

Non-profit housing associations (*Gemeinnützige Wohnungsunternehmen*) have played a traditional role in the rented housing sector. After the Second World War, they conducted the dominant part of social housing pro- grammes and of clearance policy. Housing associations were tax-exempt but had to fulfil specific obligations, for instance, to charge cost-based rents even for those dwellings which were not (or no longer) receiving direct subsidies from the social housing programmes, and to reinvest their returns into the non-profit housing sector.

While there are economic reasons supporting the case for non-profit housing associations, their political reputation became seriously damaged during the 1980s, triggered by a large-scale corruption scandal within the largest company followed by a financial collapse. The political consensus that the non-profit sector should fulfil important functions within society was broken. As a consequence, the federal government took a major dereg- ulation step by changing the associations' legal status to that of profit- oriented enterprises. The associations will not necessarily be disadvantaged, since they began the new phase in 1990 with high tax-free provisions, that is, a high potential for future depreciation allowances to set against tax. Nonetheless, that deregulation measure has fundamentally changed the structure of the rented housing market, since public authorities make only a negligible contribution to the rented housing supply.

Owner-occupation

In West Germany, in the 1970s and 1980s, the size of the owner-occupied housing sector increased markedly (Table 4.1), but was relatively small in comparison to other countries.

This was to some extent the consequence of urban planning which kept land relatively scarce and land prices high. Moreover, Germany was a main target region for post-war migration. So the demand for rented housing was comparatively high.

There are indicators, however, that the share of owner-occupation will further increase. Home-ownership rates are higher for middle-aged households (above 50 per cent) than for the young or for the elderly. In addition, more and more households buy property in condominiums rather than family homes. The tendency towards condominiums started in the mid-1970s when the emphasis of housing policy shifted from new construction towards rehabilitation and improvement of existing stock.

HOUSEBUILDING

The long-term trend in housing construction activity is mainly determined by demographic developments, migration and real income growth. In addition, changes in the real interest rate (composed of long-term rates of interest and price expectations) may induce cyclical fluctuations. A change in housing policy measures, in particular subsidies and tax relief for investors, influences the level of construction activity as well as its composition (new construction and rehabilitation or upgrading). In Germany, housing investment since 1980 reflects all these influences (Table 4.3).

During the 1980s, expectations of a decrease in population and high real interest rates induced a declining trend of housing construction activity which was only partly offset by the rise in upgrading investment. By the end of the decade, a sharp rise of in-migration in the West, falling interest rates and a substantial increase in supply subsidies caused a new housing investment boom in West and East. In East Germany, the real volume of

Table 4.3 Housing investment and housebuilding prices in Germany

	Housing investment DM billion at current prices	at 1991 prices	Price index 1991 = 100
	West Germany		
1980	117.2	169.8	69.0
1986	115.9	142.4	81.4
1987	116.7	140.5	83.1
1988	124.8	146.9	85.0
1989	136.2	154.8	88.0
1990	155.8	167.1	93.3
1991	174.3	174.3	100.0
1992	197.5	186.9	105.6
1993	215.7	195.4	110.3
	East Germany		
1991	17.5	17.5	100.0
1992	27.5	25.0	110.1
1993	39.1	32.9	118.6

Source: German Institute of Economic Research, Berlin

housing investment expanded by more than one-third per annum, triggering an inflation of housing construction prices (Bartholmai *et al.*, 1994).

HOUSING INVESTMENT, FINANCE AND SUBSIDIES

Public assistance to rented housing

Subsidies to the rented housing sector are granted by social housing programmes, by rehabilitation programmes and by tax expenditure for investors.

Social housing programmes refer to the Housing Construction Act and are provided by the federal government and state governments on equal terms. In addition, local authorities subsidise housing construction by their own programmes. The subsidy system of social housing includes a variety of schemes designed by the *Länder* with differing grant elements: either interest-free loans and a grace period for repayment or recurrent grants – which are partly repayable – plus sureties for private mortgage loans.

The general principle is that the subsidy is based on cost calculations reported by the investor and will be fixed to cover all costs exceeding the predetermined social rent (which is set below market level). The investor breaks even from the beginning if actual costs do not exceed calculated costs and if empty properties can be avoided.

On the other hand, subsidisation is conditional, that is, the investor is restrained by several conditions until all the public money is repaid: social rented housing has to fulfil minimum standards concerning facilities and size; tenancies are open to certain income groups only; social rents are essentially fixed with the exception that certain cost-based rent increases are allowed. The tenant may also be charged for a built-in reduction of subsidies over time.

In recent years, alternative schemes have been designed, mainly to overcome the poor distributional accuracy of social housing programmes. The new schemes split the subsidy into a basic investment grant and an additional means-tested grant for which only low-income households are eligible. As a consequence, the social rent depends on the tenant's income, and the subsidy can be smoothly targeted and adjusted to income changes.

Rehabilitation programmes for rented housing were introduced during the 1970s. To tackle urban decay the state governments and the federal government provided attractive subsidy schemes for upgrading in old urban areas. Investors could deduct the cost of investment in existing housing at preferential terms and/or apply for direct financial aid (grants). An energy-saving programme provided additional subsidies.

This policy coincided with and gave support to a shift of demand towards inner-city areas. As a consequence of the resulting filtering-up process which took place in the inner cities, a shortage of low-quality rental housing emerged which was not matched adequately by investment in new social housing, either in quantitative terms or value. In view of these developments, rehabilitation policy was corrected during the 1980s. The state governments' programmes expired or were reduced. However, special depreciation allowances for energy-saving measures still apply. This kind of subsidisation is open for owner-occupiers too.

As far as the tax treatment of housing investment is concerned, the German tax system, as a rule, treats housing investment like any other investment. It entails some peculiar regulations, however, providing for specific incentives:

1 There are housing-specific depreciation allowances.
2 Capital gains are tax-exempt for private investors (although certain limitations apply).
3 The government provides risk-sharing by passive-loss provisions which allow investors in rental projects to use losses from these projects to offset other income.

In particular, the combination of special depreciation allowances and tax-free capital gains exerts a strong incentive to invest in private rented housing. On the other hand, this pattern of subsidisation implies a certain short-termism of investment which tends to destabilise housing markets. The government is therefore looking for more long-term oriented subsidisation schemes.

Housing construction activity in Germany is mainly financed by special mortgage banks (*Hypothekenbanken*), communal banks (*Sparkassen*) and commercial banks. Institutions of collective housing finance (loan and building associations, *Bausparkassen*) and insurance companies hold a minor market share. The structure of finance is different for rented housing (mainly bank loans, insurance company loans) and owner-occupation (mainly building loan contracts). Moreover, investment in existing housing is to a large extent financed by the owners' own funds.

The mortgage market

The main feature of the German mortgage market is that the special housing banks follow a restrictive line in their lending activity. This results from a regulation concerning their refunding. The main method of refunding is for mortgage banks (but also communal banks) to issue securities (*Pfandbriefe*). To secure the interests of investors, restrictive legal commitments have been established, in particular restrictions on the loan-to-value ratio. As a consequence, mortgage interest rates are relatively low in Germany.

Problems of home-ownership finance

German banks practise prudential lending to owner-occupiers. Securitisation of mortgages is offered up to 60 per cent of the 'lending value', that is, less than 50 per cent of the market value of a building. About 20 per cent of value is usually refinanced by intermediation, leaving interest-rate risks with the borrower. Alternatively, German building and loan associations provide these 'second' mortgages, exposing the borrower to a higher front-load burden. As a rule, 25 to 30 per cent of value is required as a down-payment. This lending practice is enforced by applying 'burden' criteria which define the minimum consumption level as a percentage of current income. The lower current income is, the higher the required down-payment has to be. Consequently, the risk of default because of house price volatility is low in German housing finance. Bankers claim that there is no need to introduce mortgage insurance. On the other hand, a worker's household with median income is not eligible for home ownership, because they normally cannot make the high down-payments.

Housing-specific savings

Due to the banks' prudential lending practice, housing-specific savings have been a major instrument of housing finance for owner-occupiers. They gained increasing importance after the Second World War, stimulated by housing saving subsidy schemes. Subsidies for housing-specific savings are paid as grants for low-income households. Alternatively, tax deductibility may be claimed without income restriction. Grants for housing-specific savings were very important until the end of the 1970s. After that, the grant was reduced step-wise to end up as 10 per cent of the annual savings

Table 4.4 Public subsidies to the German housing sector

| | DM billion[a] | | | |
	1980	1985	1989	1994
Social housing[b]	5.3	5.1	5.4	5.3
Tax expenditure for owner-occupation	5.5	7.2	6.1	10.1
Rehabilitation[c]	1.1	0.8	0.6	1.1
Special programmes for East Germany				5.1
Total public subsidies to housing investment	12.1	13.5	12.5	21.7
Subsidies on housing contract saving	2.9	2.0	1.3	0.8
Housing allowances	1.8	2.5	3.7	6.5
Urban development	2.3	1.5	2.5	2.7
Total	19.1	19.4	19.9	31.7

Source: Expert Commission on Housing Policy
Notes: [a] 1980 to 1989: West Germany; 1994: East Germany included
[b] Federal and states expenditure for existing stock of social housing
[c] Including subsidisation of energy-saving investment

amount. Table 4.4 gives evidence of the volume of grants and tax expenditure for savings.

Contract savings and loans systems

A special branch of housing-specific savings is the contract savings and loans system (*Bausparen*) which is widely used by first-time home buyers and has for years enjoyed preferential treatment by the government. Principally, *Bauspar* banks offer loans on the basis of specific housing savings contracts in which borrowers have to make regular deposits in advance until a certain savings target is achieved. There is no connection to the capital market. Hence, the system works without any interest-rate risk for the institution. A mismatch between supply and demand of funds within the system is responded to by queuing. Hence, the waiting period for an individual borrower is uncertain in advance, depending not only on his or her personal savings schedule but also on the dynamics of the overall system. Interest rates for deposits as well as for loans are fixed and below market rates. The spread is usually two percentage points.

Depending on the volatility of market interest rates, the contract savings and loans system may have substantial redistributive effects. Mainly for this reason, and to keep waiting periods under control, repayment periods are kept short, on average ten to twelve years. Consequently, borrowers are charged with high principal payments and, hence, are confronted with a special 'front-load problem'. In addition, the cost of home-ownership is partly shifted back to the savings period. Even in a period of moderate inflation, long-term savings contracts yield 7–8 per cent interest compared to 2.5–3 per cent for housing contract savings. That certainly balances low mortgage interest rates. The real advantage seems to be, therefore, that *Bausparkassen* promise to provide a loan (at an uncertain date in the future) without any specific underwriting procedure and at fixed terms (when the loan is paid, the borrower usually has to contract a life insurance; hence, an essential part of the default risk is covered).

Public assistance to home-ownership and affordability

Home ownership in Germany has been traditionally promoted by tax concessions to owner-occupiers and grants to specialised savings institutions. Since 1957 new construction for owner-occupiers has benefited from supply subsidies for medium- and low-income groups as part of social housing programmes. Since the early 1980s the emphasis of these programmes has shifted more and more from the rented sector to owner-occupation.

Basically, social housing supply subsidies to owner-occupiers are invariable with respect to income. It is true that most states apply a two-tier scheme for low- and medium-income households, respectively. The difference is more than balanced by the system of tax releases, however.

Since 1987 the German Income Tax Law (which was modified in 1992) has treated home-ownership as consumption. Basically, home-ownership cost has to be financed out of taxed income. This is balanced by the fact that imputed income of home-ownership is tax-exempt. Apart from that, the tax system provides subsidisation in the form of tax benefits.

Until 1995, this scheme had clearly regressive distributional effects since the subsidy value of home-ownership grants to low-income households was rather low compared to the tax benefits accruing to the rich. To avoid discrimination against low-income households, the government replaced the scheme of tax benefits for home-ownership by a flat-rate scheme:

1 First-time home buyers are entitled to flat-rate tax credits (with ceilings) during the first eight years. Very generous income limits apply.
2 Families with children are entitled to additional tax credits. If a negative income tax results, the amount is not paid but may be shifted to other years.

The new scheme was enacted in 1996. Still, low-income households are disadvantaged by high down-payment requirements. It is only in East German states (*Länder*) that the federal government introduced public guarantees for private mortgages to overcome this threshold.

Housing allowances and other targeted assistance

There are two distinct reasons for problem groups emerging in the housing market. First, low-income households cannot afford housing without public assistance. Second, several social groups are discriminated against by landlords. Because of specific features which qualify them as being 'bad' tenants (for example, one-parent families, families with many children, ethnic minorities, drug addicts), they are excluded from market access even if they are willing and able to pay.

In Germany, the size of these problem groups has increased during the last decade. In 1992, approximately 1 per cent of the population were homeless people, which is four times more than ten years earlier. This coincides with the observation that the share of housing cost in disposable income increased significantly for low-income households during the 1980s. In 1988, the tenth of households with lowest incomes would have had to spend about 40 per cent of disposable income for housing, 10 percentage points more than ten years earlier, if there had been no public housing allowances.

According to specific 'poverty criteria', one-tenth of households in West Germany and one-sixth in East Germany are not sufficiently provided with housing facilities. These criteria tend to overstate the problem, however, since they take as a norm that a household should be provided with one room per person. The required standard for a family of five, for example, is a five-room dwelling. Consequently, families with several children are relatively more frequently than other households indicated as poor.

Since 1965 the federal government has increasingly used income-related housing allowances to limit the burden of housing cost for low-income households. Housing allowances are a means-tested instrument of social housing policy, depending on the size of the household, disposable income and rent. The scheme is designed to reduce the share of housing cost to approximately 20–25 per cent of disposable income. Housing allowances can be exactly targeted to low-income groups. They have as a disadvantage, however, that their subsidy value decreases over time if they are not adjusted to inflation. At present, approximately 6 per cent of private households receive housing allowances.

On the other hand, social housing programmes are still conducted to assist low-income groups, although on a reduced scale (Table 4.4). In contrast to housing allowances, the distributional effect of social housing is hard to control. As social rents are regulated, the difference from market rents increases over time, and so does the value of the subsidy. Therefore, long-standing tenants benefit the most. With rising income, they might be expected to want housing of higher quality and so to move, but often they do not, simply because the low social rent compensates them for remaining in low-standard accommodation. A large share of the social housing stock (approximately one-third) is let, therefore, to tenants whose incomes exceed the required limits. Policy towards the housing stock has attempted since the early 1980s to correct these unintended effects by charging a special fee to those tenants of social housing whose incomes have risen above the limit (*Fehlbelegungsabgabe*). Only since 1994 have social housing programmes been changed accordingly, to a scheme which provides a two-tier subsidisation consisting of a basic investment grant to the investor and income-related grants to low-income groups, similar to housing allowances.

As far as discriminated groups are concerned, local governments have practised several forms of incentives to induce landlords to provide adequate housing for these groups: grants, rent guarantees, guarantees to take over extra costs, short-term rent contracts, leased housing and, if necessary, therapeutic treatment of tenants. The economic rationale is that private landlords should receive compensation if they delegate the right to select tenants to local authorities. However, such an agreement would induce external effects into multi-family housing, as far as the sitting tenants are concerned. The landlords have preferred schemes, therefore, which maintain their right of letting but commit them to select and choose a tenant who meets the discrimination criteria. Such schemes are capable of replacing social housing construction programmes at reduced fiscal cost. Local governments have been reluctant, however, to shift to this kind of social housing policy on a wide scale. They call for social housing construction programmes to be continued, mainly because they would have to bear the cost of the new policy of accommodating discriminated groups in existing housing, whereas social housing construction programmes are funded predominantly by the federal and state governments.

DEVELOPMENTS IN HOUSING POLICY

During the last decade revitalisation of the East German housing market has been the most important challenge for German housing policy. The government chose a slow pace of transition towards a housing market and provided very generous subsidisation to local housing companies (recapitalisation), tenants (specific housing allowances), and investors (specific tax benefits for new construction and rehabilitation). Five years after German unification it seems that it will take nearly a decade to fully adjust the East German housing market and integrate it into the legal housing policy framework (*Wohnungspolitik für die neuen Länder*, 1995).

Apart from that, subsidy programmes clearly indicate a shift of support from rented housing to owner-occupation (Table 4.4). Several policy measures give evidence of this (*Wohnungspolitik auf dem Prüfstand*, 1995). First, although social housing construction programmes were reduced during the 1980s the federal government continued to support owner-occupation within these programmes. Consequently, the share of owner-occupation in social housing programmes increased. Second, tax expenditure for owner-occupation has been rising significantly since 1980. That is predominantly the consequence of higher subsidy values. Only recently the volume of home-ownership investment was rising again. Third, if it is true that subsidies on housing contract saving were reduced until recently, that was more than balanced by a general tax-exemption of interest earnings (up to DM 12,000 per person and year) which was enacted in 1994. In addition, the federal government plans to extend the narrow income limits for subsidies on housing contract saving schemes.

On the other hand, the reduction in social housing construction programmes was balanced by an expansion of housing allowances for low-income groups. That indicates a shift of emphasis from supply-side subsidies to demand-side subsidies which has two major implications. First, social housing policy is shifted towards more narrowly targeted measures and, consequently, towards more efficiency. Second, the social housing policy approach is broadened since it relies not only on social housing but on the total supply of existing housing stock. As fluctuation rates are higher in the private-rented sector than in the social-housing sector, market access for low-income groups may be improved significantly.

Subsidisation of rehabilitation and urban development programmes had been substantially increased prior to 1980. These programmes have been consolidated since then. In particular, the federal government reduced its share. The state governments and local authorities are conducting rehabilitation programmes of their own, however. Subsidies for rehabilitation are still more important, therefore, than indicated in Table 4.4 which covers the joint federal and state subsidy programmes only. In recent years, rehabilitation and urban development have been the most important subsidy issue in East Germany (Hills, 1990).

As to future prospects, two major policy changes were implemented which will fundamentally change the functioning of housing markets in the long run. First, the non-profit housing sector was extinguished. As mentioned, non-profit housing associations lost their legal tax privileges in 1990 and were transformed into profit-oriented joint-stock companies. That will enlarge the private-rented sector by more than one-third in the long run. Housing associations had been the major suppliers of the social housing stock. They were obliged to hold to specific rent regulation schemes even after their commitments to the social housing programmes had expired. Since social housing construction programmes have been conducted on a reduced scale for years, housing supply at fixed social rents will be gradually reduced to a very small market share. The supply of low-rent housing will depend more and more on the proficiency of market agents. Consequently, the fiscal cost of social housing policy will become more transparent than in the past. One might also expect that the political will to deregulate rents will be strengthened.

Second, a significant shift was implemented in land-use planning policy by entitling local authorities, in particular in East Germany, to offer public –private partnerships to developers. Traditionally, strict statutory planning procedures have been applied in Germany which assign the competency for land-use planning exclusively to the local authorities. This has the consequence that the costs of developing residential areas (including follow-up cost of building schools, recreational areas and so on) are burdened on the local authority whereas the increased property values ('planning gains') accrue to landowners. Against this background, public–private partnerships will enable local authorities to allocate planning gains more easily to finance the required infrastructure in residential development. That will certainly have an impact on the elasticity of building-land supply and will improve market adjustments to increased housing demand.

Housing policy issues in East Germany

The East German housing sector was subject to government rationing for about forty years, a situation which ended in fundamental imbalances. When Germany was unified in 1990, the existing East German housing stock was in a state of poor quality. In addition, the ownership status of housing was unclear and regulated rents were extremely low, covering less than 20 per cent of cost on average. The most urgent issues were to clarify property rights and to provide finance for rehabilitation, maintenance and operation of the housing stock (Tomann, 1992).

The Unity Treaty enacted a transfer of the former state ownership of residential property (*Volkseigentum*) to local government. To provide an appropriate institutional setting for the management of residential property, the former local housing administration units (*KWVs*) and public construction firms (*VEB Gebäudewirtschaft*) were transformed into companies with limited liability, owned by the local governments. The co-operatives, on the other hand, received the ownership status for the buildings which they had constructed, but were obliged to buy the residential estate which they had built upon.

The ownership status of private property was rather undetermined in many cases as local land registers had been closed by the East German authorities and many private owners had abandoned their property, either deliberately or under coercion.

One of the basic principles of the Unity Treaty was to restitute private-ownership rights in eastern Germany. Furthermore, whenever the rights of former owners conflicted with the rights of present users, restitution should be given priority with compensation for the former owners. Since October 1990 about 2.5 million claims on private property by former owners have been registered. About one million of these claims refer to housing, that is, one-seventh of the existing housing stock is concerned. The rest are claims on land. To settle these claims, offices for unsettled property rights (*Ämter zur Regelung offener Vermögensfragen*) have been established by the East German states. After four years, by the end of 1994, about half of the registered claims had been settled. There is still a large number of open claims which, as long as they remain unsettled, impede the process of fresh housing investment.

Restitution of old property rights as well as new construction activity will increase the share of owner-occupation substantially during the next decade. At present, however, rented housing still has a large market share (approximately 70 per cent). The Unity Treaty prescribes that in the long run the East German rented housing sector should be regulated according to western standards. That requires a radical change in rent regulation and housing finance. However, for the period of transition, the government chose a gradual strategy following social policy reasons. The system of rent control has not been abolished but will be phased out gradually.

A first raise of the general rent level was enacted in October 1991, providing for a mark-up on the basic rent and allowing local housing companies to charge operating cost (for central heating and hot-water services, with caps) on tenants. On the other hand, the federal government managed to establish in eastern Germany the western system of housing allowances for low-income households by October 1991 – starting with a simplified procedure of application – in order to avoid a cost squeeze for that group of tenants. By 1993, a second round of rent increases became effective, including additional rent adjustments in 1994. In 1995, an envisaged shift to the West German system of rent regulation was postponed until the end of 1997 and replaced by another scheme of general rent increases with only minor deviations for quality and location. Within five years, the standard rent was increased from DM 1.20 per month and square metre in 1990 to DM 6.50 (basic rent) and DM 8.20 (rent including operating cost) respectively, by the end of 1995.

The lesson is that a more market-oriented system of rent regulation can only be implemented with great political difficulties. In this case, political resistance against enforcement of the western rent-regulation scheme seems not so much to reflect the interests of the tenants, who have readily adjusted themselves to high rent increases, but the interests of local housing companies, who have accustomed themselves to cost-oriented rent increases and do not trust the vagaries of a market-oriented rent regulation. Political rent regulation, as will continue to exist until 1998, may be opportune as long as local housing markets are dominated by large companies. To change this, the federal government has bound the recapitalisation programme for local housing companies to the condition that the companies have to privatise at least 15 per cent of the housing stock, preferably by selling dwellings to tenants.

To enhance new housing construction and rehabilitation, very generous tax allowances have been temporarily enacted. Investors may deduct 50 per cent of investment expenditure from taxable incomes. That has been particularly an opportunity for West German investors who, having high taxable incomes, are looking for loss provisions. Hence, an East German boom in housing investment was triggered by West German investors; this may overshoot and induce a cyclical downswing in housing construction activity during the years to come.

CONCLUSIONS

Over the past 10 to 15 years German housing policy has implemented a general shift of emphasis from supply-side subsidies to demand-side subsidies. In addition, the federal government took this change of policy as an opportunity to gradually retreat from intervention in housing markets. Although this tendency has been interrupted in recent years, when politicians came under pressure to respond to the sharp rise of housing shortages, it will continue since there are indications that a more liberalised housing market will emerge as an outcome of political debate. Moreover, the political will to retreat from direct intervention is dictated by budget constraints. There is also a growing perception that social housing policy objectives may still be pursued by more specific targeting. Those targets are to provide affordable housing for low-income groups, to improve market access for discriminated groups and to assist first-time home buyers whose funds are limited.

The shift of emphasis pursued by the federal government is not necessarily shared by the state governments, and local authorities whose objectives are still directed more towards 'providing housing for broad strata of people' (Second Housing Construction Act). The controversy between the federal government and the state governments about the reform of the Housing Construction Act which regulates social housing programmes and about the rent regulation system is evidence of this. Nonetheless, major reform steps have been enacted towards a more targeted supply subsidisation in social housing construction and further reforms are envisaged.

As to rent regulation, there is no clear direction in which policy is moving. During the last decade a sequence of minor policy changes have been enacted, bouncing to and fro, from a squeeze of rent increases to a release and vice versa, mainly responding to changes in the actual performance of housing markets. Apart from that there is a broad political consensus that the system of rent regulation with its centrepiece of moving ceilings (*Vergleichsmiete*) should be maintained. Accordingly, the tenant's right of protection against notice (*Kündigungsschutz*) is not seriously challenged by politicians.

Part II

THE PROMOTION OF
SOCIAL HOUSING

5

INTRODUCTION TO SOCIAL HOUSING

Paul Balchin

The social rented housing stock is proportionately larger in the Netherlands, Sweden and Austria than in the EU in aggregate. Although in France the sector is marginally smaller than the EU average, the social rented stock has experienced substantial growth in recent years as a result of proactive governmental and institutional policy. Although these countries differ considerably in terms of size and demography, and in their economic, social and political backgrounds, each country experienced serious housing shortages after the Second World War, and each to a significant extent has looked to the social-rented sector to satisfy its housing needs. The social-rented sector, moreover, has (except in Sweden) performed a dominant role in dampening rent levels in the private rental market, whereas in Sweden it has been a rent leader – in both respects helping to create a 'unitary rental system' (Kemeny, 1995). Although social-democratic (or socialist) governments have particularly favoured social renting and have been in office in the Netherlands, Sweden, Austria and France for varying periods since the Second World War, there has in general been a consensus of support for maintaining or expanding this sector.

TENURE

The Netherlands has the largest proportion of social rented housing in the EU, 36 per cent in 1994 compared to an EU average of 18 per cent (Table 5.1). In contrast, the private-rented sector is relatively small (the smallest in western Europe except for the United Kingdom), 17 per cent compared to an average of 21 per cent in the EU. The owner-occupied sector, 47 per cent of the stock, is smaller than the EC average, 56 per cent, but larger than that of Germany. Since the early post-war years, both the social-rented and owner-occupied sectors have expanded, while the private rented stock has diminished.

Social rented housing was an outcome of the social reform movement of the nineteenth century, but with the introduction of state assistance under the Housing Act of 1901 the sector developed steadily until after the Second

PAUL BALCHIN

Table 5.1   Housing tenure in the Netherlands, Sweden, Austria and France, 1994

	Social rented %	Private rented %	Owner-occupied %	Co-operative housing %	Other tenure %
the Netherlands	36	17	47	–	–
Sweden	22	16	38	19	–
Austria[a]	20	25	55	–	–
France	17	21	54	–	8
EU	18	21	56	–	5

Sources: CECODHAS (1995); Österreichisches Statistiches Zentralamt (1993)
Note: [a]1991

World War when there was a surge of output. Over 75 per cent of social housing in the Netherlands was built after 1945 and most remains in good condition. There are two principal landlords: first, there are housing associations – non-profit *Woningcorporaties* – over 260 of which (under the control of local authorities) own a total of 2.1 million dwellings; and second, there are local authority housing companies – 214 of which own about 255,000 dwellings. As a consequence of the *Heerma Memorandum* (Ministrie van VROM, 1989), it has become government policy to convert these companies to housing associations. There are also other non-profit organisations catering for special needs, but these too are converting into associations (McCrone and Stephens, 1995).

Private rented housing in the Netherlands is similarly owned by two major groups of landlords: private individuals and companies or institutions. The former tend to own older and cheaper property in town centres and, because of controlled rents and security of tenure, further investment is deterred whilst the size of the stock has diminished. The latter own the greater part of this sector, the dwellings are newer and their quality is superior and rents are higher (Ghékiere, 1992; Boelhouwer and van der Heijden, 1992).

Although the demand for owner-occupied housing is boosted by more generous tax allowances than in most other countries in Europe, the expansion of the sector being an explicit aim of government, the sector remains relatively small by European standards and may not have fully recovered from the 30 per cent downturn in house prices over 1978–82.

Sweden also has a social rented housing stock proportionately larger than the EU average, 22 per cent in 1994 compared to 18 per cent (Table 5.1). But both the country's private rented and its owner-occupied stock are proportionately small compared to EU averages, respectively 16 and 43 per cent in 1994 in comparison with 21 and 56 per cent. Unlike other countries in western Europe there is a significant co-operative sector constituting 19 per cent of the Swedish stock in 1994. Since 1975 the social rented, owner-occupied and co-operative sectors have expanded while, as elsewhere in Europe, the private-rented sector has been in decline.

The provision of social rented housing is undertaken mainly by municipal housing companies. Originating in 1935, they grew rapidly after the Second World War when the government (mostly Social Democrat administrations) thought that non-profit-making companies were the best means of managing rented housing in receipt of state subsidy. In contrast to most other west European countries, to avoid social segregation, access to social housing was not subject to means-testing. Municipal housing companies in Sweden broadly fulfil the same role as housing associations in many other European countries or local authorities in the United Kingdom, and like the *habitations à loyer modéré* (HLMs) in France they are set up by the local authorities (McCrone and Stephens, 1995). The municipal housing companies have stocks ranging in size from less than 100 to 50,000 dwellings, the 14 largest owning over 21,000 units (Lundqvist, 1988; Lindecrona, 1991). Through the medium of these companies, local authorities in Sweden are able to assume a much greater degree of responsibility for the provision of social rented housing than in any other country of western Europe, with the possible exception of the United Kingdom.

Largely as a result of rent control introduced in 1942, the supply of private rented housing (confined mainly to old multi-apartment housing in urban areas) decreased substantially throughout the period to 1968, when it was replaced by a managed rental regime similar to that employed in the social rented sector. The private rental stock, however, continued to decline as more and more dwellings transferred to co-operatives formed by the tenants themselves. Thus, in an attempt to halt the continuing decline in private rented housing, there was a reversion to market rents in the early 1990s, a development which might relegate social renting from that of rent-leading to rent-influencing as in Switzerland or Germany (Kemeny, 1995).

As is common throughout western Europe, the expansion of the owner-occupied sector in recent years, albeit to a fairly modest level, has been attributable to subsidies. House purchase has been assisted by subsidised rates of mortgage interest and mortgage-interest tax relief. The deregulation and liberalisation of credit also helped to stimulate demand in the late 1980s.

Owner-occupation (and to a lesser extent renting from a social or private landlord) is, however, often less attractive than obtaining housing from a co-operative. Providing mainly multi-apartment dwellings, housing co-operatives in Sweden originated in the nineteenth century, and have since consisted of either tenant co-operatives or tenant/owner co-operatives, of which the latter are by far the more important (there is now only one notable tenant co-operative – the Stockholm Tenant Housing Co-operative (SKB)). Under the Housing Act of 1972, members of tenant/owner co-operatives are required to pay an initial fee to secure a share in the co-operative and are obliged to meet the costs of services, repairs and maintenance, but when they move out they can sell the right of occupation at a market price.

The pattern of tenure in Austria clearly demonstrates that both the social and private rented sectors are larger than the EU average, while the owner-occupied sector is marginally smaller. It might be noted, however, that whereas the social rented stock is about 2 percentage points higher than the average for the EU (20 per cent of the total stock compared to 18 per cent in the EU), the private rented stock is not only larger than the social stock but 4 percentage points higher than the EU average (25 per cent compared to 21 per cent). It could be asked, therefore, why is Austria included among the countries where social rented housing is the prime rented sector? The answer is that it is included because proactive policy in recent years has been directed at maintaining or expanding this sector, whilst the future of the older and often run-down private rented stock has in part been determined by rent regulation (originating in 1917) or by the market. More importantly, the social (cost-rental) stock has exercised a dominant role in dampening rents in the private sector, thereby furthering the development of a unitary rental system – especially in Vienna (Matznetter, 1992; Kemeny, 1995).

The social-rented sector in Austria has developed continuously since 1945, particularly in urban areas – social-rented housing constituting 39 per cent of Vienna's housing stock in 1991. There is generally an emphasis on non-profit housing associations but a few cities (particularly Vienna) have maintained municipal housing programmes – Vienna being one of the world's largest landlords, owning nearly a quarter of a million dwellings. Although some non-profit associations derive from workers' housing co-operatives of the late nineteenth and early twentieth centuries, currently most are joint-stock companies owned by cities and trade unions. The private rented sector is likewise concentrated in Vienna, comprising 40 per cent of its stock, although much of it dates from before 1918, and much of it is in bad condition and lacking in basic amenities. The owner-occupied sector in Austria, while being proportionately smaller than the EU average, nevertheless contains over half (55 per cent) of the nation's housing stock, largely because of the rural character of much of Austria, but also because housebuilding in this sector is subsidised.

In contrast with the countries considered above, the sectoral distribution of housing in France is broadly on a par with that of the EU as a whole (Table 5.1). However, although 21, 17 and 54 per cent of the French housing stock in 1994 were respectively social-rented, private rented and owner-occupied dwellings, compared to EU averages of 21, 18 and 56 per cent for these sectors, as in the Netherlands and Sweden there has been a marked increase in the proportion of both social-rented and owner-occupied housing and a decrease in private rented housing in recent years.

The number of dwellings in the social-rented sector in France increased from 784,000 to 3.5 million over 1961–88, or from 5.4 to 17.1 per cent of the total stock (Emms, 1990). The largest landlords are the HLMs, non-profit

housing associations owning about 90 per cent of the stock of social housing (Ghékiere, 1991). HLMs are either initiated by local authorities or
formed independently as non-profit companies, the former comprising
about 57 per cent of HLM housing and the latter 43 per cent. The other
social landlords include the *sociétés d'économie mixte* (SEMs) (public–
private partnerships), and the *Société Civile Immobilière de la Caisse des
Dépôts* (SCIC), one of the largest social landlords in Europe, and with
180,000 dwellings equivalent in size to Glasgow District Council (McCrone
and Stephens, 1995).

As in the Netherlands and Sweden, the supply of private rented housing
in France has decreased substantially in recent years, from 33 per cent in
1961 to 21 per cent in 1994 – in large part due to rent control and more
attractive alternative forms of investment. If this trend continues, the sector
will become very marginal (as in the United Kingdom), and a unitary rental
system will be replaced by a dualist system.

In common with other countries in western Europe, owner-occupation in
France has expanded rapidly in recent years – from 39 per cent in 1961 to 54
per cent in 1994. Demand has been substantially assisted through the
provision of a range of interest subsidies on mortgage loans, whereas, in
contrast particularly to the Netherlands, mortgage-interest tax relief and
tax-exemptions are not notably generous.

HOUSEBUILDING

After the Second World War, a very large volume of housebuilding was
undertaken in the Netherlands, Austria, Sweden and France to eliminate
serious housing shortages resulting from the Second World War or even
emanating from the 1920s and 1930s.

The Netherlands has the newest housing stock in Europe, with 75 per
cent of its dwellings having been built since 1945. Output peaked in 1972/
73 with a total of more than 150,000 completions, but subsequently fell to
about 80,000–90,000 per annum in the early 1990s (Ymkers and Kroes,
1988; Emms, 1990; VROM, 1993). Although there has been an emphasis
on housebuilding in the owner-occupied sector in recent years (amounting
to 46 per cent of total completions in 1991), 39 per cent of all completions
in the period 1960–90 were in the social-rented sector (McCrone and
Stephens, 1995). Housebuilding in the owner-occupied sector, moreover,
is particularly disadvantaged by cyclical slumps. From 1978 to 1982, for
example, the number of houses built for owner-occupation decreased from
76,000 to 58,000, but in 1982 total housing completions were higher than in
1978 because of a major programme of housebuilding in the social-rented
sector (McCrone and Stephens, 1995).

After the Netherlands, Sweden has the newest housing stock in Europe,
with 70 per cent of its dwellings built after the Second World War

(Lunqvist, 1988; Boelhouwer and van der Hejden, 1992). The number of completions increased to 110,000 in 1970 at the peak of the 'one-million-dwelling programme' of 1965–74, but thereafter the trend was downwards, to 20,000 in 1992. The municipal housing companies accounted annually for over 40 per cent of completions in the 1950s and 1960s, but during the million-dwelling programme the proportion reached 68 per cent at its peak (Lundqvist, 1988).

In Austria, to meet severe housing shortages after the Second World War (particularly in Vienna), the provision of direct object subsidies effectively guaranteed a steady volume of housebuilding in the years thereafter. In the 1980s, for example, the majority of completed multi-family blocks of flats were constructed for subsidised non-profit housing associations, a total of 700,000 having been built since 1945. In addition, the housebuilding industry has increasingly been involved in housing rehabilitation schemes – notably in Vienna.

France suffered a serious housing shortage in 1945 amounting to a deficiency of supply of 2 million dwellings resulting from inadequate investment during 1919–39, together with a loss of 1,950,000 dwellings damaged or destroyed during the Second World War (Duclaud-Williams, 1978; Emms, 1990). From 1950 a major housebuilding programme got under way with completions rising from 200,000 in 1953 to a peak of 560,000 in 1979, but thereafter falling to 260,000 in 1993 as crude housing shortages were overcome (McCrone and Stephens, 1995). Housebuilding, however, was at first confined mainly to the private sectors, but by the late 1950s the number and proportion of completions in the social-rented sector accelerated with the development of HLMs.

HOUSING INVESTMENT, FINANCE AND SUBSIDIES

Over the years, the social-rented sector in the Netherlands has increasingly relied upon state subsidy to meet a proportion of the costs of new house-building and renovation. In 1989, and in respect of 50-year amortisation periods, the prevailing system of subsidy based on the need to bridge the gap between 'dynamic cost rent' (which took account of inflation and fluctuating rates of interest) and actual rents (determined annually by government) was terminated because of the problems of forecasting future rents and inflation, and the risk of escalating deficits. As a consequence of the White Paper, *Nota Volkshuisvesting in de jaren Negentig* (Ministrie van VROM, 1989), referred to as the *Heerma Memorandum*, 1989, a fixed rate of annual subsidy was henceforth paid to local authorities which necessitated rents being adjusted if costs increased (Emms, 1990; McCrone and Stephens, 1995). In effect, whereas before the subsidy was open-ended, now rents are open-ended, indicating a shift towards the market. There has also been an attempt to target object subsidies at housing for 'Special Attention

74

Groups' – households with comparatively low incomes, 75 per cent of tenants falling within these categories (McCrone and Stephens, 1995).

In the private-rented sector, although the system of subsidies is broadly the same as in the social-rented sector, and has the same history, landlords (and particularly institutional or company landlords) normally finance private rented housing from retained profits or raised equity. Despite these financial facilities, rent control (originating from 1925, and updated by legislation in 1947 and 1950) was, however, a major cause of sectoral decline – control being particularly tight in respect of dwellings built with state subsidy. The Housing Rent Act of 1979 therefore empowered Parliament to set annually a permitted rate of rent increase, a measure that led to rents increasing by 67 per cent over 1980–91, twice the rate of inflation (McCrone and Stephens, 1995). Clearly, in the 1980s and 1990s, the dampening effect of rents in the social-rented sector must be examined in the light of a more flexible rent regime in that sector too. In addition to a freer system of rent determination, landlords benefit from a fairly generous tax regime. Although they incur either income tax or corporation tax on rent and subsidies, interest costs and depreciation are deductible from taxable income (in the latter case from whatever source), and they are exempt from capital gains tax (CGT).

The expansion of owner-occupation in the Netherlands is facilitated mainly by 30-year mortgage loans from specialist mortgage banks, insurance companies and commercial banks, but housebuilding (for both owner-occupation and private renting) is assisted by government grants made available through the local authorities. In the past they were demand-determined, but are now rationed according to budgetary criteria. As elsewhere, owner-occupiers are eligible for mortgage-interest tax relief at the mortgagor's marginal rate of tax, but in relation to the full mortgage – more generous relief than in any other EU country – and there is also exemption from CGT. These concessions, however, are partly offset by the liability to pay tax on imputed rent income (the Netherlands being the only country in the EU to have this form of taxation) and by value added tax (VAT) at 17.5 per cent on newly constructed dwellings (McCrone and Stephens, 1995).

Housing allowances were introduced in the Netherlands in 1970 as a means both of targeting assistance and of reducing government intervention in the housing market (van Weesup, 1986; Emms, 1990; Papa, 1992). From 1975/76 to 1990/91, the number of allowances increased dramatically, from 348,000 to 958,000, reflecting an increase in rents in both the social- and private-rented sectors. Unlike Germany, Sweden and France (but like the United Kingdom) allowances are not paid to owner-occupiers.

Housing policy in the Netherlands is clearly expensive. Of the total sum of public expenditure and tax transfers allocated to housing in 1990 (ƒ 16,760 million), 48 per cent was spent on object subsidies and loans, 32 per cent on mortgage-interest tax relief and 11 per cent on housing

PAUL BALCHIN

allowances. Total expenditure was equivalent to 3.2 per cent of the gross domestic product (GDP) or 9 per cent of the national budget (comparable in cost to the United Kingdom) (McCrone and Stephens, 1995).

In Sweden, and in contrast to the Netherlands, housing investment is stimulated in each of the housing sectors by a common system of subsidies. In the social-rented sector, 70 per cent of the required investment is facilitated by first mortgages from private financial institutions, and 30 per cent by second mortgages originally from the state, but since 1985 from SBAB (Statens Bostadsfinansleringsatiebolog) – a state agency. Until 1993, subsidised rates of interest started at 3.7–5.1 per cent and rose by 0.375 per cent per annum until market rates were reached. Because this was an open-ended commitment and expensive (market interest rates exceeded 12 per cent over 1980–92), since 1993 the state has subsidised 57 per cent of the market rate of interest – with the subsidy reducing annually by 4 per cent per annum to reach 25 per cent by the year 2000 (McCrone and Stephens, 1995). In the private-rented sector, a similar arrangement applied, although the second mortgage covered only 25 per cent of the required investment until 1993. Since that year, private landlords have been eligible for single 95 per cent mortgages – backed to the extent of 25 per cent by a guarantee from the Swedish National Housing Credit Guarantee Board (BKN), and with the same mortgage-interest subsidy provisions that apply to the social-rented sector. From the same sources, and until 1993, owner-occupiers likewise received 70 per cent first mortgages and 25 per cent mortgages, but at a subsidised interest rate of 4.9 per cent rising by 0.5 per cent per annum until the market rate was reached. After 1993, a subsidy of 42.67 per cent will fall by generally 5.5 per cent per annum until it is eliminated in 2000 (McCrone and Stephens, 1995). Co-operatives were able to secure 29 per cent second mortgages on top of their 70 per cent first mortgages, and like the rented sectors received an interest subsidy of 3.7–5.1 per cent rising by 0.375 per cent until market rates were reached. Since 1993, however, the second mortgage can only be obtained from a private financial institution (with 29 per cent of the loan guaranteed by BKN) (McCrone and Stephens, 1995). The subsidy provisions on this mortgage are the same as those applicable to the rented sectors. Clearly the post-1992 subsidy provisions in all sectors were designed to help curb further increases in public expenditure.

Rents in the Swedish social-rented sector are generally freely negotiated between tenants' and landlords' associations, with recourse to rent tribunals where there is a failure to agree. The municipal housing companies, nevertheless, found it difficult to cover costs, particularly in the 1970s and 1980s, and, to an extent, relied upon state grants to offset rent losses and subsidies to compensate them for empty and unlettable dwellings. But constraints on public expenditure in the early 1990s under a Conservative-led coalition resulted in rent increases of up to 50 per cent in 1991–93. In the private-rented sector, by contrast, before the early 1990s the market was highly

76

regulated. Under prevailing legislation, rents had to broadly accord to use-value, and where landlords sought to obtain an increase in rent, rent tribunals could set a fair rent based on the level of rents charged by municipal housing companies for equivalent housing (McCrone and Stephens, 1995). Through this process, the social-housing sector assumed the role of a market leader and thereby helped to maintain a unitary rental system. Since the early 1990s, however, the market-leading role of social housing has been diminished since private landlords were freed from having to base rents on social-sector equivalents. Although private landlords cannot claim mortgage-interest tax relief and are liable for CGT, they are eligible for allowances for depreciation and other costs.

Owner-occupiers in Sweden, as elsewhere, are eligible for mortgage-interest tax relief, but relief has been reduced from 50 per cent in 1982, to 40 per cent in 1990, to 30 per cent in 1991. If interest exceeds SKr 100,000 (£8,300), relief falls to 21 per cent (Lundqvist, 1988; Papa, 1992). VAT of 25 per cent on new housebuilding, together with capital gains tax of 25 per cent (on half the nominal gain) do little to stimulate demand in this sector, although the deregulation and liberalisation of credit in 1986 heralded a house-price boom in the late 1980s. Co-operative housing is generally subject to the same mortgage-interest relief eligibility and tax liability as the owner-occupied sector.

Introduced in the 1930s, housing allowances are available to all tenants in relation to income, family size and housing costs. In 1992, 30 per cent of allowances went to only 9 per cent of households (those with the lowest incomes), while 50 per cent went to the lowest 19 per cent (Petersson, 1993). Most recipients were elderly, with less than 50 per cent being families with young children. From 1980 to 1992, the total cost of allowances doubled to SKr 13,800 million (£1,200 million) (McCrone and Stephens, 1995), and because of the increase in rent and cuts in mortgage-interest tax relief, it can be expected that allowances will continue to soar. Clearly, as part of an overall package of curbing public expenditure on housing, the increase in allowances (taken together with cuts in loan subsidies) indicates the beginnings of a marked shift of emphasis from object to subject subsidies.

As in the Netherlands, housing policy in Sweden was expensive, public expenditure on this item amounting to SKr 56,000 million in 1991 – equivalent to 41 per cent of the GDP (a higher proportion than in any country in the EC). Of this sum, interest subsidies accounted for 52 per cent, tax relief for 25 per cent, and housing allowances for 12.9 per cent.

In Austria, both housing supply and housing demand are substantially facilitated by subsidies. Under the housing promotion laws, direct object and subject subsidies are available to ensure a steady flow of new dwellings; object subsidies (through the medium of subsidised loan interest) also assist housebuilding, and subject subsidies (notably tax relief on mortgage interest) facilitate demand. The former subsidy, however, is by far the more important form of assistance, accounting for three-quarters of the relevant federal budget. Of the total sum, 80 per cent is allocated in approximately equal proportions to the construction of rented flats, owner-occupied flats and owner-occupied single-family houses, while the remaining 20 per cent is spent on rehabilitation.

In total, over 60 per cent of all post-1945 housing in Austria has been financed with public subsidy – funds being derived from income tax and corporation tax (10 per cent of the revenue from each source), and from an 'earmarked' housing tax (*Wohnbauforderungsbeitrag*).

Rent regulation in Austria was first introduced in 1917. Currently, all landlords (both social and private) must freeze the income they extract from rents on new housing for a period of 10 years to allow the residue to meet the cost of maintenance. In respect of social rented housing, the Non-Profit Housing Act limits rent to the level of costs (including repayment and maintenance costs). With regard to private rented housing, all pre-1914 dwellings are subject to regulation under the Tenancy Act of 1917 (as subsequently amended); whereas rents for subsidised flats built after 1945 are regulated under special subsidy legislation, while privately financed housing built after 1945 without subsidy is not subject to rent regulation. As in other countries, rent regulation arguably has an adverse effect on investment and encourages landlords to sell off their properties to sitting tenants, or more likely to owner-occupiers.

Despite object subsidies and rent control (both intended to help households secure affordable housing), housing in Austria may still remain beyond the means of many people on low incomes. There is therefore the need for a housing allowance to be introduced to help ensure that households can obtain accommodation to suit their requirements.

Social-rented housing in France is funded by a plethora of subsidies (McCrone and Stephens, 1995). The Caisse des Dépôts et des Consignations (CDC), a public-sector institution, channels subsidised (low-interest) loans and grant aid to social landlord organisations. These include PLA loans (*prêt locatif aidé*) for HLM development – the most heavily subsidised loans; PLS loans (*prêt locatif social*) for higher-quality housing intended for higher-income tenants, and PALULOS grants (*prime a l'amélioration des logements à usage locatif et à occupation sociale*) for housing rehabilitation. CDC also draws on funds from the savings banks to supplement public expenditure, and also borrows from the market. Private rented housing is also funded by a wide range of subsidised loans. PLA loans are available from the state and via the Crédit Foncier de France (CFF), but the interest subsidy is less generous than that awarded to social landlords; PLS loans (as in the social-rented sector) are intended for higher-quality housing for higher-income tenants; PC (*prêt conventionnel*) loans are regulated rather than subsidised, and ANAH (Agence Nationale pour l'Amélioration de l'Habitat) grants are available for rehabilitation. Land-lords do, of course, obtain funds from banks and other financial institu-tions, normally in the form of fixed-interest long-term mortgages. Rents are, in general, determined in relation to the subsidised programme in operation, and are based largely on historic costs since rent-pooling is generally impracticable.

Private-sector tenants in France have for long been protected by rent control. Introduced in 1914 to minimise social unrest (Duclaud-Williams, 1978), it remained generally in force throughout the inter-war period but, under the Rent Act of 1948, rents were permitted to rise gradually to near market levels in respect of existing tenancies. By degrees, major categories of dwellings became entirely free from rent control. New tenancies, how-ever, were not subject to the 1948 Act and remained regulated. Under the Rent Act of 1982, rent levels and rent increases for newly let dwellings were brought within the jurisdiction of a national consultative body, and the 1986 and 1989 Acts enabled rents of newly let housing to be freely and mutually determined by landlords and tenants, with tenants being offered 3- or 6-year contracts (McCrone and Stephens, 1995). Private landlords were eligible for mortgage-interest tax relief (but only in relation to 10 per cent of interest paid over two years on small mortgages), but they were eligible for tax credits to assist with the cost of repairs and maintenance, and were only liable to capital gains tax on gains in excess of FFr 4.13 million (£500,000). Clearly private-sector rents were only loosely affected by social (historic cost) rents, and with the expansion of the social sector and con-tinuing contraction of the private sector there increasingly emerged a dual-ist rather than a unitary rental system, not unlike that of the United Kingdom.

Owner-occupation in France is also comprehensively subsidised (McCrone and Stephens, 1995). New housebuilding and improvements qualify for PAP loans (*prêts aidés pour l'accession à la propriété*) which are subsidised by the state and provided principally by the CFF. PAP loans are intended to enable lower-income households to buy and as such are means-tested.[1] The purchase of both new and older housing is more widely facilitated by regulated PC loans. Not means-tested and covering up to 90 per cent of the purchase price, they are repayable at a fixed rate of interest over 10–20 years. House purchase is also assisted by a subsidised saving scheme – PEL (*plans d'épargne-logement*) which is eligible for tax exemp-tion on interest received and attracts a state bonus, and renovation is subsidised by a means-tested improvement grant – PAH (*prime a l'amé-lioration de l'habitat*). Probably because of the magnitude of subsidies, owner-occupiers are eligible for only a very limited amount of mortgage-interest tax relief. At 25 per cent, relief is available for only five years and on incomes (in 1992) of no more than FFr 210,000 (£25,000) for a single person, or FFr 410,000 (£49,000) for a married couple. Whilst there is no tax on imputed rent income (as in most other EU countries), VAT is payable on new property at 18.6 per cent (1994) and owners are eligible for CGT where market values exceed FFr 4.13 million (£500,000).

Housing allowances in France are in two forms: APL allowances (*aide personalisée au logement*) and AL allowances (*allocation de logement*). Introduced in 1977, APL allowances are the more important and are available across all three sectors, provided households have been assisted by subsidised loan or grant schemes – an encouragement to investment since allowances will ensure that mortgage payments and rents will be met. AL allowances date from 1948 and are paid either to families with dependent children or dependent elderly adults, or (since 1971) are targeted at persons aged over 65, employed persons under 25, the handicapped and the long-term unemployed. The recipients of AL allowances are either tenants, or owner-occupiers who have not benefited from subsidised loans (McCrone and Stephens, 1995).

In France, as elsewhere in Europe, there has been a shift from object to subject subsidies. Whereas in 1985 object subsidies accounted for 26 per cent of state spending on housing, in 1993 their share had fallen to 12 per cent. Housing allowances, however, increased their share from 35 to 47 per cent over 1985–93. Public expenditure on housing, moreover, has decreased in real terms in recent years, amounting to 2.1 per cent of the GDP in 1985 but only 1.8 per cent in 1983, proportionately less than in the Netherlands, Sweden or the United Kingdom.

DEVELOPMENTS IN HOUSING POLICY

In the Netherlands, as a response to escalating costs and the perceived need to contain public expenditure, the Heerma Memorandum 1989 proposed that housing policy should be targeted at the needy, that there should be a greater reliance on the market and private capital, and that controls on rents should be relaxed; but in strong contrast to the United Kingdom, responsibility for housing policy should shift increasingly to the local authorities who already regulated the housing associations. In the case of housing associations, there was consequently an abolition of public loan funding, subsequently replaced by a complete reliance upon private institutions and the market. A privatisation programme has also been introduced whereby social-sector tenants are encouraged to buy their own homes, although the scale of sales at about 3,000–5,000 per annum was likely to be far less than the predicted number, 10,000 per annum by 1995 (Boelhouwer and van der Heijden, 1992). In the private-rented sector, and in response to the Heerma Memorandum, there has been a considerable liberalisation of rents. Rents on new dwellings are now at market levels, and rents for new lettings (in respect of dwellings already let at levels above that which would qualify tenants for allowances) are now market-determined.

In Sweden, in response to the disappointing economic performance of the 1980s, the coalition government embarked on major cuts in public expenditure and tax concessions in the period 1991–94, not least within the field of housing. There was a sharp cut in mortgage-interest tax relief for owner-occupiers, and a further shift of emphasis from object subsidies to targeted subject subsidies (notably housing allowances). With the return of the Social Democrat Party to government in 1994, it is highly probable, however, that the state will not abandon a policy designed to be as tenure-neutral as possible, will ensure that housing costs are approximately the same irrespective of tenure, and will endeavour to avoid policies that would result in social segregation (McCrone and Stephens, 1995).

In Austria, policy initiatives have often been concentrated in Vienna. Currently, the Vienna Land Procurement and Urban Renewal Fund (WBSF) (established in 1984) performs an important role in housing development. Since it has a near-monopoly in land purchase, it helps to keep the price of land under control and affordable for developers of low-cost housing. It also undertakes the whole process of planning from zoning to the disposal of sites to housebuilding promoters, and it co-ordinates and supervises municipally assisted housing improvement schemes.

As elsewhere in Europe since the mid-1970s, policies in France have aimed at constraining costs and targeting subsidies more effectively. The Barre Report, 1975 recommended a shift from object to subject subsidies, a recommendation implemented in legislation in 1977. Subsequently, and largely as a result of this shift, the rate of housebuilding has diminished and shortages have emerged, notably in the social-rented sector. By the early 1990s, 200,000 people were homeless and 600,000 were in temporary housing or mobile homes (Geindre, 1993). It could be argued that the introduction of right-to-buy (RTB) measures by the Juppé government in 1995 might have exacerbated the problem of homelessness – as in the United Kingdom – but it is probable that RTB in France is only having a marginal effect on homelessness since tenants of flats or houses are not eligible for discounts and they are required to have been in occupation for at least 10 years, whilst the property must be at least 10 years old (Bull, 1996). The private-rented sector, in contrast, faced rapid decline. The Lebègue Commission and the Geindre Report therefore both recommended that private landlords should qualify for more generous tax allowances (on a par with owner-occupiers), and that depreciation allowances should be deducted from income from whatever source (as in Germany) and not solely from letting (McCrone and Stephens, 1995).

CONCLUSIONS

In the EU, housing in the Netherlands is comparable only with Sweden in terms of the magnitude of subsidisation, although unlike Sweden there is little attempt to secure tenure-neutrality or to prevent a serious decrease in the size of the private-rented sector. For reasons of macroeconomic prudence, however, there is a perceived need to substantially cut public expenditure on housing. The Heerma Commission thus proposed that housing-association debt and subsidy arrangements should be cancelled (leaving the social sector with a debt-free base), and that object subsidies to owner-occupiers be abolished. Subsidies henceforth would be mainly confined to housing allowances and mortgage-interest tax relief (McCrone and Stephens, 1995). If these measures were introduced (together with others already implemented), Dutch housing policy would be among the least costly in the EU.

As in the Netherlands, there is a recognition in Sweden that public expenditure on housing might be excessive in the economic climate of the 1990s. Curbs on housing expenditure and rising rents, however, are being implemented within the context of tenure-neutrality, but this necessitates a marked increase in housing allowances to ensure that the least well-off are not disadvantaged.

The percentage of social-rented housing in Austria is still being increased, partly in response to mass immigration resulting from the opening of the country's eastern borders following the collapse of communism. But whether this growth can be sustained, or indeed whether the sector can be maintained at its present level, will depend a lot on whether or not political trends favour the weakening of the social-rented sector. Constraints on the federal budget might also result in the reduction in the size of this sector. Cuts in public expenditure might increasingly be accompanied by a shift of emphasis from object subsidies to individual housing allowances for the needy. The construction of new or rehabilitated social housing would thus be reduced, and the availability of higher allowances could prompt an increase in rents towards market levels.

In France there is a very clear need to increase the rate of housebuilding to satisfy future demand. There is a legacy of a substantial amount of problematic high-rise system-built housing, particularly in the social-rented sector, while the number of private-rented dwellings is steadily diminishing. Although the owner-occupied sector is expanding, it is still significantly smaller than in the United Kingdom. There is thus a need to: provide more resources for the social-rented sector so that it can undertake a major housebuilding programme and improve its existing stock; arrest the decline of the private-rented sector by introducing more attractive fiscal arrangements comparable to those enjoyed by owner-occupiers; and ensure that assistance to lower-income owner-occupiers is at least maintained (McCrone and Stephens, 1995).

Clearly in each country considered above there are very strong macroeconomic pressures (and sometimes political pressures) to curb public spending and to revert to market pricing across all housing sectors. There have also been major shifts of emphasis from subsidising supply to subsidising demand, and from systems of subsidy across all tenures to the provision of a welfare safety-net of housing allowances to low-income households in need. As such the corporatist or social-democratic character of several countries in Europe is under threat, with embryonic liberal welfare economies awaiting development.

NOTE

1 From 1 October 1995, PAP loans (for new housebuilding) were replaced by 0 per cent loans – the loans being determined by the composition of the household, its income and the area in which it lives or hopes to live. In Paris, for example, maximum eligible incomes were set at FFr 17,500 (net) per month for a single person or FFr 27,500 (net) for a couple, while the maximum sizes of loan were respectively FFr 100,000 and FFr 180,000. Elsewhere, eligible incomes and the maximum size of loan were notably lower.

6

THE NETHERLANDS

Peter Boelhouwer, Harry van der Heijden and Hugo Priemus

Dutch housing policy since the Second World War is in many respects comparable with that of many other west European countries. the Netherlands too had suffered a great housing shortage after the war and the government was thus obliged to intervene in the housing market via regulation and subsidisation. In contrast to many other countries, however, the housing shortage proved to be very persistent, and in fact a stable supply-and-demand situation was never reached.

Because the housing shortage was gradually eliminated, the countries in the vicinity of the Netherlands, such as Belgium, the Federal Republic of Germany, France and the United Kingdom, could allow themselves over the decades to act in a less regulatory fashion. However, despite a high level of housebuilding during those decades this situation was not attained in the Netherlands. The principal reason for this was initially the high birth-rate in the Netherlands, combined with postponed household formation. The birth-rate did not start to fall sharply until after 1972, whereas in most of the other west European countries this occurred in the 1950s and 1960s. As a result, the autonomous demand for dwellings in the Netherlands in the period 1970–87 was still very high; the number of households grew by nearly 50 per cent. In this respect the Netherlands is followed at a considerable distance by France (29 per cent), the United Kingdom and the Federal Republic (both 22 per cent) and Belgium with only 12 per cent.

To reduce the housing shortage, the production of new dwellings remained the highest priority in Dutch housing policy. During the 1950s and 1960s the housing shortage was even regarded by many as the principal cause of social concern. Thanks in part to the extensive government aid of the past decades the social-rented sector in the Netherlands was able to grow to over 40 per cent of the housing stock, unique by European standards.

Somewhat later than in other countries, more attention was given to housing condition in the Netherlands at the beginning of the 1970s as urban renewal policy was shaped and poor-quality dwellings were demolished or improved *en masse*. In the 1980s in particular urban renewal really got into its stride.

Partly because of economic setbacks, the nature of the housing shortage and the changing views of the government's role, the distribution problems in housing are receiving more attention in the 1990s. On account of shrinking government budgets, the efficiency and the effectiveness of the various housing instruments are being subjected to critical examination. Via the reduction of general building subsidies and the stress that is laid on individually linked subsidies such as the rent subsidy, the position of weak groups on the housing market is particularly receiving considerable attention in the conduct of policy.

When at the end of the 1980s a state of equilibrium in the housing market seemed within reach in the Netherlands, immigration increased again rather unexpectedly. The provisional peak was reached in 1993 with a positive migration balance of 100,000 persons, including 35,000 asylum seekers. The flow of migrants and asylum seekers led to a growing housing shortage and calls for more government intervention. As is explained in this chapter, the Netherlands government is still setting course for a market-oriented approach. The decentralisation, deregulation and privatisation of the social-rented sector are thus still being vigorously continued in 1995.

Before we review the background to new policy aimed at liberalising the role of central government in housing, the principal tenure sectors of the housing market are presented. To be able to understand the present Dutch housing policy properly, it is further of importance that the housing policy followed in the preceding decades be taken into account. Housing policy in the period 1945–90 is therefore discussed prior to a discussion of policy in the 1990s.

TENURE

As in many other countries in western Europe, three types of tenure can be distinguished in the Dutch housing market (see Table 6.1): the private-rented sector (13 per cent), the non-profit-rented sector (40 per cent) and owner-occupation, which accounted for 46 per cent of the total housing stock in 1993. The non-profit-rented sector and the private-rented sector can be further divided according to ownership. The private-rented sector is composed of individual landlords and companies. The non-profit-rented sector consists of housing corporations and local-authority housing departments.[1]

Table 6.2 presents a summary of the housing characteristics of each tenure group. The size of dwellings is given in terms of the number of rooms, not including the kitchen. Dwellings with four rooms (38 per cent of the total housing stock) account for the largest proportion of the housing stock. Large dwellings are relatively more common in the owner-occupied sector. Dwellings with six rooms or more are significantly more frequent in this sector than in the rented sector. The proportion of homes with four

Table 6.1   Tenure, the Netherlands, 1947–93

Year	Owner-occupied	Non-profit-rented sector	Private-rented sector Persons	Private-rented sector Institutions	Other/ Unknown	Total (000)
1947	28	12	54	6	–	2,117
1956	29	24	41	6	–	2,547
1967	32	35	24	9	–	3,450
1971	35	37	20	8	–	3,729
1975	39	41	13	7	–	4,281
1981	42	39	10	7	2	4,957
1985	43	41	8	6	2	5,384
1989	45	41	7	6	1	5,802
1993	46	40	8	5	1	6,304

Sources: Van der Schaar (1979); Housing Demand Surveys (WBOs), processed by OTB (Research Institute for Policy Sciences and Technology, Delft University of Technology)

Table 6.2   Housing characteristics by sector, the Netherlands, 1993 (percentages)

	Total	Owner-occupied	Rented sector Total	Rented sector Non-profit[a]	Rented sector Private[b]
No. of rooms					
⩽3	30	12	46	44	50
4	38	38	38	41	29
5	22	33	14	13	15
6	6	11	2	2	4
⩾7	3	6	1	0	2
Type of building					
Single-family house	67	91	47	49	40
Flats etc.	33	9	53	51	60
Amenities					
Bath/shower	98	99	98	98	96
Central heating	83	87	79	82	68
Cavity-wall insulation	52	53	51	56	33
Double-glazing/double windows in the living room	75	79	72	79	48

Source: Housing Demand Survey 1993/1994, processed by OTB
Notes: [a] Housing corporations and local authorities
[b] Individual landlords and companies

rooms (41 per cent) in the non-profit-rented sector is relatively high. In the private-rented sector there are relatively many small dwellings.

Almost 67 per cent of all occupied housing is single-family dwellings. In the rented sector the proportion of single-family dwellings does not differ greatly between non-profit rented housing and private-rented housing. In the former, single-family dwellings account for 49 per cent of the housing stock, and in the case of the private-rented sector the figure is 40 per cent.

Single-family dwellings are especially common in the owner-occupied sector; around 91 per cent of homes in this sector are of this form.

In the rented sector it is private-rented homes which have fewest amenities. Within the private-rented sector a distinction can be made between two types of housing: the mostly pre-war, fragmented possessions of individual landlords, and the more estate-type managed (mostly post-war) housing stock of companies, including institutional investors (Adriaansens and Priemus, 1986). The low level of amenities provided in the private-rented sector results principally from the few amenities provided in the housing rented out by individual landlords rather than by companies. Of those homes rented out by companies 86 per cent have central heating, while the corresponding percentage for rented accommodation owned by individual landlords is 57 per cent (these figures are not given in Table 6.2). The same pattern is evident in the case of insulation. Of those properties owned by companies, 53 per cent had (cavity) wall insulation and 70 per cent were either double-glazed or had double windows in the living room; the corresponding percentages for properties owned by individual landlords were 21 per cent and 35 per cent respectively. Dwellings in the owner-occupied sector have the highest level of amenities.

The non-profit-rented sector

The proportion of the total housing stock accounted for by the non-profit-rented sector rose from 12 per cent in 1947 to 40 per cent in 1993. Priemus (1995) defines social-rented dwellings in the Netherlands as dwellings owned by non-profit landlords, who manage their property within a public framework aimed at a moderate rent, an adequate quality and a focus on tenants with a below-modal income.

In the Netherlands the housing corporations (akin to housing associations in the United Kingdom) determine the identity of the social-rented sector. They are private non-profit organisations that are active solely in the interests of housing. Since the Housing Act of 1901 they have occupied a position of priority with regard to social housing. They qualify for financial aid from the state. In 1994 housing corporations owned about 34.5 per cent of the total housing stock. The first housing corporations were founded in the second half of the nineteenth century to offer housing alternatives for the low-paid. Under the Housing Act of 1901 (still in force) housing corporations may be recognised as so-called approved institutions. Many housing corporations were set up in the 1920s.

Until recently many aspects of the activities of housing corporations were regulated by the central government: housing corporations are therefore sometimes referred to as 'state private concerns'. Housing corporations are monitored primarily by the local authority in which they operate. In addition, a form of supervision is exercised by the central government.

Above all, since the 1950s housing production by housing corporations has been high, as a result of which the housing stock of housing corporations is now of relatively recent date. Housing-corporation property is mostly well maintained and the general quality of the buildings and their amenities is relatively high. The tenants come from a broad spectrum of income groups, and in terms of other characteristics too they represent a broad cross-section of society. On average their incomes are higher than those of tenants in private-rented-sector housing built before the war.

In addition to the housing corporations the Netherlands has a number of municipal housing companies that together manage 3 per cent of the housing stock. Since 1965 housing corporations have by law been alloted a more important role in the provision of non-profit rented housing than local-authority housing organisations. As a result the housing stock of local authorities is on average older and cheaper than that of housing corporations. The municipal housing companies are now undergoing a large-scale transformation into housing corporations via a process of privatisation. Between 1986 and 1992 the number of municipal housing companies decreased from 283 to 195. Before 1 January 1997 they must be transformed into housing corporations if they want to receive financial support from central government.

The private-rented sector

The share of the total housing stock accounted for by the private-rented sector fell from 60 per cent in 1947 to 13 per cent in 1993. This decline has largely been as a result of the significant fall in the number of properties owned by individual landlords (Adriaansens and Priemus, 1986).

In much of the private-rental sector there is a relation between housing and pension provision. In the pre-war stock in many cases a number of dwellings are let by private persons, who acquired their property with a view to their old age. In the post-war stock the link between housing and provision for old age is made by a separate institution: the institutional investors (pension funds or life insurance companies).

The pre-war private-rental sector consists predominantly of relatively poor, small, cheap dwellings, often in the form of medium high-rise, located in the central city districts. The occupants are either young or very old. Mobility is high. Many dwellings are sold to owner-occupiers or to local authorities or housing corporations. Management is often farmed out to estate agents.

The post-war private-rental sector, chiefly the property of institutional investors, forms a market segment completely different from the pre-war private-rental sector. The investment sector consists on average of good, large and expensive rented dwellings, concentrated in areas of the Netherlands where there is a great demand for housing. The occupants have a relatively high income and are mobile. Here too management of the dwellings is usually farmed out to estate agents.

The production of new dwellings in the private-rental sector since the Second World War has been attended to chiefly by private institutions, notably the institutional investors. The greater part of these dwellings are subsidised. Up to the end of the 1980s the subsidy regulations for private- and social-rented dwellings were the same. Since then the subsidies for new private-rented dwellings have been sharply reduced. In the period 1988–94 the share of the private-rental sector in new construction declined from about 10 per cent to 7 per cent.

The owner-occupied sector

After the Second World War the proportion of the total housing stock accounted for by the owner-occupied sector was 28 per cent (van der Schaar, 1987). After a gradual increase to 32 per cent in 1967 there was a rapid expansion in owner-occupation in the Netherlands during the 1970s, and, by 1981, 42 per cent of the total housing stock was owner-occupied. After 1981 growth in owner-occupation stagnated, but it expanded again in the second half of the 1980s and reached 46 per cent in 1993.

The promotion of owner-occupation has been an increasingly important element in government policy since the Second World War. For the direct subsidisation of the owner-occupied sector various subsidy schemes have been applied in the past decades, in the form of (income-dependent) long-term contributions or one-off grants. Direct subsidisation of the owner-occupied sector is primarily directed towards encouraging home-ownership among lower-income groups. The subsidy schemes therefore relate to relatively inexpensive dwellings or households with a below-modal income.

In addition to direct subsidisation of home-ownership, owners of a dwelling in the Netherlands qualify for mortgage-interest tax relief. In contrast to many other countries in western Europe there is no limit to the form of assistance in the Netherlands (see Haffner, 1992). Households buying existing properties are required to pay a one-off 6 per cent stamp duty, however. Furthermore, every year home-owners are obliged to include a percentage of the imputed rent of their property as part of their income. This percentage was increased at the beginning of 1990 to around 1.8 per cent of the value of the occupied property. It is expected that this rate will be increased further during the 1990s.

In 1994 the loss of income to the exchequer through mortgage-interest tax relief amounted to around ƒ 6.76 billion. In contrast to this the level of receipts from taxing the imputed rent of property was around ƒ 2.15 billion, and the receipts from stamp duty ƒ 1.87 billion. When the supply subsidies to the house-purchase sector are included (see Table 6.4) the total value of supply and demand subsidies and the cost of tax-exemptions are around ƒ 3.44 billion in the case of the owner-occupied sector.

HOUSEBUILDING

In the 1970s, a high level of housebuilding remained necessary because the housing shortage had not been eliminated. This was the result of continuing population growth, combined with an increasing thinning-out of families. The consequence of the housing policy followed in the 1960s was an increase in the average dwelling floorspace (living area, bedrooms and kitchen) from 62 m² to 67 m², with a drop in the average number of rooms per dwelling from 5.6 to 4.6 in the period 1965–75 (de Vreeze, 1993: 275).

Table 6.3 gives details of new housing completions for the period 1970–94. The principal developments that emerge from the table are:

Table 6.3 Average number of housing completions by sector, the Netherlands, 1970–94

Year	Non-profit-rented sector with operation subsidy	Private-rented sector			Owner-occupied sector			Total
		with operation subsidy	one-off grant	without subsidy	with operation subsidy	one-off grant	without subsidy	
1970	45,349	26,052	–	2,890	25,698	–	17,295	117,284
1971	50,025	33,403	–	1,925	30,917	–	20,325	136,595
1972	53,455	41,986	–	1,918	30,325	–	24,588	152,272
1973	55,765	37,626	–	2,192	30,946	–	28,883	155,412
1974	48,257	32,280	–	1,467	29,896	–	34,274	146,174
1975	40,130	23,454	–	777	31,013	–	25,400	120,774
1976	36,420	17,415	–	569	32,080	–	20,329	106,813
1977	35,315	15,122	–	734	33,362	–	26,514	111,047
1978	29,230	11,806	–	1,036	34,233	–	29,520	105,825
1979	23,596	7,208	–	940	27,812	–	27,966	87,522
1980	38,881	9,820	–	1,267	36,049	–	27,739	113,756
1981	54,979	14,538	–	1,274	30,124	–	16,844	117,759
1982	65,589	22,568	–	1,052	26,020	–	8,081	123,310
1983	52,611	21,735	–	633	30,230	–	5,918	111,127
1984	49,233	18,051	2,853		27,969	14,626		112,732
1985	34,596	16,201	4,009		25,502	17,823		98,131
1986	35,770	13,929	4,963		25,587	23,081		103,330
1987	35,851	11,443	4,418		26,297	32,082		110,091
1988	40,197	8,794	1,507	1,018	23,493	18,757	24,680	118,446
1989	35,976	6,146	2,107	1,488	20,749	15,936	28,831	111,233
1990	28,449	5,950	1,393	1,606	18,374	10,259	31,353	97,384
1991	22,514	3,985	916	1,625	15,676	8,357	29,815	82,888
1992	25,064	3,818	858	1,850	13,313	7,025	34,236	86,164
1993	22,360	3,219	482	2,220	9,995	5,359	40,054	83,689
1994	22,431	2,949	405	2,489	8,646	3,618	46,831	87,369

Source: CBS, Maandstatistiek Bouwnijverheid

- a record number of housing completions in 1973;
- a considerable drop in new building production from 1989;
- a decrease in the number of social- and private-rented dwellings;
- a considerable increase in the number of owner-occupied dwellings;
- a shift in both the private-rented sector and the owner-occupied sector from subsidised to unsubsidised housebuilding.

In the 1960's urban renewal also started. In this period clearance reconstruction occupied a central position in Dutch urban renewal. In the 1970s urban renewal underwent a change from demolition to the improvement of dwellings and the construction of new replacement housing, often in the social-rented sector (Priemus and Metselaar, 1992).

The lines for urban renewal policy in the years to come are set out in the memorandum 'Policy for urban renewal in the future' (Heerma, 1992). In this memorandum the prospect of a declining urban renewal fund is emphasised. By the year 2005 urban renewal will have largely been discontinued in the Netherlands as state subsidisation of renewal schemes will be terminated (Heerma, 1992). In the remaining years the financial share of the private sector will have to increase structurally, the share of the central government will fall from 36 per cent to 26 per cent, and 'remaining' urban renewal will be more strongly concentrated in the large cities (Priemus and Metselaar, 1992: 35).

HOUSING INVESTMENT, FINANCE AND SUBSIDIES

Although not explicitly indicated, control and reduction of government expenditure on housing is an important underlying objective of the housing policy set out in the Heerma memorandum. Cutting down the high expenditure on property subsidies plays an important role in this. Partly on account of the choice of a mixed system of supply and demand subsidies in 1974 and as a result of a significant degree of government intervention after the collapse of the owner-occupied market in 1979, by the end of the 1980s property subsidies formed the most important element of the housing budget (Table 6.4).

One important problem associated with the housing budget is that the level of expenditure is determined to a considerable extent by obligations incurred in the past (see Klunder, 1988 and Brouwer, 1988). These obligations take the form primarily of long-term annual subsidies for rented and owner-occupied housing. In the case of non-profit rented housing constructed after 1975 subsidies lasting as long as 50 years have been granted. The consequence of this was that by 1988 around 60 per cent of total expenditure on housing consisted of payments resulting from obligations incurred in the past. Only around 40 per cent of the housing budget was available for rent rebates, urban renewal and urbanisation programmes, and for subsidising new housebuilding and housing improvement. Table 6.4 gives only a partial account, however, of total expenditure on housing (further consideration of public expenditure being undertaken is given on pp. 89–90).

In order to reduce future expenditure on property subsidies the Heerma memorandum proposed the introduction of relatively high annual increases in rents and a reduction of supply-side subsidies for new (social) housing construction. Accommodation for lower-income groups should be achieved by freeing inexpensive dwellings from the existing stock for these households.

In the Netherlands, in addition to the many occupants with a below-modal income, hundreds of thousands of households with an above-modal income also live in a social-rented dwelling. This shows that social-rented dwellings in the Netherlands are also attractive to households with a relatively high income, though they are not primarily intended for this group (Priemus, 1995). Given the principles underlying the government's housing policy this situation presents a major problem. Supply subsidies are effectively being received by households which, given their level of income, do not require them, while on the other hand high levels of demand subsidies are being paid to meet the housing costs of those on lower incomes living in relatively expensive rented housing. The central government in the Netherlands speaks of a 'mismatch'. An attempt is being made to reduce this mismatch by encouraging households with an above-modal income to move on to commercial rented dwellings and (above all) owner-occupied dwellings and to allocate the social-rented dwellings freed in this way to households with a below-modal income.

Decentralisation

The decentralisation of decision-making was brought about above all by the Dwelling-linked Subsidies Order (BWS) in 1992. For social-rented dwellings and so-called social owner-occupied dwellings financial contributions are made annually. These subsidies are paid to the local authority, and since 1993 to the region, which can pass these contributions on to housing corporations with construction plans for new dwellings. With the entry of this new subsidy system the government has lost its controlling role regarding the quality of new construction. The new philosophy behind quality policy is the guaranteeing of a minimum quality by the government (via the Building Decree), while the market (residents and principals) has to weigh costs and quality against each other.

As from 1 January 1995 the Dwelling-linked Subsidies Order (BWS) 1995 has been introduced. Under this Order limited location-linked subsidies (lump-sum contributions) and 'accessibility' bonuses are paid, but generic operating subsidies disappear entirely. The age of unsubsidised building has therefore dawned in the social-rented sector. The initial rents will subsequently rise to such an extent that newly built dwellings will be beyond the means of occupants with a lower income. They will thus need to find accommodation in the cheaper parts of the existing stock. In this framework new housebuilding performs a filtering function: by building for households with a higher income, cheaper dwellings are freed for households with lower incomes. It is incidentally questionable whether housing corporations will in fact invest sufficiently in new housebuilding if the generic property subsidies are abolished. For institutional investors new construction subsidies were almost entirely eliminated in 1988. Since then they have hardly invested in new rented dwellings. Housing corporations are now entering into the same position as institutional investors with regard to investment in new dwellings for rent. Certainly if long-term interest (now about 7 per cent) were to rise somewhat, a shortage of new housebuilding initiatives threatens in the rented sector, bearing in mind that financial aid to the owner-occupied sector (unlimited deductibility of mortgage interest, subsidy on imputed rent, no capital gains tax) remains fully in effect (Priemus, 1995).

Changing position of housing corporations

For the housing corporations an important role is reserved in the accommodation of lower-income groups, despite the fact that they no longer receive generic property subsidies for new construction. Because they moreover must bear the risks of their investment decisions themselves, they will increasingly have to behave as self-supporting (social) entrepreneurs. The freedom of action that housing corporations need to shape that entrepreneurship has come about via a number of measures: for example, towards the end of the 1980s the Central Fund for Housing and the Social House-building Guarantee Fund were formed. The Central Fund is a solidarity fund, the goal of which is to reorganise financially weak housing corporations. The necessary reserves are provided by all the corporations. The Guarantee Fund is a direct consequence of the abolition of the government loans and guarantees on the capital-market loans for improvement and construction of social-rented dwellings; this fund acts as guarantor for the interest on and redemption of capital-market loans to housing corporations.

In 1993, via the Social Rented Sector Management Order (BBSH), the relationship between housing corporations and local authorities was regulated. An important change in the relationship between government and housing corporations is the shift of emphasis from preventive maintenance supervision in the form of detailed instructions to the obligation to justify oneself after the event. Next, the housing corporations have acquired the possibility, through introduction of the 'rented sum approach', of varying the rent increase within their own property. In that way they can adjust to the market situation of their own property. The most recent development with respect to privatisation of the housing corporations is what is called the 'grossing and balancing' operation. This operation, starting on 1 January 1995, entails that the current subsidy obligations of the government towards the housing corporations are bought off, largely via an accelerated re-payment of outstanding government loans. By this the ties between the government and the housing corporations are in fact broken. For social landlords this means that for the time being risks are also going to be incurred in the operation of rented dwellings and that it will be necessary to be self-reliant from the point of view of business economics, even in difficult times. The housing of low-income groups alone is no longer sufficient to achieve a satisfactory result; this is partly dependent on the income that a housing corporation manages to generate. It will be necessary to balance maintaining sufficient solvency against following a specific rent and investment policy. The social landlords will have to include these three activities as a whole in their management. The great question for the future of course remains whether the social task will live up to expectations in this process.

DEVELOPMENTS IN HOUSING POLICY

Like most other west European countries the Netherlands was confronted with large-scale housing shortages after the Second World War, and the shortages later increased as a result of a growth in the number of house-holds and the low level of housing construction. Given the manifest shortages, an unusually far-reaching level of involvement by the state in housing was temporarily accepted. The heart of government policy was formed by a rent and subsidy policy (van der Schaar, 1987). That rent and subsidy policy was supplemented by measures in the field of quality control and housing distribution.

Both non-profit rented and private-rented housing were subsidised. Government subsidies were extremely broad-ranging in scope and considerable in amount. As much as 95 per cent of all housing construction was sub-sidised (van der Schaar, 1987: 180). To be able to build large numbers of dwellings within the limited budgetary possibilities, the construction of cheap, austere dwellings in relatively large projects with a limited number of types was pursued. The social-rented sector was considered to be better able than the private sector to plan, produce and manage these large numbers of dwellings. Moreover, this sector could be better controlled by the government.

In the 1960s various points of departure for the housing policy followed (which until then had only been under discussion), such as the extensive government subsidisation of housing and the introduction of measures against the monotony and uniformity of housebuilding production. These included higher-quality standards in the subsidy conditions for social- and private-rented dwellings. At the same time liberalisation of the housing market was pursued. In practice, the extent of government involvement in the housing (construction) market did not decrease. It did, however, change in character. Housing policy became increasingly directed towards gearing building, rent and subsidy policy to consumer preferences and to concentration of subsidies on the lower income brackets, all this while maintaining a high rate of building production (van der Schaar, 1987: 377).

The enhanced quality in the new construction was increasingly motivated by 'building for the future'. From 1974 this even became the starting point of housing policy when a centre/left-wing coalition government presented the Rent and Subsidy Policy Memorandum. This argument played a particular role in the first half of the 1970s, when the national economy displayed high growth rates. In this period the owner-occupied sector underwent strong growth in the Netherlands.

For the rented sector a subsidy system selected in which properties subsidies and housing allowances were included. Through property subsidies new social-rented housing ought to remain within the financial reach of households with a modal income. To be able to keep the initial rents affordable and to avoid the occurrence of operating surpluses, a new subsidy system for the rented sector was introduced, the dynamic cost-price-based rent scheme, in which for a 50-year period the proceeds, rent and subsidy were equated with the operating and financing costs. In addition to promoting a good housing climate in general, subsidy policy had to be aimed at increasing the options for all groups of residents, including the lower-paid. The latter required additional aid in the form of housing allowance and subsidisation of dwellings for owner-occupancy. Private-sector investors financing new housing construction regarded the new subsidy scheme as an unfavourable development and they abandoned housing construction for new private-rented housing *en masse*.

The choice of a mixed subsidy system in the rented sector, the subsidisation of owner-occupied dwellings and high quality standards, in combination with a high interest rate, led to an increase in both property subsidies and housing allowances.

In 1978 a centre–right coalition government formulated new policy objectives: specifically the promotion of owner-occupancy and a reduction of government expenditure on housing. But when, as a result of an economic recession, the owner-occupied market in the Netherlands collapsed at the end of the 1970s, the social housebuilding programme was greatly expanded to guarantee the continuity in housebuilding production. The share of social-rented dwellings in building production rose from 28 per cent in 1978 to 54 per cent in 1982. While the construction of unsubsidised housing in the owner-occupied sector fell drastically, the introduction of a new system of grants for homes constructed in the owner-occupied sector prevented the level of (owner-occupied) housing construction from declining too much. The result was a sharp increase in the already not inconsiderable subsidy obligations of the central government.

After the intensification of the central government's policy from the second half of the 1960s and notably during the 1970s, from the first half of the 1980s housing policy focused (again) on reduction of government expenditure and decentralisation and deregulation of the subsidy system. The reduction of government expenses related mainly to the rented sector, involving higher rents for new rented housing, reducing expenditure on rent rebates, and significant increases in rents for existing housing. Furthermore, the housebuilding programme was cut and it was the non-profit-rented sector particularly which bore the brunt of these cuts (Boelhouwer and van der Heijden, 1992: 68). The owner-occupied sector was given an important stimulus not so much by extra subsidies as by the absence of the financial burdens introduced in the rented sector.

Within urban renewal policy a remarkable shift occurred from a centralist housing approach directed towards the lower-paid in the 1970s, to a decentralised approach in the 1980s, aimed more at the economic and cultural development of cities, in which the private sector plays an ever-greater part (Priemus and Metselaar, 1992: 5, 13). The Urban and Village Renewal Act has formed the framework for policy since 1985. Some 20 existing subsidy schemes were abolished and incorporated in a national Urban Renewal Fund of approximately 1 billion guilders a year. This sum of money is distributed among the municipalities and provinces via the 'urban renewal formula'. The central government calculates the need for urban renewal in detail, but municipalities are relatively free to determine their own priorities. Subsidies for the improvement of rented dwellings and subsidies for new construction in urban renewal areas remained centralised (until 1992, see below), like the housing allowances that play an important role in urban renewal areas and elsewhere (Priemus and Metselaar, 1992: 19–20).

Housing policy in the 1990s

There had been hardly any major change in Dutch housing policy during the period 1974–88. Central government expenditure on housing grew despite repeatedly formulated economic plans, the subsidised housing construction programme continued to be maintained at an extremely high level, and expenditure on housing allowance increased explosively (see Table 6.4).

From the end of the 1980s, however, Dutch housing policy has been going through a process of considerable change. Outlines of recent housing policy, as announced in *Volkshuisvesting in de Jaren Negentig* (Ministerie van VROM, 1989), adopted by parliament in 1990, are a reallocation of responsibilities and corresponding financial risks among the participants in

Table 6.4 Government expenditure on housing, the Netherlands, 1970–94
(in *f* million, at current prices)

Year	Property subsidies rent	purchase	Housing allowance	Urban renewal and urbanisation	Other	Total	Housing Act loans	General total
1970	275	55	0	100	20	450	1,790	2,240
1975	980	280	235	395	185	2,075	2,810	4,885
1980	1,810	430	965	805	375	4,385	4,435	8,820
1982	1,840	570	1,425	1,210	465	5,510	5,215	10,725
1985	4,050	990	1,445	1,810	740	9,035	5,355	14,390
1987	5,170	1,020	1,665	1,125	545	9,525	4,590	14,115
1994	5,072	704	2,251	1,035	1,183	10,245	417	10,662

Source: Housing Ministry budget, various years

the housing market, specific deployment of financial aid from the central government, a strong emphasis on the promotion of home-ownership and a concentration of (the remaining) financial aid on lower-income groups.

One of the starting points of the memorandum is that with the changing nature of housing (a shift from new construction to management) such central control as in the past is no longer necessary. Policy is directed towards strengthening the operation of the market. Through the process of decentralised decision-making and the privatisation of the housing corporations the central government is largely withdrawing from the housing market. Housing is becoming much less a task of the central government and much more a local and regional matter, with local authorities, housing corporations, residents and other participants in the market as the important players.

The operation of the market is strengthened via the pursuit of market rents and the reduction and phasing out of property subsidies. The housing market has to become a 'real' market. This means that in new construction an important part is reserved for unsubsidised housebuilding, notably the construction of homes for owner-occupancy. Government assistance is targeted specifically at those households that are not themselves capable of affording the market rents. The key instrument here is the housing allowance.

CONCLUSIONS

In the above analysis we have shown that for decades there was a considerable and unsatisfied housing demand in the Netherlands. Despite high production figures, a state of equilibrium is still not within reach; quite the reverse. Through the structural immigration of recent years the shortages are still growing. Not until 2030, when the Netherlands will probably have a population of 17 million, will population growth be at an end and housing demand gradually stabilise. Despite the fact that the situation in the housing market has not essentially changed, government influence has strongly declined and the future demand for dwellings will have to be met with very limited direct government aid. On account of changing ideas about the task of the government and the need to reduce and curb government expenditure, it seems that government influence will remain limited in the years to come. In this situation, the owner-occupied sector in particular will increase in importance. This is at the expense of the private-rented sector, which will probably waste away further. The big question remains how in the new structure of housing the extensive social-rented sector will develop. If we examine the combined effect of the introduction of the BBSH (1993), the BWS (1995) and the grossing and balancing operation, we see a strongly independent housing-corporation sector over which the central government has little control.

Whether the housing corporations will succeed as self-supporting social entrepreneurs in providing sufficient housing for households with a low income remains to be seen. Without property subsidies for new dwellings and the existing housing stock they will have difficulty in achieving their social tasks. Without an adequate public framework the housing corporations and municipal housing companies may gradually change colour and increasingly behave like commercial landlords.

It is, however, clear that the housing possibilities of the lower-income groups will be largely dependent on the success of filtering and therefore on the possibilities of producing and marketing more expensive (owner-occupied) dwellings. And in this way housing will become more strongly dependent on exogenous factors like demographic and economic developments.

NOTE

1 The private-rented sector is characterised as a commercial rented sector. Most of the non-profit rental sector is also privately owned. The Central Bureau of Statistics used the term 'private-rented sector' to indicate the commercial rented sector. We follow this terminology.

7

SWEDEN

Bengt Turner

This chapter will describe the housing market and housing policy in Sweden. It starts by giving basic data on the housing stock, housing consumption and changes in housing production. In another section, the structure and changes in housing subsidies are described. This is the starting point for an overview of policy concepts and policy changes in Sweden. The main emphases are on a comparison of general selective subsidies and on the future role of public housing. It is argued that general subsidies are being phased out in Sweden and are only partly being replaced by targeted subsidies. It is also argued that public housing in Sweden is becoming subject to market forces in a way that conflicts with traditional social housing policy.

TENURE

The structure of the housing stock has changed steadily since the Second World War. Table 7.1 shows the distribution of different forms of tenure at the censuses of 1945, 1960, 1970, 1980 and 1990. The share of private-rental housing, for example, has dropped from 52 per cent in 1945 to 20 per cent in 1990. That sector is at present somewhat smaller than the social-rental sector which comprises about 25 per cent of the stock. The tenant-owner (co-operative) sector has expanded and the home-ownership sector has retained its share of the housing stock.

Table 7.1 Forms of tenure, 1945-90

Year	Social-rental	Private-rental	Home-ownership	Tenant-owner (co-op)
	Forms of tenure %			
1945	6	52	38	4
1960	14	43	34	9
1970	23	30	34	13
1980	24	21	41	14
1990	25	20	40	15

Source: Housing and Rental Survey

99

Table 7.2 Dwellings per production period and form of tenure, 1989[1]

Production period	Form of tenure %				
	Social-rental	Private-rental	Home-ownership	Tenant-owner	Total
Pre-1940	5.1	19.9	23.1	9.1	19.9
1941–50	6.8	11.1	9.0	12.1	11.1
1951–60	21.2	14.0	9.9	22.6	15.2
1961–70	33.7	19.8	19.3	29.1	24.1
1971–75	17.7	5.4	14.9	9.3	12.6
1976–80	6.3	1.5	15.4	4.9	8.6
1981–85	5.8	2.7	6.5	8.4	5.8
1986–88	3.4	2.6	1.8	4.6	2.8
Share (%)	22.4	22.7	40.0	14.9	100.0

Source: Housing and Rental Survey

The differences in the composition of the housing stock at different time periods are matched by significant differences in age distribution across tenure forms (Table 7.2). The table applies to 1989 and the data are from the Housing and Rental Survey of that year. Nearly 54 per cent of dwellings in the private-rented sector are in buildings produced before 1950. The corresponding figure for public rented stock is 12 per cent. As a consequence, the social sector is heavily over-represented in the latest production periods.

It is often discussed whether private- and social-rental dwellings differ as regards household composition. For example, have private landlords smaller but more resource-intensive households than those living in socially owned dwellings? Table 7.3 shows households divided by household type and form of tenure. The share of single-person households is 63.5 per cent of all households in the private-rental sector, while in social-rentals it is 57 per cent. The share is 54 per cent in tenant-owner properties, but as low as

Table 7.3 Households according to type of household and form of tenure

Household type	Form of tenure %				
	Social-rental	Private-rental	Home-ownership	Tenant-owner	Total
1 adult without children	57.1	63.5	12.3	54.2	40.2
1 adult with children	10.8	5.7	2.4	3.9	5.2
2 adults without children	23.7	23.1	43.2	31.8	32.6
2 adults with children	8.4	7.7	42.1	10.1	22.0
Share %	22.4	22.7	40.0	14.9	100.0

Source: Housing and Rental Survey

12 per cent in privately owned properties (single-family houses). Single parents are a significantly more common household type in social-rental dwellings: less than 11 per cent in contrast to less than 6 per cent in privately rented dwellings. Similar differences, albeit of a smaller extent, exist for couples with children.

One common assertion is that households in the social-rental sector are poorer than those in the private sector. Given certain reservations this is true, as Table 7.4 shows. In the table all households have been divided into deciles according to disposable income per unit of consumption.[2] In the lower deciles, there is a large share of public-sector tenants, otherwise, the differences are slight. Admittedly, there is a certain over-representation of households in the private-rental sector in the highest deciles, but the differences are not great. Taken together, the households in the private-rental sector have somewhat more resources than do those in the social sector.

Table 7.4 Households according to disposable household income per consumption unit (in deciles), in terms of forms of tenure[3]

Income decile	Form of tenure				
	Social-rental	Private-rental	Home-ownership	Tenant-owner	Total
1	13.3	12.6	5.7	9.2	9.5
2	12.3	8.2	8.1	10.6	9.4
3	10.9	9.9	9.0	7.5	9.4
4	9.4	8.9	11.5	7.2	9.8
5	8.7	9.8	11.6	9.7	10.3
6	9.2	10.4	10.7	9.6	10.1
7	10.9	11.5	10.9	10.5	11.0
8	7.8	10.1	9.7	10.4	9.5
9	10.5	9.4	11.0	12.6	10.8
10	7.0	9.1	11.8	12.9	10.3
Total (share)	22.4	22.7	40.0	14.9	100.0

Source: Housing and Rental Survey

The distribution of households into flats of different sizes is important since it is likely that changes in the housing market will cause the greatest stress among households initially living in relatively large dwellings. Of single-person households living in one room and kitchen 29 per cent live in the private-rental sector. The corresponding figure for social-rental dwellings is 25 per cent and for tenant-owner housing, 18.5 per cent (Table 7.5). On the other hand, the share of households in larger flats is greater in the private-rental sector than in the public. The general impression, which is confirmed in other household types, is that the spread in terms of space standards is greater within the private-rental sector than in other forms of tenure. This is particularly true in comparison with the public-rental sector.

Table 7.5 Households according to type of household, size of flat, and tenure

Social-rental

Household type	Flat size					(Share)
	1 room and kitchen	2 rooms and kitchen	3 rooms and kitchen	4 rooms and kitchen	5 rooms and kitchen	
1 adult without children	25.2	51.5	20.0	2.8	0.5	57.1
1 adult with children	0.7	17.3	53.8	25.4	2.9	10.8
2 adults without children	2.4	38.0	47.7	9.4	2.5	23.7
2 adults with children	0.6	5.4	44.1	39.8	10.1	8.4
(Share)	(15.1)	(40.7)	(32.3)	(9.9)	(2.1)	

Private-rental

Household type	Flat size					(Share)
	1 room and kitchen	2 rooms and kitchen	3 rooms and kitchen	4 rooms and kitchen	5 rooms and kitchen	
1 adult without children	29.0	46.3	19.5	4.3	1.0	63.4
1 adult with children	2.3	23.1	41.2	25.0	8.3	5.7
2 adults without children	5.0	28.9	39.2	19.0	7.9	23.1
2 adults with children	–	8.8	50.3	29.2	11.7	7.7
(Share)	(19.7)	(38.0)	(27.7)	(10.8)	(3.8)	

Home-ownership

Household type	Flat size					(Share)
	1 room and kitchen	2 rooms and kitchen	3 rooms and kitchen	4 rooms and kitchen	5 rooms and kitchen	
1 adult without children	0.7	11.1	24.1	37.6	26.4	12.3
1 adult with children	–	2.3	10.6	27.0	60.1	2.4
2 adults without children	0.1	3.6	15.6	40.3	40.5	43.2
2 adults with children	–	0.9	6.3	31.6	61.2	42.1
(Share)	(0.1)	(3.3)	(12.6)	(36.0)	(48.0)	

Tenant-owner

Household type	Flat size					
	1 room and kitchen	*2 rooms and kitchen*	*3 rooms and kitchen*	*4 rooms and kitchen*	*5 rooms and kitchen*	*(Share)*
1 adult without children	18.5	49.6	26.5	4.3	1.1	54.2
1 adult with children	0.8	14.3	34.3	41.3	9.3	3.9
2 adults without children	1.5	29.4	49.4	15.8	3.9	31.8
2 adults with children	0.1	3.7	30.4	46.1	19.7	10.1
(Share)	(10.5)	(37.2)	(34.5)	(13.6)	(4.2)	

Source: Housing and Rental Survey

On average, space standards seem to be somewhat lower within the private-rental sector – if the type of household is kept constant. Since incomes are generally higher among private-rental tenants, the stress following upon the reduction of subsidies remains relatively slight within the private-rental sector.

More important is that different tenure forms in Sweden are governed by different economic regimes. The social-rented sector consists of, as a rule, one municipally owned company in each municipality. They are non-profit organisations, set up as organisations acting independently from the municipal economy.

Rents are set in negotiations between the municipal company and its tenants on a non-profit basis, with a rent structure determined on more or less market terms – a certain degree of rent pooling between different parts of the housing stock is taking place (Turner, 1979). This rent structure is then used as an upper limit when negotiations are taking place for the private sector, which is a traditional private-rented sector. The sectors are of equal size and cater for the same type of households. This is an important feature of the Swedish housing market: social and private housing sectors are competing on equal terms and are both mainstream housing market segments, open to all households.

There are, however, differences between the social- and private-rented sectors, in that low-income and immigrant households tend to be concentrated in the public sector. This is an effect of a lower average age of households in the public sector, and that the public sector cannot so easily refuse to accommodate these households, when the owner – the municipality – so demands.

The ownership sector is a traditional west European type of ownership with relatively few restrictions on property rights, nowadays. The tenant-owner sector, or co-operative sector, is a mixture of a condominium sector and a rented sector. Loans are not split up between apartments, except loans to cover down-payment or a transfer price.

The fact that household composition differs between the sectors indicates that different tenures are not perfect substitutes for each other. The present changes in economic and housing policy affect relative housing expenditure in different tenures, thereby altering the economic incentives of tenure choice.

HOUSEBUILDING

New construction reached 60,000 dwellings in 1990 and 1991. The composition mirrored the composition in the existing housing stock, but was slightly weighted towards co-opetative housing. The volume of new construction then dropped dramatically during 1992–93, because of the economic recession and the withdrawal of housing subsidies (Figure 7.1). From a peak of 70,000 dwellings in 1991 it fell by 80 per cent to reach a predicted all-time low of approximately 12,000 in 1995. The share of single-family houses out of all new construction fell from 50 per cent in the 1980s to 25 per cent in 1993 and 30 per cent in the first quarter of 1995.

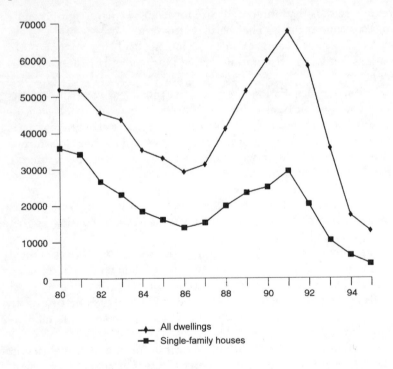

Figure 7.1 Housing completions by house type, Sweden, 1980–94

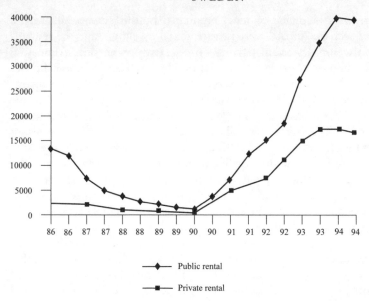

Figure 7.2 Total number of vacant flats in public and private rental, Sweden, 1986–94

After a steady decline in the 1980s to minuscule levels in 1990 the vacancy rate rose by 3.4 percentage points (from 0.2 per cent to 3.6 per cent) between March 1990 and March 1994. A large share of the vacant flats is in the recently built stock, where the vacancy rate is 5.7 per cent, and vacancy rates are higher in municipal houses than under private landlords (Figure 7.2). The aggregate numbers conceal large regional variations with high vacancy rates coinciding with more general economic problems. Given the prolonged crisis, an earlier overproduction and the further withdrawal of interest subsidies in subsequent years, it is likely that vacancy rates will stay high and the low level of production will continue for a number of years.

HOUSING INVESTMENT, FINANCE AND SUBSIDIES

There is a long tradition in Sweden of pursuing a social housing policy. Housing has for a long time been a part of the Swedish welfare system. This is clearly visible in a still high housing subsidy level with a low degree of selective targeting and detailed and far-reaching planning and regulation practice.

A few words should first be said to outline the essential characteristics of the Swedish housing market. In 1985 there were 3.9 million dwellings, of which 2.1 million were multi-family houses and 1.8 million single-family houses. Multi-family houses can further be categorised by ownership:

800,000 dwellings belonged to local municipal housing companies (e.g. a social-rented sector), 700,000 were private-rental dwellings, and the remaining 600,000 dwellings were in tenants' co-operative ownership. Almost all single-family houses were in private hands, but 100,000 of these houses belonged to other (corporate) owners. Dwellings belonging to co-operatives in practice are owned by the dweller, and if these are included in the ownership sector, this sector in Sweden is fairly large, compared to the rest of Europe: 2.6 million out of 3.9 million dwellings in all. It is also important to point out that the private- and public-rental sectors are of about the same size and roughly the same quality. There are no slums in Sweden, and the public sector is a viable alternative to the private-rental sector for most households. The rent level within the non-profit public-rental sector is also legally used as a ceiling for rent-setting in the private-rental sector.

The goal for Swedish housing policy is officially stated in a bill from 1967 as: 'the whole population shall have access to healthy, spacious, well planned and suitably equipped dwellings of good quality at affordable prices'. This goal has since then been supplemented by goals of a good environment, a right to participation for renters and a goal of an integrated household composition (from the housing policy proposition of 1974).

In terms of subsidies within the housing policy system, there are three different measures of importance: an interest subsidy, tax benefits and a housing allowance system. The first two measures are general and will be discussed in some detail below in relation to a description of fiscal aspects.

The housing allowance system is a selective measure. Housing allowances are given to families with children and to retired persons. Different regulations steer the allowances given to these different groups, and they are generally more generous to retired persons. In any case, the system does not discriminate between house types or forms of tenure , and it is given to approximately 25 per cent of households with children and to 40 per cent of all retired persons.

We have so far only discussed subsidies and economic aspects of housing policy. There are other aspects as well within the housing policy framework. One of these is the long-standing Swedish goal of integration on a neighbourhood level. Sweden has striven to prevent ethnic and economic segregation. The measures to reach this goal have included, for example, the local-authority exchequer and building a mixed housing stock, with a predominant rental subsector and avoiding the creation of a stock specially set aside for socially needy households. Sweden has also tried to prevent the creation of homogeneous old-aged communities. The idea behind preventing segregation according to age was an ambition to create a constant demand for societal facilities, such as child care and schools, on a neighbourhood level. It was also generally considered beneficial to society to have this kind of mixture; young children should meet older people on a regular basis.

This policy is now being dismantled in Sweden. There is a growing political acceptance of ethnic and socially homogeneous areas in Sweden and, for example, the co-operative housing companies have started to build apartment blocks with an age-eligibility clause. It is in fact a business idea, brought about by the fact that new generations of pensioners will be more wealthy than the present generation. In many cases, they own a house with considerable embedded equity. This will constitute a firm basis for the down-payment on a new co-operative flat in an estate that provides additional services, such as nursing facilities, specially designed recreational activities and in-house catering service.

Subsidies from a fiscal point of view

Sweden has, admittedly, one of the highest housing standards in Europe. This can be explained by a high income level and a housing subsidy system that has been most generous. Table 7.6 presents fiscal data from 1990 up to 1993.

Table 7.6 Housing subsidies and taxes in billion SKr at 1992 prices

Transfer	Year			
	1990	1991	1992	1993
Subsidies				
Interest subsidies	21.5	24.9	31.9	32.4
Tax subsidies	16.1	14.2	15.0	13.8
Investment subsidies	–	1.9	5.1	4.0
Housing allowances	3.5	5.2	5.9	6.5
Housing allowances for pensioners	6.9	8.2	8.3	9.4
Others	0.7	0.7	0.7	0.5
Sum	48.8	55.1	66.9	66.5
Taxes: real-estate tax	6.6	12.5	12.5	14.5
Sum (net subsides)	42.2	42.6	54.5	52.2

Source: Housing and Rental Survey

Subsidies have increased over time and amount to approximately 3.5 per cent of the gross national product in 1993, according to Boverket (1994: 130 ff). This makes the Swedish subsidy system one of the most generous in western Europe in the early 1990s, only paralleled by that of the Netherlands.

Most new housing units are entitled to interest-subsidised loans according to a system that has been in effect since 1975 (and was applied retroactively to houses constructed prior to 1975). In short, long-term government-guaranteed mortgage loans covering 95–99 per cent of approved building costs are granted to all new units that comply with certain government regulations with respect to a range of standards. The system acts both as a general subsidy for post-1975 houses and as a way of handling the 'tilt problem' of mortgage payments, that is, the fact that standard schedules for amortisation and interest payments imply disproportionate real payments during the first years, which may give rise to liquidity constraints (Kearl, 1979; Hansson, 1977).

Subsidised loans run at guaranteed interest rates, which start at very low levels and are increased year by year until they reach the market rate. In 1989 loans to multi-family apartment buildings started at a 2.6 per cent interest rate with a 0.25 per cent yearly increase. For one-family houses the corresponding rates were 4.9 and 0.5 per cent. The starting level was increased to 3.4 and 4.9 per cent in 1991, and a new, much less generous system was introduced in 1993.

The differences between the tenures reflect the differences in tax treatment: the lower interest rate for apartment houses aims at offsetting the favourable tax treatment of owners. With market interest rates around 12 per cent it would take around 40 and 15 years, respectively, until the guaranteed rate reached the market interest rate.

One way to gauge the impact of the subsidies is to calculate their expected present value. The results, of course, depend critically on assumed future market interest rates. Calculations by Jacobsson (1995), based on methods discussed in Hendershott *et al.* (1993), indicate subsidy values of around 2,500 Skr per square metre (assuming a fixed 11 per cent interest rate for the full life of the loan). This corresponds to around half of the average market price of a non-subsidised house. Basing the calculation on a lower interest rate of 8 per cent reduces the subsidy value to a third. There were some changes of the system in the late 1980s but they did not have a major impact on new houses. For three-year-old houses, the present value was reduced from slightly below Skr 2,000 in 1985 to less than half that in 1991.

DEVELOPMENTS IN HOUSING POLICY

Interest subsidies are now being phased out, as a result of tax reform in 1991 and housing finance reform in 1993, which set out a plan to reduce interest subsidies to new construction down to a quarter of its present value over the next decade. This is one of the reasons for the largest downfall in new construction that Sweden has experienced in modern times. There are also other explanations, such as the economic recession and increasingly real interest rates, which have decreased income and thus demand for housing.

The tax reform in 1991 decreased the marginal tax rate from 47 per cent to 30 per cent for most income earners. This led to a reduced tax benefit to home-owners, as their interest deductions on mortgages became less effective. For equity reasons, and for the sake of funding the tax cut, interest subsidies for rental and co-operative housing were reduced. At the same time indirect taxes were increased to help fund the tax reform. This also increased housing expenditure in the same way as the reduced subsidies. In all, the housing sector contributed Skr 20 billion or one-third of the losses due to the tax cut to fund the tax reform.

This brought hardship to poor households. Their housing expenditure increased, which is important to these households as housing is a large fraction of their costs. At the same time, the marginal tax cut did not help them very much, as they were not paying much tax anyway. To compensate for this, housing allowances were increased, which relieved these households to some extent, but did not give a full compensation.

Prices and rents

One important effect of reduced subsidies and an economic recession is the fall in house prices that occurred in 1990. Figure 7.3 shows the development in nominal prices of single-family and multi-family houses since 1970. The drop in prices after 1990 coincided with a sharp increase in user costs, resulting from the 1991 tax reform as well as a declining economy.

As expected, prices of rental apartment buildings and rents are closely correlated. Real rents started to increase around 1985, followed by a sharp 20 per cent increase between 1989 and 1992 (Figure 7.4).[4] The explosion in rents between 1989 and 1992 was at least partly due to the reduction in interest subsidies that was embedded in the tax reform.

A conclusion is that Sweden's attempt to reduce subsidies had an unfortunate timing, as it coincided with a decline in economic activity. New construction fell to an all-time low, rent increased and house prices fell.

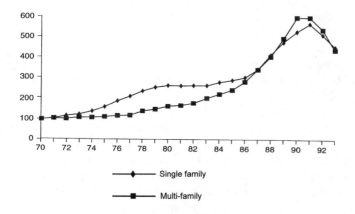

Figure 7.3 Price index for owner-occupied one-family houses and multi-family rental buildings, Sweden, 1970–93
Source: Housing and Rent Survey, Statistics Sweden, different years

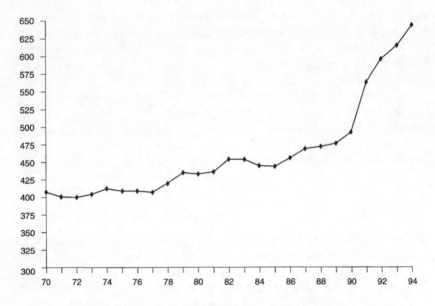

Figure 7.4   Private rents in public rental housing, Sweden, 1970–94
Source: Housing and Rent Survey, Statistics Sweden, different years
Note: Rent is measured in square meters at fixed 1994 prices

Rents as a share of disposable income have increased to 30 per cent (Eng-lund *et al.*, 1995), which creates affordability problems. It is thus likely that the adaptation time will be prolonged, and that Sweden will have to live with high housing expenditure and a low level of new construction for a long time.

CONCLUSIONS

Swedish housing policy is characterised by large non-targeted housing subsidies. This has long been seen as an important feature of Swedish housing policy and an important supplement to general social policy. As this policy is now dismantled, under political protest, it is interesting to compare Swedish housing policy with the practice in other Nordic countries, as these have the same social welfare ambitions as Sweden.

A paper on housing finance in the Nordic countries (Turner, 1990) had the apparent similarity between the Nordic countries in social, political and economic respects as a starting point. This study came to the conclusion that the housing policies were remarkably different.

The general conclusion was that Finland, Iceland and Norway have a more selective housing finance policy than Denmark and Sweden. Taking into account a somewhat smaller level of non-targeted housing subsidisation in the former countries, the result is that they have a larger share of selective housing subsidies than in Denmark and Sweden.

A selective housing policy in Finland, Iceland and Norway was also combined with housing stocks characterised by a small public or rental sectors, which may explain the policy structure.

Even if the share of selective subsidies, including finance subsidies, was large in Finland, Iceland and Norway, compared to Denmark and Sweden, it was nevertheless clear that the amount spent on subsidies was quite different. Including tax subsidies, housing subsidies in Denmark amount to 4.7 per cent of GNP. The corresponding figure in Sweden is 3.9 per cent (Boverket, 1994). This may be compared to Finland, Norway and Iceland where housing subsidies amount to 2.1, 1.5 and 1.0 per cent of GNP, respectively. These figures are from 1984. Since then Denmark has managed to reduce most of its subsidy content.

So there are two ways of reaching a housing policy goal: a selective housing policy with a low level of subsidies and a general policy with a high level of subsidies. This raises two problems: how can we explain these differences in policy and will the outcome be the same in a housing welfare perspective?

The answer to the problem is twofold. First, the structure of the housing stock offers an explanation. The small size of the public sector may be explained historically and consequently affects the structure of the subsidy system. The second explanation is that the market context is different, making a selective housing policy more feasible in Finland, Iceland and Norway because of more market-directed policies in general.

The answer to the problem of outcome is less obvious as we know very little about capitalisation of subsidies and the various housing policy measures. Even if a policy is targeted, price effects may very well change the social outcome as part of the subsidy will be lost in increasing housing prices and building costs. The long-term effects of subsidies are even less well known. The same applies to countries such as Sweden, where housing subsidies are quite evenly distributed between households in different income classes (Turner, 1988).

So, even if there is a trade-off between small selective subsidies and large general subsidies in a social-housing policy context, the outcome need not be the same as regards, for example, needy households living in deprived housing conditions. The reason for this is the capitalisation of housing subsidies and regulations.

The real difference exists when the selective policy takes another meaning. Instead of supporting selective households, the policy might instead be targeted towards a subsector of the housing market, such as the local municipal rented sector. This will, as in the United States and Australia, lead directly to access control and a stigmatised population in these subsectors.

These trends will inevitably be in conflict with increasing social demands and tensions in society. A larger share of socially deprived households must be housed by a sector undergoing a market orientation. The result may be a selective policy, where the houses and not the individuals are the targets. A subsector of the municipal housing company may, for example, be allocated to socially deprived households, but the majority of its housing stock will increasingly conform to market structures with more private-sector involvement.

Examples from the other Nordic countries do show us that a selective housing policy can be more successful. The target can be the households and not the houses. Given that a poverty trap can be avoided, and if, for example, housing allowances are combined with a more comprehensive social policy at neighbourhood level, the social outcome of a reduction in general subsidies and a more market-oriented municipal housing sector need not be devastating. This will, however, require much rethinking among policy-makers in Sweden.

NOTES

1 The table reveals fewer differences between public- and private-rental dwellings than that indicated by the Folk och bostadsräkning (Population and Housing Census). This is because the Housing and Rental Survey is a stratified sample survey and thus random and systematic errors can cause deviations. At the same time, the Population Census also has its faults.
2 This is a measurement which takes into account different household members' household costs. Two adults, for instance, compose 1.61 units and one adult and one child, 1.4 units, etc. The measurement makes it possible to compare households with different compositions in a simple way.
3 The table is based on stratified data, which explains why the decile fractions deviate slightly from 10.0 per cent.
4 The diagram refers to the stock of three-room apartments, the size and quality of which have varied slightly over time.

8

AUSTRIA

Wolfgang Förster

Housing policies and urban development cannot be considered without due regard to the social environment and, in particular, the nature of the country concerned. This is especially true for Austria – and even more for its capital, Vienna, which has a number of characteristics that distinguish it from other European cities of comparable size and character.

Austria being a federal republic (with Vienna one of the nine autonomous provinces – *Bundesländer*) enables jurisdiction for housing to be devolved to provincial government. This includes legislation on housing subsidies, on housing renewal and on allowances, but not the Tenancy Act (dating back to 1917) which until now has remained a federal law. The federal government also continues to be the most important provider of funds which are generated by means of earmarked taxes collected together with the various income taxes and subsequently distributed to the nine federal provinces. The federal constitution thus generates nine different housing policies in Austria, and nine very different policies on urban renewal as well, all within the limits of the Tenancy Act which guarantees tenants' rights to an extent unrivalled by most European legislations. Again, the situation is more complex in the capital, Vienna being by far the largest Austrian city and containing housing and renewal problems on a scale unknown to most other areas in Austria.

The other eight *Bundesländer* are subdivided into political districts. Communities with at least 20,000 inhabitants may apply for city status which gives them certain political and financing rights. At present, Austria has 15 towns with city status, 84 political districts (plus 23 districts in Vienna which have their own elected district parliaments) and 2,333 local communities (*Gemeinden*). Local community autonomy is an important feature when discussing the Austrian political system.

Recent years have seen an important shift of power – especially in the field of housing – from central to local government. In 1988 the entire legislation for new housing and for urban rehabilitation was transferred to the nine provincial governments in order to meet more precisely regional concerns and to increase the efficiency of public housing promotion.

Legislative and executive powers on the provincial level also include regional planning (with a few exceptions such as state roads, railways, forestry and water legislation) and the right to adopt their own regulations concerning the purchase and sale of land for building purposes.

TENURE

Rented housing

Since 1917, rent levels of nearly all rented flats in multi-storey buildings have been restricted by government regulation. There are rent limits for both new and existing tenancy contracts. In existing contracts, rents may only be increased to compensate for inflation or for the purpose of urgent maintenance works (cost-covering rents). All landlords – private and public – are obliged to freeze their rent income for ten years for potential maintenance works; only after that period may they freely dispose of the surplus income generated. Rents for all flats in privately owned pre-1945 buildings are regulated by the Tenancy Act. Rents for subsidised flats constructed after 1945 are regulated under special subsidy legislation.

Privately financed rental dwellings – constructed by private developers without public subsidies after 1945 – are not subject to any price restrictions. However, the number of privately financed rental flats is very low and without any significance for the housing market in general.

With regard specifically to the public-rental sector, the Non-Profit Housing Act limits rents to the level of costs (including repayment and maintenance costs).

Of course, rent regulations only work with appropriate control. Tenants may have their contracts checked by one of the tenants' organisations and in the case of an offence may apply to the courts or to a special 'Schlichtungsstelle' ('mediation office') without costs. However, regulations are rather complex, and not all tenants – especially immigrants – know their rights or dare to defend them, which in some sectors of the private-rental market leads to cases of drastic abuse.

Table 8.1 Tenure, Austria, 1991

	Social-rental sector %	Private-rental sector %	Owner-occupied sector %
Austria	20	25	55
Austria excluding Vienna	13	20	67
Vienna	39[a]	40[b]	21

Source: Österreichisches Statistisches Zentralamt (1993); Statistisches Amt der Stadt Wien (1994)
Notes: [a] 26 council housing, 13 non-profit associations
[b] 33 private-rental housing, 7 housing provided by private enterprises for their employees

The social-rental sector

The social-rental sector in Austria has been continuously developing since 1945 with a clear emphasis on promoting non-profit housing associations in the more urbanised areas. By 1991, social-rented housing amounted to 20 per cent of the national housing stock and 39 per cent of the housing stock of Vienna (Table 8.1). Only a few cities (particularly Vienna) have also carried on their council housing programmes (*Gemeindebauten*).

Whereas the roots of non-profit housing lie in the late nineteenth- and early twentieth-century workers' housing co-operatives (some of them still existing) the majority of the associations are today organised as joint-stock companies. The most important companies are owned by public bodies – cities, trade unions, etc. In the 1980s non-profit associations have been responsible for approximately two-thirds of completed dwellings in multi-family buildings, the tendency of this share still moving upward. Today they are in possession of approximately 10 per cent of the total housing stock, of approximately 20 per cent of all dwellings in multi-storey buildings and about 50 per cent of the social-rental sector (council housing and non-profit housing put together). Taking into account the owner-occupied dwellings that have been constructed by non-profit housing associations they are responsible for one-third of the multi-family housing stock. The most crucial principles of non-profit housing are:

- limited profits (by restricting the interest rate on own capital and regulating charges),
- housing costs according to costs of construction, financing and maintenance, and
- restriction of (re-)investments to the housing sector.

Up to now about 700,000 flats have been constructed by non-profit housing associations. All companies have to be members of the Austrian Union of Non-Profit Housing Associations, and are subject to comprehensive controls by the Union and by their respective provincial governments. As for the subsidised owner-occupied sector there are income limits for prospective tenants which differ between the various provinces, but in each case are high enough to guarantee a vast majority of citizens access to non-profit housing. Income limits are, however, checked only before letting a dwelling, and this has caused some discussion about the proper allocation and use of subsidised housing. Most parties agree on keeping up a high proportion of public housing and thus avoiding social ghettos, and an income-related rent system is being discussed at the moment. This would allow families whose income has significantly increased to stay in their flats, but at higher cost. At the moment, however, a differentiation of costs is only possible through additional 'subject subsidies', that is, individual allowances given to low-income tenants.

WOLFGANG FÖRSTER

The private-rental sector

As mentioned above, the strict regulations of tenure after the First World War have hampered private investment in the housing sector. Consequently most of the private-rental dwellings – with the exception of a small 'luxury' segment for temporary use by diplomats etc. – date back to the pre-1914 period. There is a significant share of such houses only in the large cities (e.g. Vienna: 33 per cent), whereas the national average is below 25 per cent. There has been a growing tendency, however, in recent years to sell off private-rental dwellings to the sitting tenants or to new owners. In many cases this has taken place in badly maintained and poorly equipped buildings, thus shifting the problems to the new owners. Immigrants seem especially to become victims of such transactions, as they depend on the private market (in most cases they are legally or practically excluded from the public-housing sector) and they are often badly informed.

As a result of the age of the buildings and of former rent restrictions (which have lately been eased to allow maintenance and improvement), but also as a consequence of speculation, many of the private-rental buildings are in bad condition. Again, this concerns mainly Vienna with its huge stock of pre-1918 houses: from a total of 738,200 permanently inhabited dwellings 58,000 (7.9 per cent) lack inside toilets or water supplies, 25,200 (3.4 per cent) do not have bathrooms. Of all (roughly 850,000) houses in Vienna, nearly 18 per cent lack inside toilets and about 7 per cent do not have bathrooms (see Table 8.2). Urban renewal programmes have also significantly improved the housing situation in the private-rental sector.

Table 8.2   Condition of housing, Austria, 1991

Category	Austria %	Vienna %
A = central heating, bath, WC	67	65
B = bath, WC	17	10
C = WC, water	5	7
D = WC and/or water outside	11	18

Sources: Österreichisches Statistisches Zentralamt (1993); Statistisches Amt der Stadt Wien (1994)

Tenant protection

The state Tenancy Act protects tenants against eviction, giving them security almost like that of home-owners. Close relatives who have lived in the same household are entitled to enter into tenancy contracts on more or less the same conditions. It is interesting to note that this feeling of security has led to vast investments by tenants in their flats, making the subsidised 'Tenants' Modernisation Programme' one of the most important means of housing renewal.

While limited tenancy contracts are possible in Austria (the limit has to be exactly three years; if the contract is extended, it becomes unlimited; for limited contracts the rent must be decreased by 20 per cent), this is not the general rule. Although the number of limited contracts has increased since 1994, about two-thirds of new contracts are still without such limits.

The owner-occupied sector

With 55 per cent of all dwellings, Austria has one of the highest rates of home-ownership in Europe (Table 8.1). Given the strong role of the public sector this may seem surprising, but it mainly results from two factors: the rural character of large parts of the country with a high share of single-family housing, and the subsidies given for the construction of owner-occupied flats. In fact, some of the nine federal provinces (*Bundesländer*) – those ruled by the Conservative Party – have concentrated on subsidising this form of dwelling, mostly by way of non-profit housing associations. Therefore only a small proportion of owner-occupied dwellings are completely privately financed; especially in rural areas the system of *Bausparkassen* (building savings banks, supported by the state through tax deductions) has proved very successful, whereas in urban areas privately financed condominiums only constitute a small 'luxury' segment of the housing market. Subsidies for the owner-occupied sector may take various forms according to the decentralised political and administrative structure, but they are in nearly all cases related to certain income limits. As there is only a small private market, however, these income limits are rather high, thus making subsidised housing – even condominiums – a middle-class programme rather than one for poor families.

The situation is entirely different, however, in Vienna, where owner-occupied flats have received only a minor share of the subsidies. Because of housing traditions and the structure of the city, moreover, single-family housing is less important than in the rest of Austria. With only 14 per cent of the stock consisting of owner-occupied flats, Vienna is a classic tenants' city.

HOUSING INVESTMENT, FINANCE AND SUBSIDIES

Since the end of the Second World War there has been a broad political consensus about the necessity of public promotion of housing, although there are, of course, different opinions about the measures to be undertaken: direct or indirect means, object or subject subsidies, public loans or bank financing.

Housing promotion has been of great importance for new housing construction. Of post-1945 housing 61 per cent has been financed with public subsidies. In multi-storey buildings this share even amounts to approximately 76 per cent, while for single-and two-family homes it is 45 per cent. No figures can be given for the owner-occupied sector, but it is important to note that in 1948 owner-occupancy was introduced, together with promotion of reconstruction after war damage and since then it has been subject to public subsidisation.

Public funds derive from income and corporation tax (approximately 10 per cent of the total amount) and by an earmarked housing tax (*Wohnbauförderungsbeitrag*: 1 per cent of incomes paid by employers and employees, see Table 8.3). The federal government is in charge of collecting this money and of its distribution to the federal provinces. The total sum amounted to some ASch 25,000 million in 1994 (1,923 million ECU; ASch 3,100 or 238 ECU per capita), which is nearly 1.9 per cent of GDP.

Table 8.3 Funds generated from taxation, Austria, 1991

Taxes	ASch billion	ECU billion	Share in %
General income tax (by employees)	10.6	0.82	53
Income tax (by independent workers)	3.1	0.24	15
Investment income tax	0.3	0.02	1
Corporation tax	1.4	0.11	7
Earmarked housing tax (1% of all salaries)	4.7	0.36	23
Total	20.1	1.55	100

Source: Köppl (1994)

Repayments of public loans serve to fund the provincial housing budgets, enlarging the disposable money by approximately ASch 4,000 million (307 million ECU) each year, which is about 15 per cent of the expenditure of direct housing subsidies.

The great majority of flats built after 1945 and used as main domiciles have been constructed with the support of government subsidies. The Austrian housing subsidy system currently consists of three main types: direct subsidies under the housing promotion laws (object subsidies and subject subsidies), object subsidies via subsidised loans, and subject subsidies via tax deductions. Frequently, these types of subsidies are combined. The major portion of government subsidies is allocated to direct object subsidies. On average, about three-quarters of federal expenditure on housing subsidies is used for this purpose (Table 8.4).

Currently about 20 per cent of the subsidy funds (16 per cent without Vienna's budget shares) are spent on the rehabilitation of old buildings; of the remainder, roughly one-third each goes into the construction of owner-occupied single houses, into owner-occupied flats and into rented flats. In Vienna, the share of subsidies for rehabilitation and for rented flats is much higher.

Table 8.4   Government spending on housing subsidies, Austria, 1991

Types of subsidies	ASch billion	ECU billion	Share (%)
Object subsidies for new housing[a]	15.2	1.17	55.6
Object subsidies for rehabilitation[a,b]	4.4	0.34	16.2
Subject subsidies[b]	1.4	0.11	5.1
Premium to savers in building banks	2.1	0.16	7.7
Tax deductions to tenants of subsidised houses	4.3	0.33	15.4
Total	27.4	2.11	100.0

Source: Köppl (1994)
Notes: [a] Via provinces
[b] Not including Vienna's budget spending, which amounts to approximately ASch 15 billion (ECU 1.15 billion)

The direct object subsidies granted in Austria guarantee a steady volume of new housing, but they also enable the subsidising authorities to influence the price, quality and allocation of flats. Since housing costs still exceed the means of low-income groups, object subsidies are supplemented by income-related individual allowances.

THE DEVELOPMENT OF POLICY: VIENNA

The strong role of the state and of the public sector in housing can only be understood by examining its historical background. Public housing had slowly been developed in the nineteenth century, the Austro-Hungarian Empire only gradually following countries like Great Britain and the Netherlands; even the German housing co-operative movement had started much earlier, whereas in Austria – with a few exceptions like the state railway housing programme and some estates built by private enterprises for their workers – the whole field of housing was left to the private market. Consequently, the housing situation in Austrian cities was amongst the worst in Europe, causing not only a high rate of homelessness in urban areas, but also the infamous *Bettgehertum* (beds in overcrowded rental dwellings which during the day were let to sub-tenants so that the tenants could afford the high rents), one of the lowest housing standards in Europe, and – not surprisingly – widespread epidemics. Tuberculosis, for some decades, was internationally known as 'Vienna disease'. These disastrous experiences with an unlimited private market may partly explain why privatisation has until now not become a major political issue.

The year 1918 marks a turning point in Austrian housing policies. Already, one year before, tenant protection measures had been introduced – meant as temporary war legislation to protect soldiers and their families against eviction; ironically, the essential parts of this first Tenant Act are still

119

in function. One of the results of this legislation – after a period of hyper-inflation had made rents practically worthless – was the complete collapse of private building. The task of housing was taken over by the state – or, more precisely, by the newly created *Bundesländer*, and especially by Vienna, which became a province only in 1923.

For social-democratic 'Red Vienna' a regulatory building and housing policy was the key to socially oriented policies at the municipal level. Such a policy was to be a compensation for the market economy that had placed housing out of reach of the socially disadvantaged. Within a decade such buildings as Karl Marx-Hof or Werkbundsiedlung provided a total of some 64,000 new, comparatively modern, flats plus a variety of community facilities. Housing construction in Red Vienna was financed through a housing construction tax (*Wohnbausteuer*) imposed mostly on wealthy citizens. At the same time numerous housing co-operatives sprang into being with blue-collar workers building flats as part of self-help initiatives. The movement was given full support by the City of Vienna, which helped to construct several estates with a total of more than 15,000 houses.

In 1934 'Austro-Fascism' brought an abrupt end to the endeavours of Red Vienna, and new housing construction more or less came to a standstill.

The destruction of the Second World War – 90,000 flats being reduced to ashes by heavy bombing – was followed by a period of reconstruction in the 1950s and 1960s. The rapid economic growth and increasing prosperity stimulated demand for more and for better-equipped dwellings. Each year more than 10,000 flats were built in Vienna, a city of about 1.5 million, housing provision being made possible primarily through substantial public subsidies. Building was commissioned mostly by the City of Vienna and by non-profit housing associations.

The role of the city in new housing

Within Austria, Vienna in many ways assumes a unique position (not only is it the capital and the largest city, by far, with a population of 1.5 million in 1991 out of a national total of 7.9 million, but it is also a *Bundesland*). It is therefore particularly interesting to examine policy with specific reference to Vienna alone, as an example of housing problems and solutions in a major city. Since the days of 'Red Vienna' the municipality has been actively participating in the construction of new dwellings to an extent virtually unrivalled by any other European city in its class. Vienna's pioneering role in council housing has made it one of the world's leading landlords, with today more than 220,000 municipal flats. Although within new public housing there has been a significant shift from council housing to subsidised non-profit associations the council housing programme is still being continued and has even been increased recently.

Currently approximately 3,000 council housing flats are constructed every year, out of a total of 8,000 to 10,000 new dwellings in the public sector (for a population of currently 1.6 million). On the other hand, only 1,000 to 2,000 dwellings per year are erected by the private, non-subsidised market; in other words, the city is actively involved in 80 to 90 per cent of new housing, which gives it a dominant role in the market and in allocation policies.

Different models of promotion have been developed, the finance coming both from federal taxes and from the city budget. As an average, 15 per cent of the city budget is used for housing purposes – including new construction, urban renewal and individual allowances. The various public sectors aim at serving different income groups – the council dwellings being the cheapest ones with the lowest income limit – but the overall objective is to create fairly mixed new communities and to avoid social segregation. Therefore there is a general political agreement to maintain a strong public sector serving large parts of the population rather than to reduce it to a poverty programme. Pressure to sell off public housing has come from parts of the Conservative Party, but this has not been continued after a rather disastrous attempt to privatise council housing in the city of Graz (where practically all tenants have refused to buy, even at below market prices).

Urban land management

In order to prevent building promoters from competing with each other in the real-estate market, applications for housing promotion have to be made in close co-operation with the 'Vienna Land Procurement and Urban Renewal Fund' (WBSF) which was founded by the city in 1984. As 80 to 90 per cent of new housing depends on subsidies, WBSF has gained a near-monopoly in purchasing land for new housing. (There is a similar organisation for the business sector.) Between 1984 and 1994, WBSF provided housing associations with more than 2.5 million square metres of real estate for a total of 32,000 flats. This helped to keep land prices under control, although a general increase could not be avoided as the Fund still has to compete with the private market. The major tasks of WBSF include not only the purchasing and selling but also the developing of land, including the requisite preparation for zoning and land-use plans, concepts for future uses, identifying the required engineering and social infrastructure, organising a planning process along competitive lines, seeing to an adequate mix regarding future uses and building promoters, as well as a variety of assistance programmes, and environmental schemes for local recreation including sports facilities and playgrounds. It is only after such planning that the land is sold to the building promoter. Other Austrian provinces are considering setting up similar institutions to provide affordable land for social housing.

WOLFGANG FÖRSTER

Allocation problems

Even with the undeniable and generally recognised success of Vienna's housing policies some problems continue to exist. These are mainly concerned with either the rise of construction costs (above the rate of general inflation), which makes new housing considerably more expensive than the existing stock, or with the city's allocation policies. While income limits exist when applying for any kind of public housing, a higher income in later years does not lead to higher rents; therefore many of the older public housing estates with costs far below the new ones have become middle-class communities whereas young families with lower incomes largely depend on the more expensive new housing projects. (It must be added, however, that because of an annual fluctuation of about 2 to 3 per cent the city can let at least 7,000 flats out of the existing stock every year – an important social argument against privatisation.)

The most severe problem, however, is the situation of foreign immigrants in the housing market. Immigrants (unless acknowledged refugees or already possessing Austrian citizenship) are by law excluded from council housing; even worse, they are in practice excluded also from the rest of the social-housing sector, as they are not eligible for individual allowances. Consequently, the vast majority of immigrants depend solely on the private market where they are crowded in the worst segment of old rental dwellings. At the same time, they often pay higher rents than Austrians (and above the legally fixed rent limits for so-called substandard flats). As these flats are mainly in a few of the inner districts immigrants concentrate in these areas, speeding up the process of segregation.

Urban renewal

Unlike most other European capitals, Vienna is still characterised by the products of massive housing construction during the age of intensive housing promotion. Nearly 40 per cent of the current housing stock dates back to the period before 1918 (Tables 8.2 and 8.5). Therefore significant sums of money from public sources are made available to maintain and improve the mostly private stock of old houses as well as the 1920s council housing estates.

Table 8.5 Housing stock: periods of construction, Austria

	Austria	Vienna
Before 1918	21	39
1919–44	10	12
After 1945	69	49

Sources: Donner (1990); Statistisches Amt der Stadt Wien (1994)

Remarkably, the city authorities have never resorted to any large-scale clearing and renewal schemes or other 'hard' measures such as have resulted in tenant protests, evictions and social ghettos in many European cities. This was due, on the one hand, to Vienna's long marginalisation at the eastern edge of free Europe and the resulting reluctance of private investors to commit large sums of money and, on the other, to the insight gained from misguided policies observed elsewhere. In addition, tenants' rights are more strongly protected by legislation and precedent than in most other countries, so that there has been a tendency to develop fairly 'soft' renewal strategies.

The mandate of the Vienna Land Procurement and Urban Renewal Fund (WBSF) therefore also includes preparing and implementing urban renewal projects, and, in particular, giving advice on, co-ordinating and supervising municipally assisted housing improvement schemes. The most significant renewal strategy is what is termed basic renewal (*Sockelsanierung*), that is, preserving, improving and modernising old housing stock without moving tenants. Such projects begin with the renovation and improvement of the house in question and the modernisation of the flats in accordance with the tenants' wishes. At the same time, preparations are made for the implementation of an overall housing improvement plan by installing the supply (electricity and/or gas) mains and disposal pipes required for future improvements and merging small flats into larger ones.

By June 1995, applications for housing improvement assistance had been filed for approximately 5,400 buildings, and applications had been approved in respect of 2,800 buildings with a total of 128,000 flats. The grants made gave rise to rehabilitation activities at approximately ASch 30.6 billion (ECU 2.35 billion). New strategies included block improvement schemes aiming at comprehensive urban renewal and at the same time actively involving all parties concerned.

CONCLUSIONS

The strong tradition of social housing has made Austria one of the countries where the proportion of this sector is clearly above the (west) European average and is still being increased. With very few exceptions – for example, the Conservative Party's demand to privatise Vienna council housing some years ago – this role of the social sector has never been questioned, and that may partly be explained by the close connections of many non-profit housing associations to one or other of the leading political parties.

Changes are on the way, however. The collapse of communism and the opening of Austria's eastern borders, together with political turmoil in neighbouring countries and new economic development as a consequence of Austria joining the EU, have led to mass immigration, to strong burdens on the budgets, but also – and even more importantly – to questioning the traditional welfare system. The so-called political '*Läger*' (societal fields dominated by one of the two great parties) are clearly diminishing; among the present opposition parties only one – the Green Party – clearly defends new municipal or non-profit rental housing, whereas the newly formed Liberals seem to support a stronger market orientation, and the growing rightist–populist party wants to eliminate more or less all pillars of the existing political system including the non-profit sector. Furthermore, the state budget situation demands new, less costly strategies. Undoubtedly, there is a growing tendency to reduce social programmes and to change the system of promotion from object subsidies to individual allowances for the most needy. As always, changes bear risks. It will, therefore, largely depend on the flexibility of the social sector, its readiness to undergo far-reaching reforms and to develop new instruments, whether Austria will either follow other European countries and gradually destroy its non-profit sector or be able to maintain its leading role.

9

FRANCE

Maurice Blanc and Laurence Bertrand

In common with most European countries, France faced a major housing crisis after 1945. During the Second World War, 400,000 dwellings were destroyed and 1.4 million were seriously damaged. But in France the housing crisis had much deeper roots than war damage. The main cause was the very poor housing situation resulting from the absence of a long-term housing policy between 1918 and 1939. At the end of the First World War, in order to protect tenants, the French government instituted a rent freeze but, because of the government's adherence to *laissez-faire*, it was not coupled with any other housing policy. Private landlords were consequently discouraged from repairing old dwellings or building new ones for rent purposes, and social landlords received hardly any help.

In the late 1950s and in the 1960s, rapid industrialisation compounded housing needs in metropolitan areas. Housing policies tried to meet this challenge but, paradoxically, slum clearance policies and urban renewal programmes did not produce many dwellings. They aggravated inner-city housing shortages, compelling former inhabitants to move out. To meet housing needs, priority was given to a vast construction programme of state-subsidised social-rented housing. Large estates were consequently developed in the suburbs for medium- and low-income households. New industrialised building techniques were experimentally used and, as a result, some of the large outer estates very quickly became dilapidated (Blanc and Stébé, forthcoming).

Three major changes occurred in the 1970s:

1 Priority was given to the construction of industrialised one-family houses for lower-middle-class households receiving state subsidies which enabled them to become home-owners in the remoter suburbs.
2 The building of large housing estates ceased, but the dilapidation of existing ones became a major concern, requiring specific housing improvement schemes in the 1980s, which were unsuccessful in preventing ethnic riots (most of them occurring in large housing estates and not in inner cities).

3 Urban renewal (involving clearance and redevelopment) was discontinued in the inner cities, being replaced by programmes of rehabilitating older housing. Although low-income tenants were eligible for specific housing benefits (see below), rent increases initiated a gentrification process.

The present situation is a complex result of various policies implemented at different periods. The state hesitates between interventionism and a market approach. Both public and private sectors are unable to provide affordable housing for the low-income groups. Homelessness is again a crucial issue.

TENURE

Some significant changes have occurred in the overall French tenure pattern over the last thirty years. Two major trends need to be emphasised: the shift towards home-ownership and the stabilisation of the social-rented sector.

The growth in home-ownership has been very strong. Only one-third of householders were owners of their homes in 1954, but 41 per cent were owners in 1964 and 50.7 per cent in 1982. Despite the slowdown of house-building, this trend has continued more slowly during the 1980s, reaching 53.8 per cent in 1993 (see Table 9.1). The demand for home-ownership remains strong. Home-owners have kept on increasing during the 1980s, both in numbers and percentage, but more slowly than previously. The high levels of unemployment, family breakdown, the slowdown of inflation and high interest rates have reduced solvency and forced households to become more careful in their residential strategies. Only the upper middle classes continued to buy houses. This is the reason why building for the social (subsidised) sector is dropping. New financial facilities to buy real estate, moreover, are contributing to a market shift from new construction to old properties.

The rented social housing sector contained 3.3 million housing units in 1993, which represented 17.1 per cent of the housing stock. In the 1960s, the

Table 9.1 Housing tenure, France, 1984–93

	Social-rental sector (HLM) (000s)	Private-rental sector (000s)	Owner-occupiers (000s)	Other (000s)	Total (000s)
1984	3,362	4,570	10,323	2,109	20,364
1988	3,622	4,292	11,386	1,957	21,257
1993	3,775	4,560	11,913	1,882	22,130
	%	%	%	%	%
1984	16.4	22.4	50.7	10.4	100.0
1988	17.0	20.2	53.6	9.2	100.0
1993	17.1	20.6	53.8	8.5	100.0

Source: INSEE Housing Surveys

development of new social housing was very important, with provision being almost exclusively in large housing estates (including tower blocks). The percentage of social housing in the total stock increased from 6 per cent in 1963 to 9.5 per cent in 1970. In the 1970s, priority was given to the rehabilitation of existing social housing within large estates, and the social housing stock was increasing slowly but steadily, housing one household in six. The private-rented sector declined until 1988, both in numbers and percentage, and a slight increase appeared afterwards (see Table 9.1).

HOUSEBUILDING

New dwellings construction: a dramatic fall

In the 1980s, the level of housebuilding declined until 1986 (Figure 9.1 and Table 9.2). In 1980, housing starts reached 400,000 units. In 1984, it decreased to 295,000 units and remained stable until 1986. In the late 1980s, the construction of dwellings recovered to reach a peak in 1989 but a more important drop occurred afterwards. The worst year was 1993, with only 257,000 starts, the lowest level since the 1950s. In 1994, signs of a slight recovery were emerging.

Table 9.2   New housebuilding starts, France 1980–94

	Subsidised rented sector (PLA)		Subsidised home-ownership sector (PAP) with housing benefit		Home-ownership sector (PC) with housing benefit		Private sector		Total	
	(000s)	*%*	*(000s)*	*%*	*(000s)*	*%*	*(000s)*	*%*	*(000s)*	*%*
1980	60	15.1	120	30.2	100	25.2	117	29.5	397	100.0
1981	56	14.0	126	31.5	82	20.5	136	33.9	400	100.0
1982	64	18.6	127	36.9	81	23.6	72	20.9	344	100.0
1983	58	17.4	115	34.6	90	27.1	70	21.0	333	100.0
1984	55	18.7	113	38.3	92	31.2	35	11.8	295	100.0
1985	65	22.0	93	31.5	105	35.5	33	11.0	296	100.0
1986	60	20.3	86	29.1	99	33.5	51	17.1	296	100.0
1987	54	17.4	78	25.2	114	36.8	64	20.6	310	100.0
1988	54	16.5	60	18.3	108	33.0	105	32.1	327	100.0
1989	50	14.7	48	14.2	105	31.0	136	40.1	339	100.0
1990	47	15.2	38	12.3	102	33.0	123	39.6	310	100.0
1991	60	19.8	33	10.9	90	29.7	120	39.6	303	100.0
1992	63	22.7	30	10.8	70	25.3	114	41.2	277	100.0
1993	72	28.1	32	12.5	50	19.3	103	40.2	257	100.0
1994[a]	78	27.4	47	16.5	42	14.7	118	41.4	285	100.0

Source: French Ministry of Housing and Infrastructures
Note: [a] Provisional

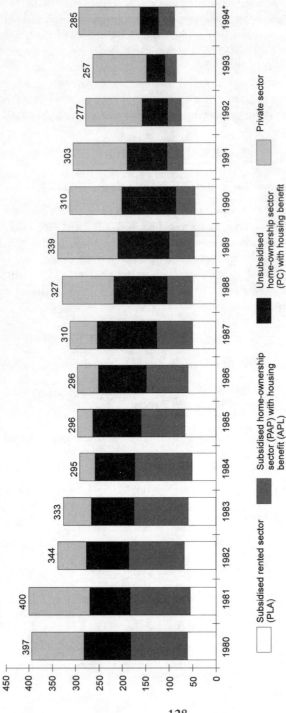

Figure 9.1 Trend in new housebuilding starts in France, 1980–94 (000 units)
Source: French Ministry of Housing
*Ministry assessment

Despite the fact that individual housing was predominant from 1975 to the early 1990s, its volume has regularly declined. It has halved since the early 1980s. The recovery of housebuilding between 1986 and 1989 was because of the construction of flats. The drop in individual housing was particularly steep in the social sector. The proportion of the private-rented sector grew from 11 per cent in 1985 to 40 per cent of the output in 1993 (35,000 units in 1984 to 103,000 units in 1993). The proportion of subsidised dwellings started in the rented sector remained stable, between 55,000 and 63,000 units per year (1984–92), that is, between 19 and 23 per cent of the output, before expanding to 78,000 units (or 27 per cent) in 1994.

Social housing improvement schemes

As early as 1977, a specific regeneration programme (*Habitat et vie sociale* or HVS) was launched on some dilapidated HLM estates, with state partnership. In 1981, after riots in a suburb of Lyon, the new socialist government reviewed the scheme and found that housing had been modernised but often with very little citizens' participation. This was one of the reasons for implementing a new and more ambitious programme in 1982 (*développement social des quartiers* or DSQ – the social development of neighbourhoods and/or communities).

One major idea was that modernisation of housing was not enough, creating job opportunities for unemployed tenants being the crucial issue. There is no evidence of significant results in this field. The controversial programme *régie de quartier* is encouraging self-help groups at a community level to take charge of the good image and the maintenance of the neighbourhood. Some see in it a first step towards professional integration, others look suspiciously at it as a source of cheap labour for cleaning and repairing large housing estates.

Even improved social housing estates are retaining their bad reputation and single-family houses are largely preferred (and idealised) by potential home-owners. In some circumstances, modernisation strategies centred on the community have produced the unexpected result of reinforcing isolation and stigmatisation of social housing estates. For this reason, the programme was renamed *développement social 'urbain'* or DSU in 1988, indicating that a neighbourhood is an element of an urban system and a revitalisation strategy should act at both community and city levels. But renaming the programme is not sufficient to solve the problems of social marginalisation.

Housing conditions

In 1945, of a total stock of 13.4 million housing units, more than 750,000 were substandard (unhealthy, overcrowded, etc.) and only 1.2 million (hardly 9 per cent) were meeting 'modern comfort' standards, specifically running water, inside WC, a shower or a bath, electricity and central heating. Housing, however, was not a priority on the French agenda, which focused on rebuilding (on the same site) dwellings damaged during the war. Compared with its neighbours, France has experienced a longer and more severe housing crisis. According to the 1954 census, more than a third (36 per cent) of the population lived in overcrowded dwellings. The next census in 1962 was more detailed and showed 19 per cent of the dwellings with no running water, 39 per cent with running water only in the kitchen, and only 28 per cent containing either a shower or a bath.

One of the most notable features during the 1980s was the reduction in size of the average household (from 2.7 persons in 1982 to 2.57 in 1990), as a consequence of the increasing numbers of persons living alone (the elderly, divorcees and young adults leaving their parents). The main causes are ageing and new ways of life, creating the diversification of housing needs during the life cycle. The quality of the French housing stock has improved and almost every dwelling now has basic amenities. The number of rooms per dwelling has increased from 3.65 in 1982 to 3.8 rooms in 1990 and the occupancy has decreased from 0.74 person per room in 1982 to 0.68 in 1990.

Homelessness

Strictly speaking, the homeless are sleeping rough on the streets. However, homelessness also includes households experiencing various precarious and/ or substandard housing conditions (shelters, furnished rooms, etc.). According to different definitions, estimates of the number of homeless greatly differ. The third report of the European Observatory on Homelessness (Daly, 1994) suggests that, as a realistic estimate, there were 627,000 homeless in 1992, slightly above 1 per cent of the French population. Homelessness is changing; the same report emphasises two worrying trends: the increasing number of young homeless under 20 and female homelessness.

Forty years ago, in a period of rapid economic growth, homelessness resulted from a shortage on the supply side. Today, in a period of economic crisis and long-term unemployment, homelessness is increasing while un-affordable dwellings remain empty. Good-quality housing, even in the social-rented sector, is too expensive and unaffordable for those in greatest need. Housing benefits are not enough to make the rent affordable (see below). The government is cautiously considering requisitioning empty dwellings for the homeless, but the need for new and inexpensive housing is obvious.

The 1990 Act on the implementation of the right to decent housing (*Loi Besson*) compels local authorities to implement a programme meeting the housing needs of 'disadvantaged persons'. In 1994, there were hardly

10,000 additional social dwellings supplied, compared to a minimum of 30,000 per year required over a period of five years to meet the needs (Raillard, 1995). Accommodating the homeless is not enough. There is the need to help them find their own way into the community. Social workers dealing with the homeless are very concerned. Homelessness is a newsworthy issue and people unanimously agree something must be done for the homeless, but 'not in their back yard'.

HOUSING INVESTMENT, FINANCE AND SUBSIDIES

Housing investment, finance and subsidies is becoming a scientific field; it has created its own jargon and is rich in acronyms! Some of the most frequently used are explained below and there is an overview of how the state is operating in the housing market.

Loans and grants for the construction of dwellings for rent purposes: PLA and PLATS

Before 1977, the state subsidised the HLM sector with long-term grants (40 years and over) at a low interest rate (between 1 and 3.5 per cent). It was replaced in 1977 by loans for rent purposes (*prêts locatifs aidés* or PLA), including a grant (12.7 per cent of the amount) and a loan at a higher interest rate (5.8 per cent). These loans are mainly, but not exclusively, provided to HLM organisations, either for new buildings or for improvement schemes. Private developers are eligible as long as they respect standards of amenity, occupancy levels and, last but not least, rent regulations in respect of the repayment period (32 years). Tenants of these dwellings are eligible for APL (*aide personalisée au logement*) housing benefit. Some of these loans (*prêts locatifs aidés 'très sociaux'*, i.e. 'very social', or PLATS), are targeted on new housing for very low-income households. The grant is higher (20 per cent) but consequently rents must be lower (under 80 per cent of the PLA rent).

Loans and grants for home-ownership: PAP, PAS and PEL

The 1977 reform also introduced a new loan for home-ownership (*prêt d'accession à la propriété* or PAP), with an interest rate between 6 and 7 per cent and a repayment period of from 15 to 20 years. This loan is restricted to new dwellings, or to old dwellings refurbished by professionals (self-help rehabilitation being excluded). Eligible households must be under a certain income level and they are eligible for an APL housing benefit.

A new kind of loan was implemented in 1993 for low-income households willing to refurbish an old dwelling by themselves (*prêt pour l'accession sociale* or PAS). A special savings scheme, created in the early 1970s, is very popular (*Plan épargne logement* or PEL). After a saving period of five years, households can receive a grant and a loan at a low rate (6 per cent), irrespective of their incomes.

Aids for improvements in the public sector: PALULOS

A new grant (*prime à l'amélioration des logements à usage locatif et à occupation sociale*) was also created to improve existing HLM rented stock more than fifteen years old. It covers 20 to 25 per cent of improvement costs, up to a guideline unit cost limit of FFr 85,000.

Aids for improvement in the private sector: ANAH and PAH

The reform also aimed at the improvement of the existing private stock and special grants have been implemented. The ANAH grants (*Agence nationale pour l'amélioration de l'habitat*) are reserved for landlords accepting rent regulations. Home-owners under a certain income level are eligible for a grant (*prime pour l'amélioration de l'habitat* or PAH). These grants are available to provide basic amenities and repairs for substandard dwellings. The government-backed agency, ANAH, has been set up to manage subsidies to the private sector. The subsidies reach 25 to 35 per cent of the repair costs, up to a guideline unit cost limit. But the amount of a PAH grant per dwelling is greatly inferior to the amount of the ANAH grant, because of differences in the cost limits for work aided by these grants.

Grants in public and private sectors are increased in specific schemes such as Social Development of Neighbourhood and/or Communities (*développement social des quartiers*), or Housing Improvement Areas (*opérations programmées d'amélioration de l'habitat*).

Tax incentives

Housing taxation is a very complex field; the state budget needs to reconcile contradictory aims: more resources for the state and fewer taxes as a stimulus for the housing market, either on the supply or the demand side. Annual interest paid for the repayment of a loan for the purchase, or for the improvement, of a main residence are partially tax-free under certain income limits (FFr 459,429 for a household with two children in 1992). Interest from PEL savings and grants are tax-free and new home-ownership is exempt from land taxes.

Private landlords can also deduct from their fiscal income part of their expenses for their rented housing. Construction for the private-rented sector is free of land tax for two years. HLM organisations pay lower VAT, lower duties on the sales of dwellings, and lower land taxes. They do not pay corporate taxation (*impôt sur les sociétés*).

State housing expenditure: 1980–93

Housing expenditure has risen from FFr 38 billion in 1980 to FFr 53.6 billion in 1993, at constant 1993 prices (see Tables 9.3 and 9.4). Taking inflation into account, this increase reached 41 per cent in 1993 (Figure 9.2). Expenditure mostly increased before 1984 and reached a peak in 1986, followed by a reduction to its 1988 level. There was a new increase after 1992, mainly due to housing benefits. Housing benefits currently amount to two-thirds of state housing investment. The level of bricks and mortar expenditure has been reduced from FFr 30 to 25 billion in constant prices between 1980 and 1993, despite a peak of FFr 37.5 billion in 1984.

Table 9.3 Public expenditure on housing, France, 1980–93 (million francs at 1993 value)

	1980	1982	1984	1986	1988	1990	1991	1992	1993
Bricks and mortar aid	30,181	33,139	37,488	35,615	26,788	27,662	25,983	27,181	25,188
Subsidised home-ownership sector (PAP)	2,611	3,437	7,106	9,325	7,667	6,844	6,917	6,269	4,416
Public-rented sector (PLA + PALULOS)	4,585	10,313	12,409	11,519	6,684	5,078	5,060	5,545	6,022
Repairs and improvement in rented sector (PALULOS[a])				2,233	2,042	2,283	2,275	2,581	
Repairs and improvement in private sector (ANAH + PAH)			1,791	2,337	2,731	2,502	2,246	2,445	2,874
Personal housing benefit	7,785	13,286	16,120	18,639	19,715	22,112	20,255	19,826	28,428
New housing benefit (APL)	666	4,633	7,317	11,289	12,954	14,117	13,159	13,429	13,628
Old housing benefit (ALS)	7,119	8,545	8,793	7,332	6,730	7,918	7,096	6,398	14,800
Total	37,966	46,424	53,608	54,254	46,504	49,774	46,237	47,007	53,616

Source: Cour des Comptes
Note: [a] Million constant francs 1992

Table 9.4 State housing expenditure, France 1980–93 (percentages)

	1980	1982	1984	1986	1988	1990	1991	1992	1993
Bricks and mortar aids	79.5	71.4	69.9	65.6	57.6	55.6	56.2	57.8	47.0
Subsidised home-ownership sector (PAP)	6.9	7.4	13.3	17.2	16.5	13.8	15.0	13.3	8.2
Public-rented sector (PLA + PALULOS)	12.1	22.2	23.1	21.2	14.4	10.2	10.9	11.8	11.2
Repairs and improvement in rented sector (PALULOS)				4.1	4.4	4.6	4.9	5.5	
Repairs and improvement in private sector (ANAH + PAH)			3.3	4.3	5.9	5.0	4.9	5.2	5.4
Personal housing benefit	20.5	28.6	30.1	34.4	42.4	44.4	43.8	42.2	53.0
New housing benefit (APL)	1.8	10.0	13.6	20.8	27.9	28.4	28.5	28.6	25.4
Old housing benefit (ALS)	18.8	18.6	16.4	13.5	14.5	15.9	15.3	13.6	27.6
Total[a]	37,966	46,424	53,608	54,254	46,504	49,774	46,237	47,007	53,616

Source: Cour des Comptes
Note: [a] Million francs at 1993 value

Housing benefits: APL and *allocation logement* (AL)

The proportion of housing benefits has increased from 26 per cent in 1980 to 62 per cent of the housing expenditure in 1993. The growth of housing benefits in housing expenditure is coherent with the 1977 reform. However, housing benefits have increased from 8 to 28 billion francs between 1980 and 1993 (i.e. 3.6 times more in constant francs). This is more than expected; the 1977 reform had also aimed at a progressive reduction of housing benefits with the increase of household resources.

The rise was particularly strong between 1980 and 1986, as a result of the new APL housing benefit (*aide personnalisée au logement*) in the public-rented sector, but the existing housing benefit (*allocation logement*) has remained roughly constant except between 1992 and 1993 (when students

FRANCE

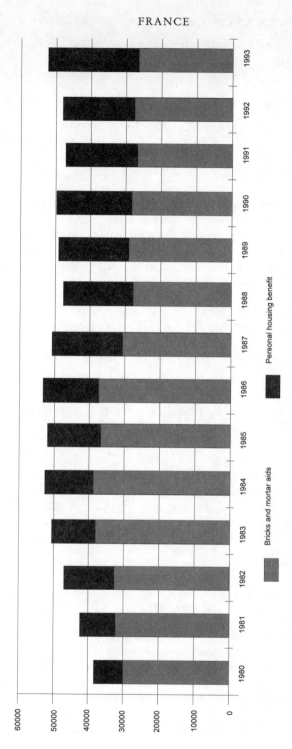

Figure 9.2 Trend in state housing expenditure in France, 1980–93 (million francs at 1993 values)
Source: Cour des comptes, 1994

135

became eligible for housing benefits). Housing benefit is currently received by 5.9 million (one in four) households.

Public loans and grants for construction in the rented sector

Public investment in the rented sector (PLA) increased at the beginning of the 1980s and (together with PALULOS grants) reached a peak in 1983 of over FFr 12 billion (at 1993 values), before decreasing dramatically to about FFr 5 billion in 1991 (Figure 9.3). A slight rise occurred in the early 1990s.

Despite the reduction in expenditure, the number of rented dwellings started has remained roughly constant (between 55,000 and 65,000 units per annum). The cost of resources on the market has also decreased. The state grant per dwelling was reduced in 1987 and social builders had to collect supplementary subsidies from local authorities and firms (through what is called '1 per cent patronal'). The state made a special effort in the early 1990s to revive the construction of social-rented dwellings (47,000 in 1990, 78,000 in 1993).

Public loans and grants for home-ownership

State investment in social home-ownership has steadily risen from 2.66 billion in 1978 to 9.6 billion in 1986 (at 1993 prices). But a dramatic fall has followed (4.4 billion in 1993) (Figure 9.4). The number of financed dwellings (PAP) has declined from 180,000 in 1978 to 42,300 units in 1993 (−76 per cent). The rate of the PAP mortage loans is progressive and has more and more placed a heavy burden on the borrowers, especially for those who became owners in the mid-1980s. With the slowdown in the rate of inflation and the economic crisis, many borrowers could not afford repayment (7.5 per cent of the whole in 1988). At the same time the state has reduced the income limits for eligibility for ownership subsidies. All these factors contributed to the drop of social ownership in the late 1980s and the early 1990s.

Public loans and grants for rehabilitation in the public sector: PALULOS

State investment decreased from FFr 2.5 billion in 1980 to 1.8 billion in 1987 at constant (1993) prices. In the early 1990s, a new improvement scheme was launched (1990–94) and expenditure increased to 2.6 billion in 1992. No reliable data on improved dwellings are available but the number is estimated to be between 160,000 and 170,000 dwellings per year since 1988 (some dwellings might already have been improved before the availability of PALULOS assistance).

FRANCE

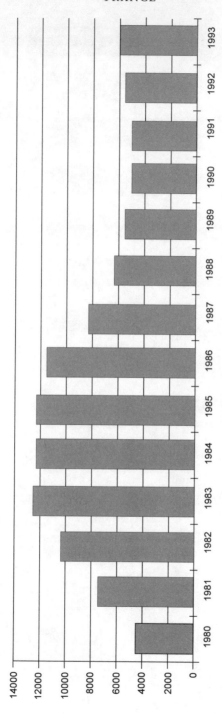

Figure 9.3 Trend in state housing expenditure for subsidised rented sector (PLA + PALULOS), France, 1980–93 (million francs at 1993 values)

Source: Cour des Comptes, 1994

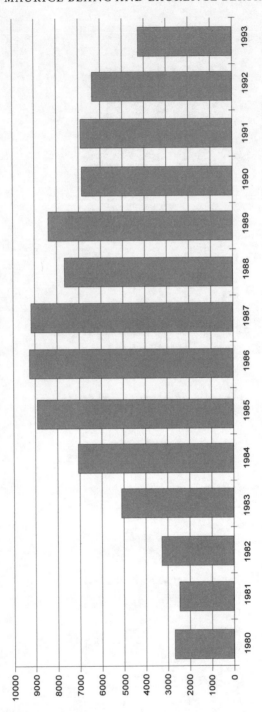

Figure 9.4 Trend in state housing expenditure for subsidised home-ownership (PAP), France, 1980–93 (million constant Francs 1993)
Source: Cour des Comptes, 1994

Public loans and grants for rehabilitation in the private sector: PAH, ANAH

In 1990, basic amenities (bathroom or shower, central heating, indoor WC) were lacking in 24 per cent of dwellings compared to 65 per cent in 1970 and 37 per cent in 1982. State investment in ANAH improvement grants increased steadily from FFr 1.4 billion in 1984 to 2.2 billion (at 1993 prices). PAH subsidies have remained more stable since 1985. But investment in the private stock has remained below the expenditures on the public-rented stock. The gap reached a third in 1992.

The number of dwellings repaired with ANAH grants increased from 34,000 in 1982 to 50,000 in 1989. A decline followed, but a new peak was reached in 1992 when 56,200 dwellings were repaired. With regard to PAH, the number of dwellings was higher at the beginning of the 1980s (55,000 units in 1982) than hitherto, but declined in the mid-1980s (to 43,000 units in 1986 and 1987). It subsequently increased to reach a peak in 1992 (52,200 units).

DEVELOPMENTS IN HOUSING POLICY: THE 1970s AND 1980s

Housing policies are particularly necessary when the market is unable to house the poor and the homeless. Any policy with such an objective has to make a choice between two strategies The first is subsidising the social sector and allowing it to provide housing cheaper than market levels. It is a form of state intervention on the supply side, called in French *aide à la pierre* ('bricks and mortar aid'). The other strategy is distributing housing benefits widely enough to allow the poor to obtain a dwelling on the market. This neo-liberal approach on the demand side is called *aide à la personne* (personal housing benefit).

Most housing policies combine these two strategies, but historically they have emphasised one or the other. State grants supporting ownership of newly built single-family houses emerged in France at the end of the nineteenth century, but they played a marginal part before the 1970s. As previously noted, subsidies for the construction of social-rented dwellings were dominant in the late 1950s and in the 1960s.

Social housing and the state before 1977

French social housing is referred to as *habitations à loyers modérés* or HLMs (housing with moderate rents). The most important HLM organisations are public organisations sponsored by local authorities (city or department), others are either non-profit private companies or co-operatives (Blanc, 1993). In the 1960s, the usual form of central-government financial incentives were long-term loans (over 40 years) at lower interest rates than free-market rates (between 1 and 3.5 per cent, according to the social targets of the scheme). HLM organisations had to repay their loans with the rents they collected. Thanks to the low cost of money provided by the state, HLMs could afford lower rents than the private sector. In order not to disturb the market and to prevent distortions in targeting the poor, access to HLM housing has been restricted. Applicants must prove that their income is under a certain level. But here lies a first contradiction: if HLMs admit the very poor, there is a high risk that they will not be able to pay their rent. Good management implies selecting tenants who will presumably pay their rents regularly. HLM tenants were not rich, but they tended to be steady workers, employees or civil servants with a low to middle income.

The shift towards the market in the 1970s

In the early 1970s, central government moved towards a housing policy favourable to home-ownership. It encouraged the building industry to use industrial processes in order to develop cheap one-family houses on suburban estates. It also began to give financial incentives which appealed to middle-class and skilled working-class families in order to help them become owner-occupiers. Many of these families were formerly HLM tenants. When better housing conditions are available, people in social-rented housing who can afford to buy tend to move out. For that reason, solvent tenants have been leaving dilapidated HLMs.

The 1977 reform of housing subsidies

Household mobility was reinforced by the 1977 reform of housing subsidies. A critical review (Barre, 1976) pointed out that the social housing system was failing to target the very poor. It suggested a market approach giving poor tenants more choice and it inspired the 1977 reform. The introduction of a new and more attractive housing benefit (*aide personnalisée au logement* or APL) was the cornerstone of this reform. APL had two main objectives:

1 Permitting the rehabilitation of substandard housing, in both the private and social-rented sectors, but without excluding poor tenants who were unable to afford the rent increase. They receive an APL which is given directly to the landlord and is deducted from the rent.
2 Encouraging low-income tenants to become owners of new-built suburban single-family houses. In this case, APL reduces the repayable loan.

140

APL: some discrepancies between aims and results

In housing improvement schemes, tenants are eligible for APL only when their landlords, either public or private, have a contract with the state. Because of the state aid they receive, landlords have two obligations: accepting the state regulations of rent increases in improved housing and retaining existing tenants who do not wish to move. Should the rent increase, however limited, be unaffordable, APL should allow tenants to meet the increase. The APL is given selectively in the areas where local and central government have decided to concentrate their action. It is thus a tool in urban policy. But the landlords are free to accept or refuse the contract proposed by the state. In the public sector, almost every HLM organisation has adopted this contract. In the private sector, in order not to have their hands tied, certain landlords prefer to improve their properties without state subsidies and thus be free to follow market prices. Their low-income tenants are not eligible for APL and have no other choice left than to move out.

After some 15 years, the results are paradoxical. In the short term, low-income families live in modernised dwellings and, thanks to APL, pay a lower rent. However, large families who have a right to APL because of their numerous children will in the long term, when their children get older, lose this right, making the rent unaffordable. APL is slowing down (but not preventing) the gentrification process of improved inner-city areas. In the social-housing estates, households with incomes slightly above the official poverty level have no right to APL, so their rents increase greatly (often by 40 to 50 per cent) and they attempt to move out. In contradiction with its aims, APL produces a new segregation, concentrating the poor in modernised (but still socially depressed) HLM estates.

APL has clearly been a strong incentive for social-housing tenants to become home-owners, but many of them underestimated the costs of home-ownership and found themselves in great difficulties with the repayment of their debts, because of high interest rates in a context of diminishing inflation. In the mid-1980s, some measures were introduced for protecting borrowers and helping them to renegotiate interest rates with banks and building societies. It gave birth to an act called the *loi sur le sur-endettement* (act on 'over-debt').

Limits of the 1977 housing reform

The radical housing reform that occurred in 1977 aimed at facilitating housing market restructuring for reasons of greater efficiency, and to progressively reduce the level of state investment. It blended the roles of state and market more closely, in an attempt to integrate the various housing sectors, new construction and old existing stock, private and public sectors, and ownership and rent tenures. The law introduced new tools such as loans for the construction of social-rented dwellings, *prêt locatif aidé* or PLA, or for social home-ownership, *prêt d'accession à la propriété* or PAP. Subsidies have been granted for improving or modernising both public and private existing stock.

In 1977, APL housing benefit was specifically targeted at two-parent families with two children, with low but regularly increasing incomes, and it was expected to cover the costs of improved housing quality. This reform had social and redistributive effects and it was supposedly tenure-neutral. The cost of this benefit to the government was expected to quickly diminish with the general improvement of housing conditions and the resumption of economic growth.

However, the social and economic changes during the 1980s were in the opposite direction. The increase of unemployment, stagnant incomes, and divorces, the decrease in the rate of inflation, the increase in interest rates, and the growing role of banks in housing finance, have reduced investments, both in new construction and in existing stock modernisation. The results of housing reform were far from expectations. APL undoubtedly affected the housing market since low-income households received more help through APL for becoming home-owners of new housing and thereby sustained the building industry during the economic crisis. But while public expenditure on 'bricks and mortar' assistance has been reduced, expenditure on APL housing benefit has hugely increased (Tables 9.3 and 9.4).

The housing market has become more and more rigid and segmented. For example, the social-rented sector is no longer the first housing step in the residential ladder to home-ownership. Since new housebuilding in the social-rented sector or in social home-ownership has dropped, the turn-over in the existing public-rented stock of dwellings has slowed down since 1980, hardly reaching 10 per cent in 1992. Housing market unification is far from being achieved; a complete state withdrawal from the housing field would have dramatic consequences but it can hardly be expected in the foreseeable future.

Adjustment policies

In the 1980s, some adjustment policies followed the 1977 housing reforms to widen the scope of the market, but also to protect social cohesion and to create balance between urban areas. Despite more and more sophisticated public procedures, the government faced growing difficulties in harmonising its social goals with the extension of market regulations. It is now attempting to transfer part of these problems to local authorities through numerous local agreements. However, despite decentralisation laws in the 1980s, financial incentives remain at central-government level. 'Partnerships' are allowing some control over local-authority housing policies.

In the late 1980s, tax relief was intended to revive housebuilding in the private-rented sector. The 1992 INSEE housing survey showed a slight increase in the private-rented stock after contraction in the early 1980s. Most of these new dwellings were small and located in attractive urban areas. Tax relief for housing is obviously popular with a well-off population, but currently the government is considering its reduction. It is, clearly, a very sensitive political issue.

DEVELOPMENTS IN HOUSING POLICY: THE 1990s

In the early 1990s, housing was towards the top of the political agenda. In 1990, the creation of the short-lived Department of Urban Affairs (*Ministère de la ville*) gave a strong impetus and some important laws were adopted, although not necessarily efficiently implemented. In 1990 came the law implementing the right to a dwelling (*Loi Besson*) which introduced the legal recognition of the right to decent housing. To reduce homelessness, local authorities (at both town and department levels) must statutorily implement programmes for housing and integrate policies in respect of 'disadvantaged' persons. Many authorities are reluctant and implementation is not very effective. In 1994, central-government funds for creating rented dwellings for the most needy were not entirely spent (Raillard, 1995).

In 1991 came the law (*Loi d'orientation sur la ville* – guidance law on towns) aimed at introducing a variety of good-quality dwellings as a way to preserve social cohesion and to reduce the imbalance between urban areas. Towns of over 200,000 inhabitants were required to produce a local housing plan assessing their housing needs, to organise social housing construction and to help the local housing market to operate in a better way through public–private partnerships. The municipalities with few social dwellings have to choose between building new social dwellings and paying a special tax for the increase of the whole social-rented stock in the town. These local housing plans were slowly being implemented but, after 1992, the regulations were relaxed.

New construction was at its lowest in 1993 (Table 9.2 and Figure 9.2) and central government bent its market approach, again being active on the supply side. Priority was now given to the construction of new housing instead of to improvement grants. A special effort has been made in recent years for the construction of new social-rented housing (72,000 units in 1993 and 78,000 estimated units in 1994). However, social housing has dramatically dropped, from 120,000 units in 1980 to 32,000 in 1993. Its proportion has fallen from 30 per cent at the beginning of the 1980s to 12 per cent in the early 1990s.

Social landlords remain major partners in the implementation of local housing policies. But they are in a paradoxical situation, between market forces and public service. More and more, they have to house the poor and at the same time they have to incur higher taxes and other financial burdens (loan repayments, unpaid rents and cost of repairs). They complain that their role is more and more difficult and they demand a reform of finance. They also ask for better partnership arrangements during the preparation of local housing plans. At the same time, private charity organisations have developed responsibilities towards the homeless in partnership with the public authorities, and sometimes they find it difficult to work with HLM organisations.

Future housing needs

Assessing future housing needs is always a risky task. The INSEE housing-need assessment model (Lacroix, 1992) takes into account: (a) demographic projections, calculating housing requirements for new households; (b) housing need with the renewal of the existing stock; (c) the need for leisure dwellings. The model estimated that housing needs would be 365,000 dwellings per year between 1990 and 1995 and 316,000 dwellings per year between 1995 and 2025.

Housing needs related to demographic trends form some 60 per cent of the whole and are not expected to decrease in the short term. Housing requirements should decline by the end of the period (2025) as a result of the ageing of the population and of the fertility rate decreasing to 1.8, which is not enough to maintain population size. The reduction in the average household size (from 2.88 in 1975 to 2.57 in 1990) should continue, to reach 2.3 in 2020. This is a result of various factors, such as staying longer in the parental home for many young adults, longer life expectancy of aged couples, and increases in the number of one-person households and one-parent families.

The growth of the population over 15 should amount to 17 per cent between 1990 and 2020, and the increase in the number of households should be as much as 23 per cent over the same period.

Demand might change, becoming more sensitive to housing quality. Households might be more attracted by inner-city rented flats with shopping and leisure facilities in the neighbourhood. They might also require more space, especially the elderly. This concern for quality appears in the regional distribution of housing requirements. Housing demand should be higher in the South-East (the Mediterranean coast) and the Ile-de-France (Paris region). The pattern of single-family houses in ownership in the suburbs is unfitted to these new aspirations. Economic and employment uncertainties will influence the tenure choice of households, with perhaps a greater interest in renting.

As in the previous years, the renewal of the housing stock is assessed at 60,000 dwellings per year (conversion, demolition, merging). Despite the recent recovery in new construction (277,000 units in 1992, 257,000 units in 1993, and 285,000 units in 1994), the foreseeable needs will largely be in excess of the actual level of supply. The housing market deficit will clearly mean continuing housing problems for low-income households.

CONCLUSION

Since the 1977 reforms, and even before, French housing policy has aimed at market enlargement and at a good partnership between the state and the private sector. Nowadays housing issues divide the government. It hesitates whether to act on the supply side to fulfil specific social goals or to maintain its faith in the market. Inevitably, therefore, partnership between private and public sectors remains firmly on the central-government agenda.

Previous housing policies have indeed increased the level of housing quality and at the same time the costs of building and the levels of rents, particularly in the social-rented sector. The development of home-ownership has led to a reduction of the private rented stock. But recent trends in demand show the need for affordable rented dwellings, not only for the poor but also for the elderly, divorcees and the young.

It is also more difficult to accommodate in the public rented stock the growing numbers of people who are in an unstable situation as a result of economic circumstances and/or family problems. This shows the ambiguous role of the social-rented stock. The old and cheap private stock of dwellings, where the poor traditionally received shelter, is also disappearing with the general trend of improvement activity, followed by rent increases. This emphasises the need for public intervention in the private-rented housing sector.

Despite the fact that central-government is retaining its control of housing finance, local authorities are more and more expected to manage the urgent needs of dwellings. This trend strengthens the legitimacy of local housing policies which are still incurring great difficulty in their development and their implementation. The relationship between central and local authorities is increasingly complex. At the local level, contradictions also remain between community development and economic development, and between competing towns in a context of decentralisation. Nowadays, housing issues are a sensitive electoral issue, at both local and national levels. This might lead to new reforms in housing finance.

Part III

THE DOMINANCE OF OWNER-OCCUPATION

10

INTRODUCTION TO OWNER-OCCUPATION

Paul Balchin

Ireland, Spain, Italy and the United Kingdom are a very diverse group of countries – demographically, economically, socially and politically, although curiously they are all located around the western or southern fringes of Europe. In each country, and for different reasons, owner-occupation is the dominant tenure. (It is perhaps interesting to note that several other countries in Europe with very large owner-occupied sectors are also on the geographical periphery of the continent: Finland, Norway, Portugal and Greece.)

Ireland is probably the only country in western Europe to have experienced a long-term reduction in the size of its housing stock. Whereas the total number of dwellings amounted to 941,000 in 1911, the total housing stock in 1961 had fallen to 676,000 – a result of emigration, abandonment and clearance. The condition of its housing, moreover, was very poor: in 1946 only one-sixth of its dwellings had all basic amenities, and one-fifth were intensely overcrowded (Power, 1993).

Spain, although experiencing a population increase of 1 per cent per annum throughout the 1960s and 1970s, incurred large-scale net emigration to the rest of western Europe until about 1973. There has since been net immigration, particularly from north Africa, and continuing large-scale migration from rural to urban areas – 19 per cent of the country's population living in cities of more than 100,000 in 1940, and 42 per cent living in cities of this size in 1981 (McCrone and Stephens, 1995). Although the 1950 Census indicated that there was currently a housing shortage of 1 million dwellings (a legacy of the destruction of the Civil War of 1936–39 and a period of slow recovery in the 1940s and 1950s), by the 1990s 15 per cent of the housing stock was vacant and 15 per cent of dwellings were second homes (McCrone and Stephens, 1995).

Following the damage inflicted upon the built environment by the Second World War, Italy suffered a severe housing shortage well into the 1950s, but large-scale housebuilding over two decades produced a crude surplus of dwellings by the late 1960s. There was still, however, much overcrowding – particularly in areas of net in-migration, while poverty

149

increased as segments of the population – the unemployed, the elderly and one-parent families – were disadvantaged by rapid economic change. The decline in the rented sectors also had an adverse effect on the welfare of low-income households and young families. There was also concern that rapid suburban development failed to take environmental considerations fully into account, while marked quantitative and qualitative differences in housing provision between the north and centre of Italy and the less-developed south remained.

The United Kingdom also suffered from the wartime destruction of a proportion of its housing stock. Over 200,000 dwellings were lost during the Second World War and a further 250,000 were severely damaged. This, together with very little new housebuilding in the war years, resulted in an overall shortage of 2 million dwellings in 1945 (Holmans, 1987). Much of the existing stock, moreover, was old and in a very poor condition, yet the housing shortage in the post-war period should not have been an insuperable problem. Whereas in the years 1945–90, the population of the Netherlands increased by 60 per cent and that of France, Germany and Spain by 40 per cent, the population of the United Kingdom grew by a modest 16 per cent (McCrone and Stephens, 1995), making it possible for a crude housing surplus to be obtained without a rapidly accelerating demand on resources.

TENURE

Ireland has the largest proportion of owner-occupied housing in the EU, an estimated 80 per cent in 1995 compared to an EU average of 56 per cent. The social-rented sector is relatively small – 11 per cent compared to 18 per cent in the EU, while (except for Italy) Ireland has the smallest private-rented sector in the EU – only 9 per cent compared to 21 per cent in the Union (Table 10.1). Since the 1950s, the owner-occupied sector has grown substantially, the local-authority stock has grown marginally, the number of housing-association dwellings has been negligible, and the private-rented sector has declined dramatically – largely because even low-income households could afford to buy and marginal households had their needs satisfied in an adequate supply of low-rent local-authority housing. Because the private-rented sector has shrunk due to its inability to compete with other sectors and functions in tenurial isolation, there is clearly a 'dualist' system of renting (Kemeny, 1995).

The high proportion of owner-occupied housing in Ireland (although a legacy of tenure reform after independence and of the country's overwhelmingly rural economy), is now largely attributable to the impact of the Consolidated Housing Act of 1966. The Act resulted in a substantial volume of housebuilding leading, virtually, to a total replacement of the rural stock by a modern owner-occupied sector (Power, 1993). Demand for

INTRODUCTION TO OWNER-OCCUPATION

Table 10.1   Housing tenure in Ireland, Spain, Italy and the United Kingdom, 1995

	Owner-occupied %	Social-rented %	Private-rented %	Other tenure %
Ireland	80	11	9	–
Spain	76	2	16	6
Italy	67	6	8	19
United Kingdom	66	24	10	–

Source: CECODHAS

new housing was facilitated by the availability of loans, grants and subsidies, while sitting tenants of local-authority stock were encouraged to purchase their cottage dwellings by the provision of financial assistance.

Although the 1966 Act reaffirmed the responsibility of the local authorities in the provision and management of social housing (Power, 1993), the sector remained very marginal. It housed larger-than-average households (there were 4.4 persons per household in social housing in the 1980s compared to an average of 3 per household in Ireland as a whole) (Blackwell, 1988); over half the residents in local-authority estates were 16 or under, compared to 30 per cent in Ireland overall; over 70 per cent of heads of households were unemployed (compared to one-third nationally); and estates accommodated four times the national level of one-parent families (Power, 1993). The polarity between owner-occupiers and renters was clearly evident.

Spain also has a proportionately large owner-occupied sector – an estimated 76 per cent of its housing stock in 1995 compared to an average of 56 per cent in the EU (Table 10.1). The social-rented sector (originally owned mainly by the government organisation, Instituto Nacional de la Vivienda, but since 1984 largely under the ownership of the autonomous communities or the municipalities) is, however, the smallest in the EU – only 2 per cent of the stock in Spain compared to 18 per cent in the EU, while the private-rented sector (although smaller than the EU average of 21 per cent) is, at 16 per cent, comparable in size to that of the Netherlands or Sweden. The remainder of the stock is owned by non-profit independent organisations that retain links with devolved government. Since the 1950s, the substantial expansion of owner-occupation has been mainly at the expense of a declining private-rented sector – the social-rented sector never having really emerged on any scale.

The growth of owner-occupation was a combined result of the absolute decrease in the number of private-rented dwellings under rent control, and the failure of social-rented housing to emerge after the Civil War (McCrone and Stephens, 1995). The owner-occupied sector was, and is, the principal recipient of public expenditure on housing. Under Franco, and by means of a highly regressive system of subsidy, the financial institutions were

151

empowered to provide 'official credit' at sub-market interest rates to developers to build houses for sale. From the late 1950s, there was a shift of emphasis from object to subject subsidies. In the allocation of the remaining subsidies, there was a further shift of emphasis from facilitating new house-building to assisting rehabilitation.

In Spain in the 1980s (as in the United Kingdom), there was increasingly a transition from a mixed economy to a free market, not least within the field of housing. One of the least regulated financial systems in Europe emerged and (among other effects) resulted in an explosion of mortgage credit – producing a house-price boom and serious problems of affordability by the end of the decade. The government therefore reintroduced various forms of subsidy to assist house purchase, related progressively to individual income and household circumstances. By the mid-1990s, however, it was unclear whether or not these were just temporary. Also, in the 1980s, deregulation was extended to private-rented housing (in respect of new lettings), in an attempt to revive the sector; but this was followed by a reversion to intervention when, in 1992, the government introduced subsidies for the development of private-rented housing, at a level more generous than those available for owner-occupation (McCrone and Stephens, 1995). Clearly, the very small-social rented stock (amounting to only 12.5 per cent of the rented stock and 2 per cent of the total stock) had very little impact on the private-rented sector and, as such, demonstrated that Spain had a 'dualist' rather than a 'unitary' system of renting (Kemeny, 1995).

The pattern of tenure in Italy not only shows the dominance of owner-occupation, but the very small scale of the rented sectors. Italy has an owner-occupied sector significantly larger than the EU average – an estimated 67 per cent in 1995 compared to 56 per cent in the Union, but the social- and private-rented sectors respectively contain only 6 and 8 per cent of Italy's housing compared to 18 and 21 per cent in the EU (Table 10.1).

As in Ireland and Spain, home-ownership in Italy has expanded substantially in recent years – almost exclusively through new housebuilding in the 1950s and 1960s, but since the 1970s the conversion of rented housing to owner-occupation has also increased the size of the sector. The decrease in the private-rental stock was, in large part, associated with large-scale tenant eviction, while the social-rented sector has frequently been a victim of cuts in public investment and remains very small and in a deteriorating condition. Clearly, the development of a 'unitary' rental system is not practicable since the atrophied condition of both the social-rental and private-rental sectors render them uncompetitive – a 'dualist' rental system of a dominant owner-occupied sector and weak social sector prevailing.

In the United Kingdom, owner-occupied housing constituted an estimated 66 per cent of the total housing stock in 1995, notably higher than the EU average of 56 per cent. In contrast to Ireland, Spain and Italy, the size of the social-rented stock was also proportionately large – larger even than the EU average, being 24 per cent in the United Kingdom compared to 18 per cent in the Union. The private-rented sector, however, was relatively small, only 10 per cent compared to 21 per cent in the EU (Table 10.1). Nevertheless, the decline of this sector is (with the notable exception of Germany) a feature of most European countries, but because of rent control and regulation, slum clearance and the absence of tenure-neutral policy, its contraction in the United Kingdom is particularly marked (McCrone and Stephens, 1995).

The social-rented sector in the United Kingdom differs greatly from that of other European countries: it consists (or consisted) of several different types of landlords – local authorities (by far the largest owners), New Town Development Corporations (until they were largely wound up after 1979), housing associations, Scottish Homes, and Tai Cymru – its Welsh counterpart. But since 1979, the Conservative government, believing that the social sector was too large, employed policies that changed the pattern of tenure in the United Kingdom to a greater extent than similar changes elsewhere in western Europe (McCrone and Stephens, 1995). Policies were adopted which substantially increased the size of the owner-occupied sector (it had been growing steadily throughout most of the post-war period, but expanded dramatically in size in the 1980s), while the social-rented sector was intentionally and substantially reduced in scale (after growing continuously until the late 1970s).

Policy also aimed at reviving the private-rented sector, but generally failed to arrest the long decline from its 1914 peak (Department of the Environment, 1987; Scottish Development Department, 1987). It was clear, however, that whereas in the past the social-rented sector (by virtue of its size and comparatively low rents) might have had an impact on private-sector renting and might thus have almost functioned within a 'unitary' rental system, this was becoming less and less likely as the size of the social-rented sector was reduced and rents increased. The United Kingdom system of renting was thus becoming increasingly dualistic (Kemeny, 1995).

In contrast to several other countries within the EU, there has never been any attempt in the United Kingdom to formulate tenure-neutral policies, indeed policies designed to expand owner-occupation often appeared very much more ideological and contentious than elsewhere (McCrone and Stephens, 1995). Within the social-rented sector, housing provision was dominated by the local authorities – a major difference between the United Kingdom and the rest of the EU. Motivated, in part, by the very problematic nature of this sector, policies were aimed at reducing the scale of local-authority housing (where possible by privatisation). For a long time, local-authority rents had been low, which resulted in a poor level of maintenance and consequential problems of management (Grieve, 1986); there were serious structural problems associated with novel methods of construction employed in the 1960s and early 1970s, and there was often an absence of a social mix on many estates and lack of social amenities, situations which together with below-average incomes were exacerbated by rising unemployment in the 1980s and 1990s (McCrone and Stephens, 1995). Policy towards private-rented housing likewise differed from that generally adopted elsewhere. Rent control and regulation had lasted longer and was more rigid than in any other west European country, except for Spain.

HOUSEBUILDING

By the mid-1990s, almost 40 per cent of Ireland's housing stock dated from the boom years of the 1970s and 1980s when about a quarter of a million new dwellings were completed (Power, 1993). But by the 1980s, a balance between the number of dwellings and number of households had been achieved and this heralded a reduction in the rate of housebuilding – particularly in the social (local-authority) sector. Whereas there were 5,500 local-authority dwellings completed out of the total of 28,000 completions in 1981, by 1988 there were only 1,450 local-authority completions out of a total of 15,650.

In the 1970s and 1980s in Ireland, there was an accelerating shift of emphasis from new housebuilding to rehabilitation. There had already been a marked improvement in the condition of housing: for example, whereas in 1946 only 36 per cent of dwellings had inside piped water, 23 per cent had an inside toilet and 15 per cent had a fixed bath, by 1981 the respective percentages were 100, 85 and 82 (Blackwell, 1988). In 1982, however, a task force was set up by the government to target improvement grants at the worst slum housing in the inner cities – and specifically housing occupied by the elderly; and the Urban Renewal Act of 1986 subsequently targeted (50 per cent) grants to owner-occupiers to build or renovate housing in 14 towns and cities, in addition to providing tax relief to owners for improvement works anywhere in the republic (Power, 1993). In the social sector, a residential works scheme (RWS) was set up in 1985 to improve 80 run-down local-authority estates. Grants of 100 per cent (totalling £63 million over 1986–91) were made available to local authorities and targeted at 20,000 dwellings in 80 of the most run-down estates.

Housebuilding in Spain has been undertaken largely within the owner-occupied sector and has been substantially assisted over the years – first by object subsidies to developers and latterly and indirectly by subject subsidies to housebuyers. The number of subsidised housing starts in the owner-occupied sector increased from respective totals of 200,000 to 465,000 under the Housing Plans of 1951–55 and 1956–60, and a further 65,000 to 100,000 dwellings were completed over the decade without government assistance. Under the subsequent Housing Plan of 1961–76, 190,000 subsidised dwellings were completed in the period 1961–64, rising to 350,000 in 1973–76, but subsidised housebuilding as a proportion of total completions decreased from 60 per cent in 1971 to 52 per cent in 1975 (McCrone and Stephens, 1995). With the shift from object subsidies to subject subsidies in 1978, the total number of (indirectly) subsidised completions increased from 402,500 in 1978–80 to 484,300 in 1984–87, but thereafter declined to 221,700 in 1988–91, or from 49 per cent of total completions in 1978–80 to 54 per cent in 1984–87 falling to 22 per cent in 1988–91 (Banco Hipotecario, *Nota*, various issues).

In the social-rented sector in Spain, almost all the stock was built in the 1950s and 1960s as part of large-scale slum-clearance and renewal programmes – and often with the use of low-standard, low-cost materials. To further the revival of the private-rented sector, subsidised interest rates on loan capital became available to investors. It was anticipated that under the Housing Plan of 1992–95, an additional 30,000 units would thereby become available, adding 17 per cent to the 1991 stock of rented housing (McCrone and Stephens, 1995).

Large-scale housebuilding in Italy in the 1960s and 1970s was on a par with that of France and Germany, with approximately 500,000 dwellings being built per annum. In the 1980s and 1990s, although output diminished to 250,000 to 300,000 dwellings per annum, the scale of housebuilding still remained at least as high as in France and higher than in Germany or the United Kingdom. Whereas in Italy, 5.3 dwellings were constructed per 1,000 population in 1986, in Germany only 4.1 per 1,000 were built, and in the United Kingdom the proportion was as low as 3.8 per 1,000.

A little later than elsewhere in Europe, rehabilitation in Italy began to supplement new housebuilding in the 1970s, and it expanded thereafter. Rehabilitation had accounted for as little as 15 per cent of total housing investment in the 1960s, but by the 1990s the proportion had grown to nearly 50 per cent. Since 1978, legislation has provided local authorities with the financial means and planning powers to undertake large-scale rehabilitation. Local authorities (including the major ones) have consequently produced and implemented plans for extensive housing improvement.

In the 1950s and 1960s, housebuilding in the United Kingdom was substantial, output reaching a peak by 1967 with over 400,000 completions. Both Conservative and Labour administrations gave priority to local-authority housebuilding in the first ten years after the Second World War, but with the ending of material shortages and building licences in the 1950s, the owner-occupied sector soon accounted for the largest share of completions. Whereas in England and Wales alone there had been a housing shortage of 800,000 in 1951, by 1971 a housing surplus of half a million dwellings was recorded. With cuts in public expenditure, total completions decreased substantially thereafter, mainly through dramatic reductions in the rate of housebuilding in the social sector. In 1991, for example, of the total number of 161,500 housing starts only 26,500 dwellings were built for the housing associations and local authorities.

In common with Ireland and Spain, there has been a fairly recent shift of emphasis from new housebuilding to rehabilitation. Under the Housing Acts of 1969, 1974 and 1980, and the Local Government and Housing Act of 1989 improvement grants and a wide range of other assistance have been available for the renovation of housing and the installation of basic amenities, although grants became largely means-tested after 1989 and targeted mainly at needy households rather than primarily at housing in poor condition. Although rehabilitation policy since the 1960s has been instrumental in reducing the proportion of unfit housing and housing without basic amenities, there remains, however, the very major problem of housing in serious disrepair.

HOUSING INVESTMENT, FINANCE AND SUBSIDIES

For several decades, governments in Ireland promoted the expansion of owner-occupation with the use of object and subject subsidies – with a view to ultimately establishing universal home-ownership. Grants were provided to help households build replacement or new housing in rural areas, special grants were available for first-time buyers, and owner-occupiers were generally eligible for mortgage-interest tax relief. After 1985, however, in order to make maximum use of the existing housing stock and to respond to macroeconomic pressures, the rate of housebuilding was decelerated by means of a 'gradual reduction in government incentives to build or buy housing for owner-occupation' (Power, 1993). Mortgage-interest tax relief was cut from 100 to 90 per cent of interest payments, with the maximum amount of relief reduced from £4,000 to £3,000 per annum.

Local-authority housing in Ireland was made available under the Consolidated Housing Act of 1966 to all households unable to provide accommodation for themselves – subject to an income ceiling of £12,000 per annum and, in Dublin, a residence qualification of up to two years. Since local authorities charged differential rents (depending on household income and ability to pay), there was no need to introduce a system of housing allowances – in marked contrast to elsewhere in Europe. The result was often unrealistically low rents, completely unrelated to the cost of provision. Since rent shortfalls were met by the government, there has been an attempt in recent years to limit these payments to provide an incentive for local authorities to charge more realistic rents, but instead reduced funding was often an excuse for poorer services (Power, 1993).

Non-profit housing associations in Ireland, from 1984, were eligible for direct government assistance in the form of subsidised loans for up to 80 per cent of the cost of building, but only in respect of housing for special categories of people such as the disabled and elderly – and subject to a

156

cash limit of £20,000. Because the repayment of the unsubsidised 20 per cent of the loan necessitated high rents, the scheme failed to help those most in need. The government consequently raised the subsidy to 95 per cent but only in respect of housing for the homeless. Overall, the scheme promoted very little housebuilding, the housing-association stock reaching only 1,600 dwellings by 1990 (Power, 1993).

The small private-rented sector in Ireland is still required to provide accommodation for households denied access to local-authority housing (notably single people and low-income childless couples), and to help house the homeless. But although preventing further decline, attempts to revive the sector (through the abolition of rent control on most lettings and rent assistance for households on welfare benefits) failed to expand the sector. Tax incentives were therefore introduced for new-build rented housing in 1984, but they had little impact and were soon discontinued. From 1988, however, landlords became eligible for generous tax relief – a measure which had some success in increasing the supply of private rented housing (Power, 1993), but the sector remains in isolation unequal to compete on equal terms with social-rented housing.

Owner-occupation in Spain has been facilitated in recent years by the deregulation of mortgage lending in 1982 and a subsequent increase in credit from private-sector financial institutions and the state mortgage bank (Banco Hipotecario de España) culminating in a house-price boom in the late 1980s. Home-ownership has also been promoted by a wide range of object and subject subsidies. Until the late 1960s, the national Housing Plans concentrated on the former mode of assistance as a direct means of reducing the housing shortage emanating from the Civil War. Subsidies in the form of grants, loans and tax-exemptions were available to developers largely through the Instituto Nacional de la Vivienda (INV) and the Banco de Crédito a la Construcción, and were associated with a high volume of output. However, a disproportionate number of subsidies helped to fund the development of expensive rather than more widely affordable housing, supply began to exceed demand despite housing shortages, there were many incomplete dwellings, while comparatively poor-quality high-rise development was undertaken within the urban areas – the result of low planning standards (McCrone and Stephens, 1995). There was consequently a shift of emphasis to subject subsidies. To lower the cost of mortgages, interest and grant subsidies became available in respect of 'officially protected housing' (VPO) set up initially by the 1978–80 Housing Plan, and subsequently, low-income purchasers of 'special regime' housing (designated under the 1988–91 Housing Plan) became eligible for assistance. Although, until 1994, only new housing qualified for VPO subsidies, thereafter rehabilitated housing became eligible, and the 1992–95 Housing Plan extended assistance to second-hand dwellings, while under this Plan, and in response to the preceding house-price boom, middle-income buyers were eligible for subsidised loans in respect of 'housing under controlled prices' (VPT) (McCrone and Stephens, 1995).

Owner-occupation in Spain is very greatly assisted by some of the most generous tax incentives in the EU. Mortgage-interest tax relief is highly regressive and capital allowances can be deducted from tax liability. Owner-occupiers are, however, liable to a nominal level of tax on imputed rent income (2 per cent in 1995), limited capital gains tax (CGT) even on their principal dwelling, and value added tax (VAT) of 3–6 per cent on new housing.

In the small social-rented sector there is a minimal degree of policy initiative. As in Ireland, rents are fixed at a subsidised sub-market level, but very low-income households are eligible for further subsidies akin to housing allowances. Because of its very small size, however, the social sector is clearly unable to influence the larger private-rented sector in terms of rents, permitting thereby the maintenance of a dualist system of renting.

In the comparatively large private-rented sector, however, governments have for long intervened in the market. Rent control in the larger urban areas was introduced in 1920, and extended to the whole of Spain in 1931 – the level of regulation remaining high until the 1980s (McCrone and Stephens, 1995). As a result of the Boyer Decree, 1985, new tenancies were deregulated to deter landlords from keeping their dwellings empty during the house-price boom, while older tenancies were subject to either frozen rents or rents index-linked to inflation – frozen rents being applicable to 75 per cent of all tenancies. Legislation in 1994, however, aimed at deregulating the vast majority of tenancies dating from before 1985, and although there is no system of housing allowances for private-sector tenants to compensate them for rising rents, under the 1992–95 Housing Plan they were eligible for grants towards rent deposits, and, as part of the same plan, investors qualified for subsidised loan rates – both initiatives being aimed at expanding the sector (McCrone and Stephens, 1995).

As in Ireland and Spain, the development of a large owner-occupied sector in Italy is supported by government subsidy. Grants and loans (in the form of low-interest mortages) are available to public authorities, builders and households, while mortgagors are eligible for tax relief. Tenants of housing produced and managed by the Istituti Autonomi de Case Populare (IACP) and the municipalities are also recipients of subsidies – funds originating from the national budget or from employee and employer contribution; and the cost of housebuilding land is – in effect – subsidised through the exercise of local-authority compulsory-purchase powers on the basis of sub-market valuation.

In Italy, government intervention in housing markets has fluctuated wildly. In the 1950s and 1960s, coalition governments showed a reluctance to invest in the social-rental sector. Although a total of 800,000 social-rental dwellings were built between 1951 and 1970, 850,000 dwellings were privatised in the social sector in the same period. In the 1970s, however, governments were more favourably disposed to intervention. Under the 1978 Ten Year Plan for Public Housebuilding, 100,000 social-rented houses were to be built each year, while a strong emphasis was placed on rehabilitation.

In the United Kingdom, public policy has for many decades promoted the development of the owner-occupied sector and in a way very different from that employed elsewhere in the EU. There are no low-interest loans to households to facilitate house purchase (unlike in Spain, Italy or France), and there are no housing allowances available to housebuyers (in contrast to France, Germany and Sweden) (McCrone and Stephens, 1995). There is, however, a substantial amount of mortgage-interest tax relief available on mortgages up to £30,000 (in 1995), although relief was progressively lowered in stages from a maximum of 45 per cent in 1992 to 15 per cent three years later. Owner-occupiers are further assisted by an absence of tax on imputed rent income, exemption from CGT on their principal or only house, and by VAT not being levied on new construction.

Social housing in the United Kingdom has also been promoted in a fundamentally different way from that undertaken elsewhere in Europe. As part of a programme of planned cuts in public expenditure since 1979, local-authority borrowing was subject to the very tight direct control of central government through a system of 'credit approvals', yet local authorities were generally not eligible for capital subsidy, in contrast to their counterparts elsewhere in western Europe (McCrone and Stephens, 1995). Instead, under the Local Government and Housing Act of 1989, they were assisted by a recurrent housing revenue account subsidy, which was intended to bridge the gap between costs and guideline rents set by central government, taking into account local-authority responsibility for the payment of housing benefit (a housing allowance). But since housing benefits were often not fully covered by the subsidy, local authorities were obliged to raise rents above guideline levels. In contrast to local authorities, housing associations received a capital grant – the Housing Association Grant (HAG) – but no direct revenue subsidy. Allocated by the Housing Corporation since 1975, HAGs originally provided most of the capital funding required by the associations (supplemented by loans from local authorities), but under the Housing Act of 1988 the HAG contribution to association funding was substantially reduced (falling to 58 per cent by 1985/86), with funding from private financial institutions increasing reciprocally. Rents consequently rose to near-market levels to provide an acceptable return on private investment, while more and more tenants became eligible for housing benefits. Thus with reductions in government investment in both local-authority and housing-association housing, and with rising rents, the social-rented sector is less and less able to exert any downward pressure on pricing in the private-rented sector – the rental market becoming more and more divided into a dualist system (Kemeny, 1995).

Throughout much of the twentieth century, however, private rented housing in the United Kingdom was subject to some form of rent control or regulation. As in France and Germany, it was introduced at the beginning of the First World War as a temporary measure, but in the United Kingdom was not substantially dismantled until the Housing Act of 1988 introduced market rents for new tenancies under assured tenancy and assured shorthold tenancy arrangements – increasing the need for housing benefits. As 'rent allowances', they were not introduced in the private-rented sector until 1972, whereas in France they have been available since 1948, and, in contrast to housing allowances elsewhere in Europe, they cover only a proportion (and not all) of any increase in rent (McCrone and Stephens, 1995). The very rapid decline in the size of the private-rented sector in the United Kingdom since 1913 (when it accounted for 90 per cent of the housing stock compared to only 10 per cent in 1995), was not only attributable to rent control. The very unfavourable tax regime in the United Kingdom puts the sector at a very great disadvantage compared to its counterparts elsewhere in western Europe and compared to owner-occupation. There is an absence of any form of depreciation allowance, incomes from rents are taxed at the landlord's marginal rate of tax (whereas owner-occupiers are exempt from tax on imputed rent income), and rented dwellings are subject to CGT (McCrone and Stephens, 1995).

The cost of housing policy varies from country to country, but in Spain (allowing for tax relief) it amounted to only 1.09 per cent of the gross domestic product (GDP) in 1987 and 0.98 per cent in 1990. In the United Kingdom, however, it rose from 2.7 to 3.3 per cent of the GDP from 1988/89 to 1992/93 – broadly equivalent in scale to Ireland and the Netherlands but not as much as in Sweden where public spending on housing exceeded 4 per cent of the GDP in the early 1990s. Of the £19,946 million of public expenditure on housing in the United Kingdom in 1992/93, 39 per cent was absorbed by housing benefits and 26 per cent by mortgage-interest tax relief (both largely subject subsidies), yet only 12 per cent was directed at housing corporations and 5 per cent was government subsidy to local-authority housing (both essentially object subsidies).

DEVELOPMENTS IN HOUSING POLICY

In Ireland, Spain, Italy and Britain, high levels of owner-occupation were increased further by the privatisation of social-rented housing. In Ireland in 1973, sales of local-authority housing to sitting tenants (previously on a small scale) were extended by the introduction of discounted prices, subsidised loans, mortgage guarantees and grants – a total of 202,000 dwellings being sold off in the period 1973–88. New and more favourable incentives for sitting tenants to buy were introduced in 1984, but many tenants were reluctant to purchase their existing homes. A £5,000 surrender grant was therefore introduced (in 1984) to encourage tenants to buy in the open market. Although intended to make local-authority housing available for the homeless, the grant was taken up by tenants of the worst housing and resulted in a massive problem of empty and difficult-to-let housing (Power, 1993). Surrender grants were thus discontinued in 1987, and in 1988 larger discounts were introduced to encourage tenants (regardless of their length of residence) to buy their existing homes. Thus, despite a large building programme in the 1970s and 1980s, the local-authority stock decreased from about 16 per cent in 1971 to 11 per cent in 1995. To diminish the size of the stock further, the Irish Department of the Environment (1991) thus recommended that estates of new local-authority housing should no longer be developed, that low-income house-buyers should receive special help including loan guarantees from local authorities, and that long-standing tenants should continue to be encouraged to buy their own homes – a proposal facilitated in 1992 in respect of flats (Power, 1993).

In Spain, an active policy was employed in the 1980s to sell off the social sector to its tenants at discounted prices – in large part to enable the Autonomous Communities and municipalities to avoid the cost of repairs and renovation (McCrone and Stephens, 1995); and in Italy, the 1978 Ten Year Plan was soon eroded by subsequent legislation, and (in 1993) the government decided to privatise half of the public housing stock in order to reduce the budget deficit of the IACP and to cut public expenditure.

It was in Great Britain, however, where the extent of privatisation was, in volume terms, the greatest. A large proportion of the local-authority stock has *either* been sold off to its tenants under the Right to Buy (RTB) provisions of the Housing Act of 1980 (and subsequent legislation), *or* transferred to housing associations, *or* sold to private developers and private landlords. From 1980 to 1992, a total of 1.6 million dwellings were transferred to the private sector in England and Wales and an additional 292,000 sales took place in Scotland (in each case most dwellings being sold to their tenants under RTB arrangements) (McCrone and Stephens, 1995). Together with the decrease in the rate of housebuilding in the social-rented sector, sales led to a reduction in the proportion of this sector from 32 per cent of the total United Kingdom housing stock in 1979 to 24 per cent in 1995. Under the Local Government and Housing Act of 1989, moreover, local authorities ceased being 'providers' of new social housing (that role was transferred to the housing associations) and instead became 'enablers' – a shift of responsibility coinciding with large-scale voluntary transfers of social housing from the local authorities to housing associations. By 1995, however, housing associations owned little more than 3 per cent of England's stock of housing – a proportion which was destined to grow rapidly, given policy continuity. A major outcome of privatisation and the decreased

PAUL BALCHIN

rate of housebuilding in the social-rented sector has been a dramatic increase in the level of homelessness. Between 1979 and 1991, England
experienced a 165 per cent rise in homelessness, bringing the number of
'statutory' homeless households to 152,000 (or about 400,000 people). Increases in Scotland and Wales by 1991 brought their numbers to 18,000 and
10,000 households respectively. In many urban areas, the number of statutory homeless exceeded the number of vacant local-authority dwellings – a
problem that can only be solved by a higher rate of new building.

CONCLUSIONS

In each of the countries considered above, owner-occupied housing has
been vigorously promoted by government through the provision of subsidies (which in some cases have shifted in emphasis from stimulating
supply to facilitating demand). At the same time, there has generally been
less and less support for social-rented housing, and particularly for municipal housing. Apart from reduced public funding and the increase in rents
towards market levels (necessitating housing allowances or the equivalent),
the social-rented sector has been disadvantaged by extensive programmes of
privatisation.

Because of the absence of tenure-neutrality in policy, there has been an
increased degree of polarisation between owner-occupation and social renting, and the problem of how to deal with the more deprived components of
housing demand remains largely unresolved. The private-rented stock,
meanwhile, has also decreased in scale – in part because of rent control
but also through unfavourable tax treatment and low investment returns.
Only through the introduction of tenure-neutral policies in respect of
owning and renting are both rented sectors likely to revive, but a strong
and unified rental market would only emerge if tenure-neutrality were also
a feature of policy relating to both parts of the rented sector – a situation far
removed from the prevailing dualist system of renting characteristic of
European countries with a dominant owner-occupied sector.

11

IRELAND

Patrick McAllister

Analysis of the structure of housing finance and tenure in the Republic of Ireland provides some interesting insights into the interaction of history, culture and policy with the evolution of housing provision. It would be inappropriate to examine Irish housing trends without reference to the influence of British policy developments throughout the period of study. Indeed, this does not confine itself to housing policy: the influence of British theory and practice can be seen in urban and other policy areas. In this context, comparison with Northern Ireland highlights some of the main distinctions and similarities between the British and Irish experiences. However, with regard to this chapter, concentrating on the Republic of Ireland the main areas of analysis will be tenure trends, systems of housing finance (including subsidies), the extent and nature of privatisation in the housing process, and current housing issues and problems. The issues of homelessness and race will not be discussed. This is because they are much less problematic in terms of housing policy in Ireland with its low immigration and high emigration.

Any attempt to explain the current tenure pattern in Ireland must begin in the second half of the nineteenth century. At this time it was the province of Ulster that was the most industrialised and urbanised. The rest of Ireland was dominated by Dublin with a largely rural hinterland. It is in the rural areas that the pattern of owner-occupation originates. In the second half of the nineteenth century 'The Land Question' in Ireland became inextricably bound with the national struggle. A combination of absentee landlords, rural poverty and political unrest led to a highly politicised land debate. In order to attempt to pacify Ireland in a political sense, a succession of Land Acts was introduced by the British government, enabling the vast majority of tenants to buy the freehold interest in their properties. The result is that rural Ireland is now dominated by a large number of owner-occupiers. This is in contrast to the United Kingdom where agricultural land is owned by a relatively small number of large landowners. However, the political struggle over land was a factor in producing a national predisposition towards owner-occupation rather than renting.

During this period Dublin was experiencing similar urban and housing problems to those of Victorian London. Similar attempted solutions were also being implemented. The philanthropic movement can be seen in the activities of the Guinness Trust and the Artisans' Dwelling Company. Moreover, the severity of the urban crisis in Dublin forced state intervention in housing provision in Dublin before it had occurred in London. The Artisans' and Labourers' Dwelling Act of 1876 enabled Dublin Corporation to provide housing at rents lower than comparable philanthropic housing.

TENURE

The large-scale sale of agricultural land and buildings to tenants at the beginning of the twentieth century provided a firm foundation for the continued growth of owner-occupation as the dominant tenure. This century has seen steady growth in the proportion of houses owned outright or with mortgage. Table 11.1 illustrates the growth of home-ownership since 1951. This growth has largely been achieved at the expense of the private-rented sector which has declined dramatically following the British pattern. At present the owner-occupied sector accounts for approximately 80 per cent of the total housing stock. In 1987, approximately 45 per cent of all households were outright owners and 34 per cent had a mortgage (Blackwell, 1990). The private-rented sector experienced relative decline until 1971 when the proportion of households in this sector began to stabilise at approximately 11 per cent. Trends in the public-sector housing stock are interesting. Its relative importance peaked in the early 1960s, followed by relative decline until the present. This reflects government policy of selling state housing to tenants at discounted prices. The sale of local-authority housing occurred in the Republic of Ireland before it became a significant feature of British housing policy. Another notable feature is the very limited contribution the voluntary/housing-association sector has made to housing provision in Ireland.

Table 11.1 Tenure, Ireland, 1951–90

	Owner-occupation	Local-authority rented	Private-rented	Housing associations and others
	%	%	%	%
1951	54	11	32	0.5
1961	60	18	17	0.5
1976	71	16	11	0.5
1981	74	12	10	0.5
1990	78	13	9	0.5

Source: Power (1993)

Owner-occupation

Given the common political and ideological support for home-ownership in the Republic of Ireland and the United Kingdom, it is not surprising that they have many of the same systems of subsidy for owner-occupiers. In common with the UK, tax relief is available in respect of 80 per cent of the interest payable up to certain limits. The level of tax relief has been linked to the rate of tax payable. However, the government is now moving towards tax relief available only at the standard rate. There are specific incentives to encourage first-time buyers whereby tax relief is available at 100 per cent of the mortgage interest for the first £5,000 (married) and £2,500 (single) for the first five years in which they claim mortgage interest relief. In common with the United Kingdom, the imputed rental income and capital gains from owner-occupied dwellings are excluded from the tax base.

Traditionally the main source of funds for households wishing to buy their own home has been, as in the United Kingdom, the building society movement. Until the 1980s the building societies dominated the savings and mortgage market, largely through tax advantages afforded to savers. The relatively small number of major societies limited competition and produced a "queuing" system for mortgages. The life-assurance companies played a significant role in the 1960s but became less important players in the 1970s. With financial deregulation in the mid-1980s the banks quickly established themselves as a major force in this market. By 1990 the banks accounted for 34 per cent of the value of all new mortgage loans advanced (Quinlan, 1995).

Until 1987 the public sector was a significant provider of mortgages to low-income home buyers who wished to purchase their local-authority home. This was either directly through the local authority or through the Housing Finance Agency (HFA). The Housing Finance Agency was established in 1981 to raise additional funds by debt issues for lending to low-income purchasers, largely supplanting local authorities in this role (Murphy, 1995). Throughout the 1980s, local authorities and the HFA accounted for approximately 20 per cent of the annual mortgage payments. However, in 1987 this role was effectively privatised. Banks and building societies were encouraged to lend to the typically lower-income groups with limited guarantees from the state.

The system of financing owner-occupation has been biased towards new-build rather than rehabilitation of the existing stock. Purchasers of new houses are exempt from stamp duty whilst purchasers of previously occupied houses have to pay relatively high rates compared to Northern Ireland where the rate for all property is 1 per cent. During the 1970s the building societies tended to favour new-build, often arranging block mortgage schemes with large speculative housebuilders (Murphy, 1995). The first half of the 1980s saw a substantial increase in state subsidies for owner-occupation. Blackwell (1990) estimates that, between 1980 and 1986, tax expenditures on subsidies to owner-occupiers grew by 150 per cent in real terms. However, during the second half of the 1980s the state started to remove this bias towards new-build. In 1986 the £3,000 mortgage subsidy was abolished, to be replaced by a £2,250 builders' grant, which was in turn abolished in 1987. That year marked a turning point in all sectors of the housing market. The growth of the current account deficit forced the government to cut back on public expenditure and the housing sector was to experience marked change in its financing. The continuing cutback on expenditure on tax relief for mortgage interest payments also reflects the government's move away from fiscal subsidies to owner-occupation.

At roughly 80 per cent of the housing stock, this sector can now be viewed as saturated in the sense that everyone who can and wants to buy his or her own home does so. However, the government has been trying to encourage more marginal groups to enter this sector. This is through shared ownership, which involves the house being part owned by the occupier and part owned by the state. The occupier pays rent on the part owned by the state and mortgage repayments on the interest if appropriate. Tenants have the opportunity to buy part or all of the remaining freehold interest in the property as soon as their income allows. This type of arrangement provides an interesting example of overlap between two major tenures.

Local-authority housing

In the Republic of Ireland public-sector housing is administered by local authorities. It is funded by a combination of internal capital receipts (generated by the sale of existing stock) and by non-repayable grants from central government. Local authorities are required to examine regularly housing conditions and requirements, to formulate comprehensive housing programmes, and provide information and advice. They are responsible for the building, allocation and maintenance of all publicly owned housing.

Although the local-authority-owned housing accounts for a relatively small proportion (13 per cent in 1990) of the households by tenure (Table 11.1), local authorities have been responsible for the construction of approximately 30 per cent of the total stock (Blackwell, 1990). The reason for the apparent discrepancy lies in the large-scale sale of state housing to tenants. Given the political consensus about the desirability of owner-occupation as the preferred form of tenure, over 200,000 houses have been sold to sitting tenants. Tenants have been encouraged to buy by a range of incentives, including subsidised loans, discounts, grants and guarantees, and favourable valuations (Power, 1993). In contrast to the United Kingdom, apart from 1989 the peak years of selling were in the 1970s rather

Table 11.2   Sale of local-authority dwellings, Ireland, 1983–93

Year	No. of dwellings sold
1983	3,492
1984	2,732
1985	1,550
1986	533
1987	2,000
1988	4,816
1989	18,166
1990	5,600
1991	3,143
1992	1,332
1993	3,942
Total	47,306

Source: Various Annual Housing Statistics Bulletins

than the 1980s. Table 11.2 illustrates the uneven pattern of sales during the period 1983–93.

At present the tenant is entitled to a 3 per cent discount for each year of tenancy subject to a minimum of £3,000 and a maximum of 30 per cent. Additional incentives include the fact that no stamp duty is payable on the sale of the dwelling and that the local authority bears all legal costs associated with the sale. In 1989 an increased incentive was introduced for a limited period whereby tenants could obtain a 40 per cent discount in price. The reason for the relatively low number of sales in the period 1985–87 lies in the introduction of the surrender grant scheme.

In 1984 the Irish government introduced a cash grant (£5,000) to induce local-authority tenants to leave their existing local-authority house and enter the owner-occupied sector. It was viewed at the time as a relatively cheap way of making local-authority housing available to people on the waiting list. However, the surrender grant scheme had variable impacts on local-authority estates. The majority of grants made tended to be to tenants on the less desirable estates. Moreover, it was the more prosperous tenants who took the incentive. The result was that the cycle of deprivation in the so-called sink estates was made worse. Vacancy rates in the most stigmatised and disadvantaged estates increased dramatically (Powers, 1993). The surrender grant ceased in 1987 following widespread condemnation from groups involved in housing.

The surrender grant scheme accelerated a much broader trend in the local-authority sector – residualisation. It has already been illustrated that the relative importance of local-authority housing has been decreasing since the 1970s. This has meant that state housing is tending to become part of the social welfare net rather than a mainstream housing provider. Between 1983 and 1987 the proportion of local-authority tenants who were (wholly or mainly) dependent on welfare payments increased from 51 per cent to 70 per cent. Local-authority housing has increasingly become the preserve of the unemployed, one-parent families and the working poor.

The private-rented sector

In common with the UK the Irish government has been trying to stimulate this sector because of its role in promoting labour-market flexibility and the lack of public funds for housing investment.The private-rented sector in Ireland has had a similar experience to that in the UK. Its importance as a provider of mass housing peaked in the early part of this century and declined until the 1970s when it stabilised at around 11 per cent (approximately 124,000 units) of the housing stock. However, this includes 22,000 dwellings which are let rent-free, usually as part of the tenant's employment. In order to encourage increased provision of private-rented housing, tax incentives have been introduced. Tax relief is available against rental income for the costs incurred in provision of privately rented housing by new-build or conversion. Initially the tax incentives were only expected to apply until 1991 but the success of the scheme has resulted in the government continuing incentives until 1997. Another fiscal incentive is that the elderly (over 55) are given tax relief on rent payable by them in the previous tax year.

The voluntary sector

In the Republic of Ireland the voluntary housing sector has been less important than in the UK. The National Economic and Social Research Council commented in 1988 that 'Ireland stands in marked contrast to many European countries by its absence of significant voluntary and co-operative housing sectors' (NESRC members quoted in Blackwell, 1988: 53). However, since 1984 capital assistance has been available to approved voluntary organisations that provide accommodation for qualified people. In contrast to the UK, grants are made available through the local authorities to approved organisations. The *Plan for Social Housing* categorises the elderly, homeless, handicapped, victims of violence or desertion, and lone parents as qualified for voluntary housing. Grants can amount to 90 per cent of the cost of a project within certain limits. Interestingly, the Irish government has made funds from the national lottery available for the provision of communal facilities in voluntary housing schemes. The government has also tried to encourage the voluntary sector to become involved in the provision of housing for low-income groups by the introduction of rent subsidies. This is a good example of a shift away from directing subsidies at buildings rather than people. The voluntary sector now accounts for 10 per cent of total capital expenditure on housing (Quinlan, 1995).

DEVELOPMENTS IN HOUSING POLICY

A key document determining the strategic aims and objectives in the Republic of Ireland is the *Plan for Social Housing* (Department of the Environment, 1991). This document has been a fundamental part of a shift in attitudes towards the role of the state in providing social housing needs. The document states that 'the overall approach will be a broader more diverse one [which] entails the introduction of a range of complementary and innovative measures that will reduce the dependence on local-authority housing' (DoE, 1991: 10). The plan sets out what it hopes to achieve. Aims include:

- improved opportunities for community and voluntary housing,
- more choice in housing,
- greater prominence for housing in urban renewal.

There has been a shift towards encouraging rehabilitation rather than new-build. The plan suggests that local authorities should move towards buying existing dwellings with a view to refurbishment. It is also suggested that they should avoid building large estates of new buildings. Associated with a move away from new-build has been the refurbishment and improvement of existing local-authority stock. Resources are now being directed at improving local-authority estates. In this process great emphasis is being placed on tenant consultation and the devolution of management responsibility to local housing offices.

CONCLUSIONS

The 1980s and 1990s have seen some significant changes in housing policy in the Republic of Ireland. It is possible to identify a number of broad strands. The national predisposition towards owner-occupation has been encouraged. Shared ownership has been introduced in order to encourage more marginal groups to enter this sector. However, at the same time, the state has been prepared to decrease the fiscal subsidies available to home buyers. The dominance of owner-occupation has been reinforced by the large-scale sale of local-authority housing to tenants. This policy has exacerbated the residualisation of this sector. The majority of resources are now directed at improving the existing local-authority stock rather than building new houses. However, the government has recognised the importance of diversity in the housing sector and has placed particular emphasis on encouraging the voluntary and private-rented sectors. Moreover, there has been an explicit policy to eliminate the bias towards new-build and to encourage the rehabilitation of the existing stock. It is expected that these broad trends will continue towards the end of this century with policy adapting to changing circumstances.

12

SPAIN

Baralides Alberdí and Gustavo Levenfeld

This chapter primarily analyses access to housing in Spain, and illustrates how the almost total absence of rented accommodation has meant that property ownership is the only realistic option for most. After a consideration of tenure and a brief review of housebuilding, the chapter focuses on investment, finance and subsidies, and analyses recent developments in housing policy. It particularly examines the current controversy and debate about economic policy for the years 1996–99 which aims to reduce the public debt, contain inflation and refine Spain's approach to the European Union. Economic developments will have a considerable effect on housing policy even though the latter is not one of the policies to be harmonised and is the independent domestic responsibility of Spain and each of the other members of the European Union.

TENURE

One of the features that distinguishes Spain from other countries in the European Union is the high percentage of home-ownership. This has not,

Table 12.1   Tenure in Spain, 1960–91

Year	Owner-occupation (000s)	Rented (000s)	Other (000s)	Total (000s)
1960	3,558.5	2,988.2	482.0	7,028.7
1970	5,394.3	2,555.1	554.9	8,500.3
1981	7,629.7	2,168.7	632.5	10,430.9
1991	9,166.1	1,757.5	901.2	11,824.8
	%	%	%	%
1960	50.5	42.5	6.9	100.0
1970	63.4	30.1	6.5	100.0
1981	73.1	20.8	6.1	100.0
1991	77.5	14.9	7.6	100.0

Sources: Instituto Nacional de Estadística:
'Censo de Viviendas de 1960, 1970 y 1981. Resúmenes Nacionales';
'Censo de Población y Viviendas 1991. Muestra Avance. Principales Resultados Madrid, 1992'

however, always been the case, as can be seen from Table 12.1 which shows the historical development of tenancy conditions. Rather, it is the direct consequence of the housing policies followed from the end of the Civil War up until the present day.

Owner-occupation

Figure 12.1 shows the development of access conditions to private home-ownership. It shows the relationship between the mean cost of housing, at a national level, and the mean disposable income in industry and services, prepared by means of a periodic survey by the *Instituto Nacional de Estadística* (National Statistics Institute). It also shows the burden of the first year's repayments of a loan covering 80 per cent of the house price, with each year's prevailing market conditions, both including and excluding any tax relief available.

The figure does not show information prior to 1982, because of lack of data. It can, however, be assumed that the data for previous years would be very similar to 1982, although slightly less favourable to the purchaser.

The following overview can therefore be offered:

1 From the late 1970s up until 1984 house prices remained reasonable, by European standards. Nevertheless, unfavourable financing conditions led

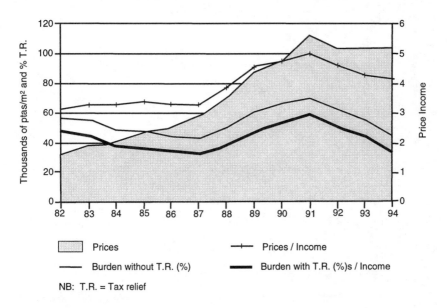

Figure 12.1 Mean house prices/mean salary ratio: level of burden, with and without tax relief, Spain

Source: Ministry of Public Works, Instituto Nacional de Estadística and own sources

to burdens which were virtually unsustainable, even after application of the generous tax relief described later in this chapter.

2 Financing conditions improved in 1984 and by 1987 the burden had become considerably smaller and more manageable. During this period the economy was still deep in the depression which began in the early 1970s and which restricted the demand for housing, allowing house prices to remain stable.

3 The whole of Europe began to emerge from the recession round about 1986. As a result the demand for accommodation, held back for many years, was unleashed. The sections of the population with incomes higher than those considered here greatly increased the demand for housing, although availability did not grow at the same rate. The result was a rare increase in prices which peaked in 1991 (Figure 12.2), the year in which demand began to show signs of waning. During this period the sections

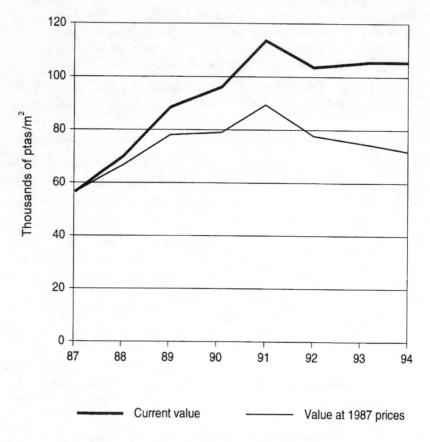

Figure 12.2 Mean house prices in Spain, 1987–94
Source: Ministry of Public Works, Instituto Nacional de Estadística

of the population on lower incomes could not find anything within their means. As a result, the burden incurred that year was more than 70 per cent compared to income (Figure 12.1).

The rise in house prices pushed up the price of building land which had serious effects on the policies put in place, as will be seen in the section on housebuilding.

In 1991 another cycle of recession began and in current terms house prices stopped rising, which means that in real terms prices began to fall slowly (Figure 12.2). Once again, however, the cycle of recession greatly restricted demand, despite the fact that better purchase conditions were seen during the period in question. The housing market was greatly weakened. The recession appeared to come to an end in 1994 and the housing market started to show positive prospects, although prices remained a barrier for the vast majority of the population.

In the period from 1960 to the present, the availability of rented accommodation has seriously diminished and for those needing housing virtually the only option was that of home-ownership. For this reason, measures adopted by successive governments have sought to facilitate this form of access to housing. At the same time, and with the long-term view in mind, legislation has tried to improve options on renting.

Public-rented housing

One of the other features differentiating Spain from other European countries has been the almost total absence of public-rented accommodation. Historically, public-rented accommodation has been promoted, with public help, by private initiative and after a certain period, usually 25 years, it took on the status of privately leased housing. Investment in private-rented accommodation and the promotion of public-rented accommodation progressively disappeared, given its scant appeal to owners.

A survey on rented accommodation carried out in the late 1980s (Ministerio de Obras Públicas, 1989) showed that public-rented accommodation in public property accounted for only 1 per cent of main residences.

Private-rented housing

With regard to the private-rented accommodation policy, it should be pointed out that after the end of the Civil War a freeze on rents was decreed, which, as the result of various rulings, continued until 1964 when the Ley de Arrendamientos Urbanos (Law of Urban Leasing) was announced; this remained in force until 31 December 1994. In 1960 rented accommodation accounted for 45.5 per cent of the total number of main residences.

The 1964 legislation affected only leases signed after that date, whilst the status of accommodation with frozen rents or rent subject to minimal review was unaffected. The new legislation covered free contracting in terms of prices, to be reviewed annually based on the retail price index, but failed to specify any time period for the tenant, enabling up to three members of the family to take over the rent after the death of the original tenant. As a result a property was often unavailable to the owner for more than 50 years. The business of renting was further complicated by the existence of a long and complex eviction process in cases of non-payment.

Slowly, property owners began to avoid renting and the availability of this type of housing reached its lowest levels in the mid-1980s, at which point this type of accommodation virtually disappeared.

In 1985, a ruling without legal force allowed tenancy agreements after that date to be established without restriction on price or duration. There was an immediate positive effect on availability, with a high number of vacant properties coming onto the market.

Notwithstanding the precarious nature of the ruling (because it was not enforceable in law), the coincidental simultaneous boom in house prices and the short-term contracts imposed by owners, which were widely contested by public opinion, meant that this new situation was a temporary one, even though it was clear that greater freedom promoted availability of housing for rent.

According to the survey referred to above (Ministerio de Obras Públicas, 1989), the percentage of rented accommodation reached 11.8 per cent that year, of which 26 per cent were at a frozen rent, with contracts signed before 1964 and of indefinite duration; 52 per cent were under contracts signed after 1964, with the possibility of a rent review in accordance with the retail price index (RPI) but of indefinite duration; and the remainder (22 per cent) were leases after 1985, with price and duration freely agreed by both parties.

In 1991, rented accommodation accounted for 15 per cent of main residences. The 1985 rule led, therefore, to a partial recovery of renting as an accommodation option and halted its spectacular decline.

Finally, in November 1994 a new Law of Urban Leasing (which will be explained later) was approved, and it came into force on 1 January 1995. This law foresees a new era for access to rented accommodation.

HOUSEBUILDING

Table 12.2 shows that the annual level of housebuilding fell slightly over the period 1978–91, but Table 12.3 reveals that there has been an increase in family and main residences which far exceeds that of the population, as well as a remarkable increase in the number of second and unoccupied residences. Likewise, a progressive restructuring of tenancy conditions has favoured home-ownership.

Table 12.2   Housebuildings starts in Spain, 1978–91

	Annual average (000s)	Total (000s)
1978–80	273.5	820.6
1981–83	235.5	706.6
1984–87	222.1	888.5
1988–91	249.0	996.0

Source: Ministry of Public Works

Table 12.3   Housing supply in Spain, 1960–91

		1960 (000s)	1970 (000s)	1981 (000s)	1991 (000s)	Increase 1981–91 %
1	Family residences (1 = 1.1 + 1.2)	7,726.4	10,655.8	14,726.9	17,160.7	16.5
	1.1 Occupied	7,359.7	9,300.1	12,330.7	14,453.7	17.2
	Main	7,028.7	8,504.3	10,430.9	11,824.9	13.4
	Secondary	331.0	795.8	1,899.8	2,628.8	38.4
	1.2 Unoccupied and others	336.7	1,335.7	2,396.2	2,707.0	13.0
2	Lodgings	132.8	53.8	22.9	12.0	−47.6
3	Co-operatives	15.9	23.8	21.9	24.8	26.9
4	Total (4 = 1 to 3)	7,875.1	10,733.4	14,771.7	17,197.5	16.4
	Population	30,583.5	33,956.4	37,746.3	39,433.9	4.5
	Number of households	–	–	10,660.9	11,830.0	11.0

Sources: Instituto Nacional de Estadística:
'Censo de Viviendas de 1960, 1970 y 1981. Resúmenes Nacionales'
'Censo de Población y Viviendas 1991. Muestra Avance Principales Resultados Madrid, 1992'

The most important events between 1981 and 1991 were a large increase in the number of households (11 per cent) and the number of main residences (13.4 per cent) but only a modest increase in the population (4.5 per cent). This led to a significant decrease in the shortage of main residences, of which there are now virtually as many as there are households (see Table 12.3). In view of this fact, since 1991 it could be said that theoretically there has been no housing shortage in Spain. There is, however, a hidden shortage on account of access problems resulting from high housing prices compared with family income.

Likewise, housing conditions have improved considerably. According to the 1991 INE housing census (Instituto Nacional de Estadística, 1991), the mean number of people per residence was 3.3. The average surface area was approximately 75 m^2, with an average of 4–5 rooms per house: 99 per cent of houses had electricity, running water and a WC, 95.5 per cent a bath or shower, 83 per cent central heating and 76 per cent a telephone.

The estimated net change in housing that will take place during the 1990s and up to the year 2009 indicates a downward national trend, beginning towards the end of the 1990s (San Martín, 1993). Nevertheless, there is a significant potential demand for housing, particularly in large cities, which has not been met in the early 1990s, as a result of the economic recession, and which will be fuelled by considerable changes in the building of new houses, especially with the reduction in their size. For example, it is expected that in Madrid the average size of households will decrease from 3.3 in 1991 to 2.6 over the period under analysis.

From a macroeconomic point of view, housing in Spain accounts for 40 per cent of investment in construction and 5 per cent of GDP. Real-estate debt is about 25 per cent of GDP.

One of the more noteworthy, and unfortunate, factors with regard to housing in Spain has been the strong revaluation of the net worth of real estate. In 1994, Spaniards had a real-estate wealth of between 5 and 6 times their disposable income (Naredo, 1993). Housing policy has greatly contributed to the overvaluation of these assets over the last decade, especially with regard to urban and land regulations and the tax position on acquisition. These factors have forced Spaniards to spend a vastly inflated percentage of their savings on this investment.

HOUSING INVESTMENT, FINANCE AND SUBSIDIES

Since owner-occupation was the dominant tenure available, policy has been steered towards improving access to this form of housing.

Public expenditure

From the mid-1970s onwards, governments have been aware of the need to apply tax benefits to house buyers. With very few changes since then, the purchase of property has had the following tax advantages:

- Since it was introduced in the mid-1980s, Value Added Tax has been typically reduced by 6 per cent in the case of new houses. There is another tax, of equivalent value, applicable on the purchase of an existing house.
- House buyers can deduct 15 per cent of the purchase investment from their tax bill annually, with the limitation that the sum to which the 15 per cent applies does not exceed 30 per cent of the base taxable amount. The repayment sums made annually on the loaned capital are considered as investment in housing.
- Interest paid annually on repayments of loans for house purchase can be deducted from the base taxable amount, by way of expenses, up to a maximum of 800,000 Ptas.

The result of these last two measures is in effect a decrease of 3 or 4 percentage points for the borrower compared with the nominal rates of interest. The noticeable improvement that this tax relief has on the level of burden can be clearly seen in Figure 12.1. Table 12.4 shows the reduction in tax which results from tax benefits granted for investment in housing and for interest on borrowed capital used for purchase. It does not include fiscal benefits gained from the application of a reduced rate of VAT. The totals amount to about 0.5 per cent of GDP.

These tax benefits are currently being questioned by some sectors of government, given their regressive nature, whereby the higher the purchaser's income, the more favourable the benefit.

Table 12.4 Tax allowances for housing, Spain, 1990–94 (in billions of pesetas)

	1990	1991	1992	1993	1994
Deductions from the taxable base for interest on invested private capital[a]	100.1	106.1	124.1	136.8	150.8
Deduction from the quota for house purchase or restoration	111.3	128.4	159.4	185.5	183.0
Total	211.4	234.5	283.5	322.3	333.8
As a percentage of GDP	0.42	0.43	0.48	0.53	0.52

Sources: Fiscal expense budgets. Ministry of Economy and Finance and Argentaria
Note: [a] Own estimate

Financing of housing

In the mid-1970s, the Spanish financing system was totally controlled by the Banco de España, with fixed savings and borrowings interest rates. In addition, it compartmentalised its functions. Mortgage financing was the responsibility of the public bank and the savings bank system, the latter being subject to high quotas of obligatory investment. The commercial bank, although it was not *de facto* excluded from this type of financing, in practice rarely used it, restricting itself to short-term financing.

The reform was started in the mid-1970s but was not completed until the mid-1980s, giving rise to a fully deregulated system which resulted in a huge rush into mortgage financing by the banks.

The following milestones stand out in relation to property financing:

1 The obligatory investment quotas were gradually eliminated by all bodies and disappeared in the mid-1980s; this did not lead to the withdrawal feared from the financing of housing of the *Cajas de Ahorro* (savings banks).
2 Until 1982 mortgage financing carried fixed interest rates, which meant that the finance bodies would not grant repayment terms exceeding 8–10

years. In 1982 the practice of applying variable indexed rates was introduced and this became widespread by 1984. The most important consequence of this was the offering of longer repayment terms. If these did not generally exceed 15 years, it was because of the high interest rates that persuaded borrowers to turn down longer terms.

3 The *Ley de regulación del Mercado Hipotecario* (mortgage market regulation law) was approved in 1981 and improved organisation of the market. The most far-reaching new development in the case of housing was the permitting of loan/value ratios of up to 80 per cent, the effect of which is shown above.

Nowadays, house financing is satisfactorily served by the financing bodies without any credit restriction. Banks and savings banks competed hard to attract borrowers. At the end of 1994, the savings banks had 55 per cent of the market and the banks the remainder, but disadvantageously: real-estate debt has increased 4.7 times since 1983 and increased from 14.9 per cent of GDP in 1983 to 25 per cent in 1994.

The introduction in Spain of variable interest rates, which are revised on the basis of objective reference indices, allowed for interest rates which are competitive, in real terms, with those of the major countries of the European Union, although they are pegged to an upswing, in nominal terms, because of the slightly higher inflation rate in Spain than in other countries. The terms offered are of up to 20 to 25 years, although loans are usually taken on for 15 years (for reasons already explained) with interest rates of around 11 per cent. The loan/value ratio normally used is about 80 per cent. Figure 12.3 shows the course of the interest rates in force in Spain since 1987, the year in which data became available, while Figure 12.4 shows the comparative course of the mortgage interest rates, adjusted for inflation, of various European countries, including Spain. These show from the end of 1993 onwards a progressive convergence.

In 1994 a law was introduced in Spain which fostered competitiveness between the credit institutions. This law reduced the high costs which in practice prevented the transfer of a mortgage loan from one institution to another offering lower interest rates, by reducing the notarial and registration rates and the penalty for early repayment and exempting this type of operation from tax.

Figure 12.1 shows the difficulty of access to home-ownership at market prices. Approximately 40 per cent of families have a disposable income no higher than the mean industry and services salary used as the point of reference in this figure. Consequently, direct public assistance became necessary and in 1978 the VPO (officially protected housing) rule was approved. This has remained in force since then with slight amendments designed to channel this assistance more directly to the purchaser and facilitate financing for buying older housing.

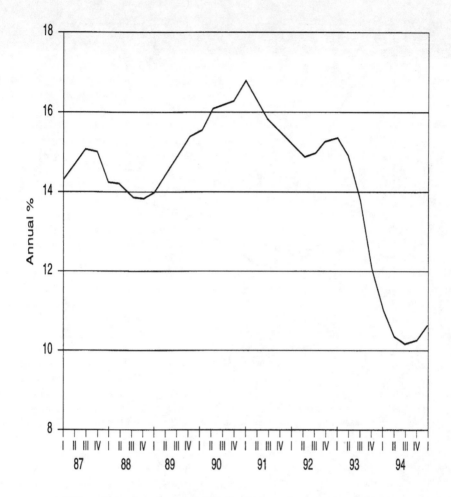

Figure 12.3 Development of real mortgage interest rates in Spain
Source: Banco de España

The objectives of this policy were fundamentally based on promoting demand by creating conditions favourable to purchasers, primarily though interest-rate subsidies and, in the most extreme cases, subsidies reducing the initial contribution. Such offers were encouraged, though to a lesser extent, by granting loans to developers at rates slightly below market rates. To avoid all gains going directly to developers as a consequence of their setting sale prices, it was necessary, logically, to limit the sale price of housing.

Basically, the VPO system was designed as follows:

1 Only the construction and sale of newly built houses with a usable surface area no greater than 90 m^2 and with a capped sale price were to be covered.

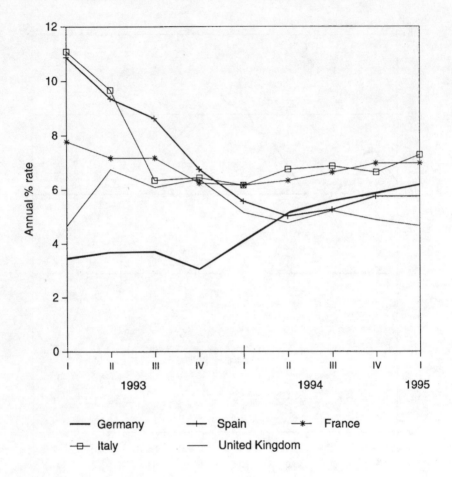

Figure 12.4 Inflation-adjusted mortgage interest rates in various EU countries
Source: Federación Hipotecaria Europea (European Mortgage Federation)

2 Construction of this type of housing was entrusted to private enterprise.
3 Financing was granted to the developer at below-market interest rates ('agreed rates'), with interest being paid only on the invested capital until the work was completed. After handover to the purchaser, which was immediate because of the great demand, the purchaser took over the loan repayments at an interest rate reduced by the state subsidies.
4 The financing system furnished the financing, committing itself, by means of an agreement, to grant loans at a rate slightly below market rates ('agreement rate').

5 Negotiation of the precise amount of annual financing, the level of assistance, the loan conditions and, of course, the payment of subsidies and grants, are covered by central government. The autonomies are fundamentally limited to the administration (i.e the granting of the right to assistance) and the management of the policy.

The outline given is a basic one. Over time and with a certain amount of success, assistance was extended to restoration (1984) and to the purchase of older housing (1989). Assistance was also granted to developers in the rented sector, but they have only really ever had a token presence in this sector.

Likewise, in 1978 a public-development VPO system was created, paid for out of state budgets and later out of those of the autonomous communities, which by and large also allocated property. In 1989 a private-development VPO category was added, endowed with greater assistance and aimed at the population with the lowest level of income. This last category is increasingly replacing the former public VPO system.

Figure 12.5 shows the housing projects started, both private and VPO housing (with the latter being differentiated between publicly and privately developed) since 1980. Table 12.5 shows the direct cost to the state and the autonomous communities in subsidies and grants of VPO interest rates, as well as the direct investment in the construction of council housing, a cost of about half of GDP.

Three periods during the development of the VPO policy are worthy of note:

1   Until about 1986 more than 50 per cent of the new housing developed came under VPO, and the general level of housing construction, which until 1978 had been falling alarmingly, levelled out. In quantitative terms

Table 12.5   Direct housing expenditure, Spain 1990–94 (in billions of pesetas)

	1990	1991	1992	1993	1994
Central state					
Investment	12.6	5.5	10.8	15.5	13.4
Grants for purchase and restoration	12.8	12.7	16.1	18.7	22.7
Subsidy of interest rates	55.9	54.8	44.3	60.5	56.1
Autonomous communities					
Investment	112.2	117.6	134.7	143.0	143.0
Grants for purchase and restoration	26.0	24.4	31.9	47.3	71.0
Total	219.5	215.0	237.8	285.0	306.2
As a percentage of GDP	0.44	0.39	0.4	0.47	0.47

Source: General state and autonomous community budgets and Banco Hipotecario de España

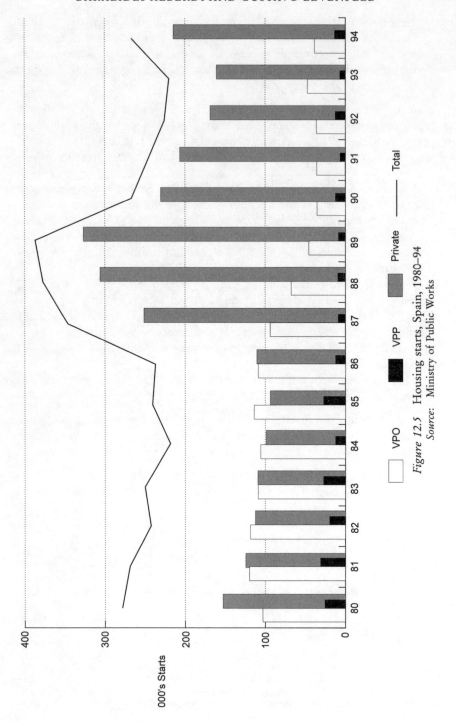

Figure 12.5 Housing starts, Spain, 1980–94
Source: Ministry of Public Works

this policy can be judged a success, in spite of the fact that it deviated somewhat from its social intention.

2 From 1987 onwards, an important event took place. The high demand for housing, which occurred as the result of the crisis of 1985–86 being overcome, pushed development towards private housing, which was much more attractive for developers. VPO construction decreased significantly, leading to a shortage for those sections of the population for which it was intended. However, the problem was worse than that: the rise in the cost of private housing led to an increase in the cost of building land. As a result, in vast areas of Spain VPO was not possible as the high land prices could not be passed on because of the maximum prices set for this type of housing. This led to a serious availability crisis for those with lower incomes.

3 In the most recent Housing Plan (1992–95) measures were taken to alleviate this. Maximum sale prices for VPO were increased, primarily in the large cities, with purchasers being compensated with larger loans and increased assistance. Even more important is the transfer of land by the public bodies at prices compatible with VPO sale transfer. This transfer was inadequate on all counts and the development of this type of housing has not appreciably recovered. Furthermore, if land continues to be expensive for private housing, the price difference for land transferred for VPO will be covered in other ways or passed on to VPO in second transfers.

DEVELOPMENTS IN HOUSING POLICY

The 1978 Constitution shaped Spain into a state of autonomous regions. The autonomies have a legislative body and an autonomous government, elected by means of universal suffrage within each territory, with the autonomy to legislate and carry out certain policies. In theory, the authority for housing policy and the organisation of territory have been virtually entirely transferred to the autonomous governments. The autonomies are basically funded by taxes and transfers received annually from central government and from other organisations; these contributions make up 16 per cent and 73 per cent respectively of the total.

Nevertheless, central government still plays an active role in social housing policy, through its general design of such policy and its financial control, and town halls thus have specific competencies relating to housing.

We are going to specify which are the competent bodies for the different means by which housing policy in Spain is implemented and how these are often the shared responsibility of various administrations. The fiscal policy related to housing falls within the responsibility of the central government, as does the legislation governing rents. These regulate the mortgage market and the financing of housing. The 1992 *Ley del Suelo* (Land Law) was approved by the central government, but approval of town planning falls within the responsibility of the autonomous regions and design within that of the town halls. The governments of the communities of Valencia and Madrid, however, approved their own land laws in 1994. Finally, social housing policy is designed by the central government and implemented by the autonomous regions.

However, the autonomous regions do carry out self-financed policies specific to their own areas. Examples of this are the construction of housing that is later to be allocated for purchase or for rental, the restoration of historic urban areas and public housing, and the generation of public land resources and of policies specifically for particular communities. Town halls also develop specific policies for which they have budgetary responsibility.

Given the numerous means employed and the existence of various decision-making centres, the implementation of housing policy in Spain is not always coherent. In fact on many occasions the results are in conflict.

Rented housing policy

Since the arrival of democracy in 1978, governments have been conscious of the need to develop the rented housing market, in order to satisfy both the housing needs of the lower-income population and entrance to the market, and mobility of the population.

After the above-mentioned 1985 ruling a new *Ley de Arrendamientos Urbanos* was drawn up, as was absolutely necessary. Nevertheless, despite the fact that various drafts were prepared, no government managed to present a Bill to parliament until December 1992.

Two years after being presented in parliament, the new *Ley de Arrendamientos Urbanos* was passed at the end of 1994, after difficult debates both in parliament and by public opinion in the media. The 1994 law establishes, with regard to housing, free agreement between parties relating to the fixing of the amount of rent and the duration. Nevertheless, if the duration is for less than five years, the tenant has the unilateral right to extend it to the said five years. Prices are to be reviewed annually and in accordance with the RPI. Recovery of the property by the owner is thus guaranteed within a reasonable period of time. The Law eases eviction proceedings resulting from non-payment, which resolves another of the points of conflict of earlier legislation.

The new leasing Law also covers contracts already in existence on that date. First it deals with the review of the amount, setting a maximum period of 10 years so that, by means of scaled increments, the updated rent incorporates inflation accumulated from the date of the initial contract, and is thereafter reviewed in accordance with the RPI. The possibility of substitution by family members is restricted, thereby shortening the period of unavailability of an owner's property. As the Law has only recently been passed, it is not possible to confirm a revitalisation in the availability of rented accommodation. Nevertheless, given the reasonable nature of the new law, an increase can be expected in the availability of this type of accommodation for which there is undoubtedly a demand.

Restoration policy

In VPO policy the first measures to encourage restoration were taken in the mid-1980s. The response was very poor and was limited to isolated cases of restoration of housing and buildings. Private enterprise has also been involved in restoration, basically through the restoration of buildings situated in better districts for subsequent sale.

Total restoration of districts has only been carried out in a few cities and tourist centres. It is not really possible to speak of a serious restoration policy, rather on the contrary the policy has remained at a secondary level. In 1994 restored housing accounted for 9 per cent of the total number of houses available for occupation and 1.9 per cent of the housing budget.

Land policy

The system for designating land for development has shown itself to be inefficient throughout the period under analysis. The land development system starts with a planning phase, to be carried out by the town halls and approved by the relevant autonomous region. Planning must comply with a basic standard (the *Ley del Suelo* (Land Law)) approved by parliament. Responsibility for the application of this standard falls, however, entirely with the town hall.

According to the standard, all municipal territory must be classified in three basic categories of land: *urban* land, land for *development*, and land *not for development*. Once the plan comes into effect, it is only possible to build immediately on *urban* land and on *land for development* once it has been developed (including the construction of infrastructures and the redistribution of ownership in an equitable manner according to authorised uses). The potential availability of land therefore remains reduced in the medium term (4 to 6 years), without there being any efficient way of avoiding the owners of this type of land retaining it. Land classified as *not for development* is not classified as such in perpetuity, unless it is specially protected. It can become eligible for development after the plan is revised, which happens every five years.

The individual criteria of each town hall in drawing up its general plan determine the availability of medium-term building land and, therefore, the evolution of its price. It should be pointed out that from 1986 onwards the cost of housing increased spectacularly, mainly as a result of the increase in land prices, bearing in mind that the price indices for construction have moved at a rate slightly below the RPI, with the fundamental cause of the increase in land price being the artificial scarcity brought about by the planning. In this regard it is clearly dangerous when the taxes received through the transfer of land and housing are a good source of income for the town halls, when the amounts are greater, and so is the number of transfers.

CONCLUSIONS

Housing policy in Spain during the period 1980–95 has been centred fundamentally on the availability of and access to home-ownership. In practice, there has been no real reason for owners to be interested in offering accommodation for rent due to a *Ley de Arrendamientos Urbanos* which went directly against their interests. Access to home-ownership has been encouraged from the demand side via a relaxation of the financing system and improvement in the functioning of the mortgage market, making it more competitive, and the granting of more generous tax benefits to home buyers.

In order to meet the demands of those with lower incomes, the VPO policy was designed. As a result of this policy, financing particularly favourable to purchasers of so-called VPOs was put together, thanks to grants and interest subsidies accorded by the state. These houses had to be new and meet certain floorspace and price limitations. Developers of this type of accommodation, which was private, also benefited from privileged loans, although to a lesser extent. The VPO policy also supported the construction of housing for rent, although the response from developers was minimal. It also covered restoration and the purchase of older housing, although there has been little achieved in these fields. The policy for public construction of housing has been particularly restricted during the period under analysis, and has been covered since 1989 by a special VPO set of rules.

Likewise, governments have continued, unsuccessfully, to search for a solution to the lack of rented accommodation on offer, for which an amendment to the leasing Law that was in force was necessary. In 1985, a completely free ruling did not find acceptance with tenants but was accepted by landlords, and immediately a large number of houses became available for rental. Finally, in November 1994 a new *Ley de Arrendamientos Urbanos* was passed; this could be said to constitute the start of a new era for rented accommodation.

The promotion of restoration has not yielded good results, as it has not been widely carried out. Private enterprise did not become sufficiently involved despite the assistance offered. The total restoration of some districts, with state help, has occurred thanks to direct intervention by the town halls, but lack of resources has not allowed this practice to develop to the levels desired.

During the period under analysis, low yields on urban land often led to a scarcity of this type of land and a consequent increase in its price. These increases have had a very negative effect on council housing policy, leading to a serious shortage of the type of accommodation that is aimed at the poorest strata of the population. During 1994 there was intense debate about the liberalisation of land, with the aim of drawing up the necessary reforms.

Currently, in the context of the drafting of the next VPO plan (1996–99), various points of housing policy are being reviewed:

- tax deductions for the purchase of a main residence to be replaced by others which promote renting of homes;
- the trend towards council housing being offered for rent;
- incentives for the restoration of whole districts;
- changes to the system for designating building land.

Finally, it should be pointed out that policies which differentiate between specific population groups such as the elderly, women, immigrants, shanty town dwellers, the young, and other groups, such as have been developed in other European countries, have not, as yet, been adopted in Spain. For this reason we have not made particular reference to them.

13

ITALY

Liliana Padovani

It is a common belief that Italy like many other European countries in the 1990s has attained a good 'average' level of housing provision. The most recent (1991) Census reported that there was a total of 25 million dwellings, occupied or unoccupied, for 19.9 million households, and 104.1 million rooms for 56.8 million inhabitants. As a result, in Italy today, nearly 2 rooms are available for each inhabitant and 1.26 dwellings for each household. Dwellings are spacious, with an average of 4.3 rooms, and are equipped with basic facilities. The drive for rehabilitation quickly gained momentum in the 1970s and 1980s and has accounted for almost 50 per cent of the total investment in housing in recent years. The proportion of owner-occupiers is quite high and accounted for 70 per cent of all occupied dwellings in 1991 (Table 13.1).

These figures are just averages and have little real significance in the evaluation of housing need, but they do give an idea of the huge development of the housing stock in Italy since the end of the Second World War. From the situation of serious crisis and of dire housing shortage in the immediate post-war years, the country has now reached a point where housing stock surpasses, in quantitative terms, standards generally considered

Table 13.1 Housing, selected indicators, Italy, 1951–91

	1951	1961	1971	1981	1991
Total dwellings					
Rooms per inhabitant	0.79	0.94	1.18	1.53	1.83
Dwellings per household	0.97	1.03	1.09	1.18	1.26
Occupied dwellings					
Persons per room	1.4	1.2	1.0	0.8	0.7
Dwelling size	3.3	3.3	3.7	4.2	4.3
Household size	4.0	3.7	3.4	3.0	2.9
% of dwellings with bathroom	11.1	–	–	–	99.7
% owner-occupied	40.0	45.8	50.8	58.9	68.0

Sources: ISTAT (National Institute of Statistics), CENSIS

188

optimum. At the same time, the expectations of a considerable number of Italian households have expanded to the point of what could be defined as housing 'opulence'.

This change of scenario is the outcome of intensive private and public investment in new housing and of a strong orientation of public policies in favour of the development of the housing stock and of owner-occupation in particular. Housebuilding has gone on, at a sustained pace, throughout the past four decades despite the fact that, by the end of the 1960s, the number of dwellings was greater than the number of households and the standard of one room per inhabitant had been attained. Throughout this period, housing stock, as well as land in residential use, has grown much faster than population or number of households (Table 13.2).

Table 13.2 Population and housing stock, Italy, 1951–91

	1951	1961	1971	1981	1991
No. (in thousands)					
Population	47,516	50,624	54,137	56,557	56,778
Households	11,814	13,747	15,981	18,632	19,909
Rooms	37,342	47,528	63,834	86,618	104,152
Total dwellings	11,411	14,214	17,434	21,937	25,029
Index (1951 = 100)					
Population	100.0	106.5	113.9	119.0	119.5
Households	100.0	116.4	135.3	157.7	168.5
Rooms	100.0	127.3	170.9	232.0	278.9
Total dwellings	100.0	124.6	152.8	192.3	219.3
Land in residential use	100.0	–	–	320.0	358.0

Source: ISTAT, Censis

However, this massive amount of housebuilding and the continuous expansion of the housing stock have not produced equally satisfactory effects, either (a) on the more traditional front of satisfying crude housing deficit, or (b) on the more general front of addressing, diversifying and evolving housing needs expressed by the different social groups. It would probably be true to maintain, as we will see later on, that many of the current housing problems are, if not the direct consequence, at least highly interconnected, with the way the housing construction industry has developed, as well as the way urban growth processes have been managed throughout this period.

Indeed, an evaluation of the present housing situation takes on a different aspect when it is analysed from a point of view which examines not only the evolution of the overall housing stock and of the average housing conditions, but also takes into account the spatial and social features of dwellings and households and their trends. Sharp, and sometimes widening, differences in housing conditions may thus be observed between urban and non-urban areas, between the developed North and Centre or the new regions of the 'diffuse economy' of the 'Third Italy' (Bagnasco 1977) and the less developed South, and between the various social groups.

With regard to what was previously referred to as crude housing deficit, while it is true that, over this long period, the housing conditions of a considerable number of households did improve, other groups are still experiencing difficulties. We observe that:

• part of the population is still living in overcrowded dwellings.[1] A proportion of 3.5 per cent of the stock (about 700,000 households) according to the 1991 Census can be considered overcrowded, with more than 1.5 persons per room;

• forms of house-sharing persist (some 200,000 families are affected according to the 1991 Census);

• the number of families on the waiting list for public housing is very high. It has been estimated that in 1991 there was a total of 700,000 outstanding applications pending at the Istituti Autonomi per le Case Populari (IACP).[2] This figure is dramatically high in a country where the public-rental sector amounts to less than 1 million dwellings (4.2 per cent of total occupied stock, 16.5 per cent of the rented sector);

• the number of families subject to eviction orders from the private-rented sector is also very high. Some 800,000 households have been affected in the last decade and the number of eviction orders grew from 73,300 in 1989 to 129,200 in 1992 (Table 13.3).

Table 13.3 Evictions from the private-rental sector, Italy, 1984–93

	1984–85	1986–87	1988–89	1990–91	1992	1993
No. (average per year)						
Eviction orders	113,837	110,109	76,936	86,188	78,342	69,400
Implementations	52,102	41,085	61,200	98,884	109,426	129,169
Forced evictions	18,508	21,559	13,759	16,556	17,788	19,598

Source: CENSIS, 28° Rapporto sulla situazione sociale del paese 1994, F. Angeli, Milan, 1995

With regard to the more qualitative aspects of housing needs, new problems emerged during the 1980s. The most important of these include:

• the creation of new forms of severe 'housing stress', which encompasses a variety of situations. These include the new areas of poverty caused by the ongoing processes of economic change, as well as the new phenomenon of immigration from third world countries which has heavily affected Italy since the mid-1980s, or the problems of specific population groups hit during the 1980s by changes in housing policies, in housing markets, or by the sharp decline of the rented sector (the elderly, one-parent families, new households);

- increasing dissatisfaction on the part of the inhabitants with the 'quality of life' in many of the suburban housing developments built in the 1960s and 1970s, which are poor in urban quality and in levels of services and communications. Such developments make up a large proportion of the housing stock;
- the problem of providing basic services and infrastructure to the 'illegal city' built during the 1970s and 1980s, through informal and illegal construction practices, in some regions of the centre and south of Italy. In 1984, the number of illegal dwellings built since 1971 was estimated (CENSIS, 1985) to be about 2,723,000 or 12.3 per cent of the total stock.

As we can see from this outline, the situation is fairly contradictory. On the one hand, new housing production as well as rehabilitation are proceeding apace, albeit with the backing of illegal forms of construction; larger numbers of families are becoming home-owners; and a remarkable proportion of households have more than one dwelling. On the other hand, though, despite zero population growth and the expansion of the stock, areas of housing deficit persist, while processes of change, both inside and outside the housing sector, are making various social groups highly vulnerable. Furthermore, illegal housing construction is leading to serious urban problems and changing perceptions of the quality of life are making a large part of the built environment undesirable.

This contradictory situation, the first signs of which were already emerging in the early 1980s, would have required the start of a wide-ranging process aimed at recasting housing policies and refocusing the public sector's role in housing provision. In fact, throughout the 1980s the basic concepts underpinning housing policies were still within the frame of the model developed in previous decades, highly oriented towards home-ownership and the expansion of the size of the stock. Little attention was paid to the fact that this model was progressively losing its capacity to satisfy the area of social demand. When there was a serious housing deficit in the 1950s and 1960s, this model was able to ensure an improvement in the housing conditions of a considerable number of people, albeit with social, environmental and urban quality costs. During the 1970s, however, when a better balance between housing availability and population had been achieved, this capacity was considerably reduced. The households that, in the previous decades, had invested in the private-rented sector, so contributing to a diversified provision of dwellings, shifted their attention towards other investments or other sectors of the housing market, mainly second homes, either in tourist areas or in the cities (Coppo, 1994a). Thus, the main outcome of this policy of indirect and non-targeted subsidies to housing was a change in the structure of the housing stock, with a further growth of home-ownership, a decline of the private-rental sector and a sharp expansion of second homes (dwellings not used as primary residences accounted for 50 per cent of the total increase of housing stock between 1971 and 1981). Since the 1970s, this model of housing policy has tended to work more as a tool in support of the upper layers of housing demand, rather than in support of social needs. These were the reasons at the start of the 1980s for housing policy in Italy to be refocused.

TENURE

In 1991, 68 per cent of the housing stock was owner-occupied, 25.3 per cent rented (19.9 per cent private, 5.4 per cent social), and a residual 6.7 per cent was in other forms of tenure.

The proportion attained by the owner-occupied sector is very large; this sector has expanded sharply from 40 per cent of dwellings in 1951 to 59 per cent in 1981 and 68 per cent in 1991. The pace of growth has sharpened in the last two decades, particularly during the 1980s. This trend, according to ISTAT data, has continued in more recent years with the proportion of home-owners among all households growing from 63.1 per cent in 1984 to 70 per cent in 1991 and then stabilising after that (Table 13.4).

Table 13.4   Tenure, Italy, 1951–91

	Owner-occupation	Rental sector	Other	Total
	(000s)	(000s)	(000s)	(000s)
1951	4,301	5,241	1,214	10,756
1961	5,972	6,076	984	13,032
1971	7,767	6,769	766	15,301
1981	10,333	6,225	984	17,542
1991	13,419	5,000	1,317	19,736
	%	%	%	%
1951	40.0	48.7	11.3	100.0
1961	45.8	46.6	7.6	100.0
1971	50.8	44.2	5.0	100.0
1981	58.9	35.5	5.6	100.0
1991	68.0	25.3	6.7	100.0

	Owner-occupation				
	North-west	North-east	Central	South	Italy
	%	%	%	%	%
1951	32.5	38.3	36.9	48.5	40.0
1961	38.7	46.5	43.8	53.0	45.8
1971	43.1	53.1	49.4	57.7	50.8
1981	52.8	61.1	59.2	63.3	58.9
1991	64.2	70.2	69.5	69.1	68.0

Source: ISTAT, CENSIS

The change in the method of access to ownership in the more recent periods is worth noting. In the early post-war years, new housebuilding played a dominant role in the move to owner-occupation, but since the 1970s, and particularly during the 1980s, transfer from private renting has also become very important. As a matter of fact, the growth of ownership was accompanied by a sharp decrease in the private-rented sector. From 1981 to 1991 the owner-occupied sector grew by more than 3 million dwellings, while the rented stock lost 1.2 million units. It seems reasonable to argue that, in this decade, the withdrawal of investors from the private-rented sector and the consequent decrease of the supply in this sector has largely contributed to shift households towards ownership. The pressure on households to enter the owner-occupied sector was high, as proved by the fact that the incidence of housing expenditure on household revenues was higher in the owned sector than in the rented (in 1989, 19.6 per cent against 16 per cent) and increasing. Another form of pressure came from the increasing number of eviction orders from the rented sector (Table 13.3). In some areas of high housing stress, like the Milan metropolitan area, 47 per cent of dwellings were bought by the previous inhabitants, and while 50 per cent of housing bought contained only 1–2 rooms, as few as 30 per cent of households were composed of one or two persons.

The problem of eviction may be considered an unexpected consequence of the Fair Rent Act of 1978. The aim of this law was to overcome the excessive segmentation of the rented market (partly free, partly under control) and to act as a regulator of rent prices. To this end the law introduced two forms of regulation: the first was on the cost of rent, which was determined through an institutional procedure (3.5 per cent of the 'rental value' of the dwelling, based on the construction cost of public housing multiplied by a set of coefficients), the second was on the length of the lease, established as four years, after which the landlord (or the tenant) was free not to renew. The first unforeseen result – at least in the numbers involved – was that at the end of the first four-year period a large number of landlords did not renew rent contracts. This started a very complex and long judiciary procedure that developed in four stages: a first communication of contract rescission, followed by an eviction order, then by a judiciary order, and finally by forced eviction with the intervention of the police. From Table 13.3, which shows the evolution over 1984–93 of this sequence of steps, the disproportion between the number of announcement orders and actual evictions is clear. Evicting a household was a difficult and unpopular measure. Furthermore, public housing had to face the problem of this new and large demand for rented housing. Thus the political answer to this problem was to try to postpone its implementation by establishing a scale for evictions. This explains the large number of households under eviction order. Of this large group, some of those enjoying better resources, under the pressure of eviction, tried and succeeded to switch to home-ownership. A second weaker group accepted this 'sword of Damocles' situation and did not leave the house. A proportion are evicted and their number was growing by about 20,000 per year in 1993.

Many local governments, especially those in the larger urban areas with high incidences of evictions, had to face acute problems and ended up utilising almost all the public-sector resources in housing families evicted from the rented sector. This further weakened the already low capacity of the public sector to satisfy the growing demand from the areas of severe housing stress.

The public housing stock in Italy, whose size has changed little since the 1960s, is composed of 826,000 dwellings managed – and largely owned – by IACP, and another 200,000 dwellings owned and let by local authorities.

The size of this stock is very small and the public-sector development plan proposed at the end of the 1970s was progressively abandoned in the following decades. During the 1980s, mainly in the second half, resources allocated to public housing declined from 52,775 billion lire in the first phase (1978–81) of the Ten Year Plan to the 13,490 of the period 1992–95. This trend may be perceived also in the number of completions, which fell from 18,500 dwellings in 1980 to 12,200 in 1992 (Table 13.5).

Table 13.5 Public housing, new housing construction, Italy, 1980–92

	1980	1985	1990	1991	1992
No. of dwellings					
Fully subsidised (*sovvenzionata*)	18,480	14,600	12,514	12,335	12,204
Partially subsidised (*agevolata*)	86,999	49,200	58,909	58,909	57,452
% of total new housing					
Fully subsidised (*sovvenzionata*)	4.4	4.6	4.4	4.4	4.4
Partially subsidised (*agevolata*)	20.5	15.4	20.5	20.5	20.5

Source: ANIACAP, Lavori in corso, Rapporto Terzo anno, CNR, Roma, 1993

During the 1980s the central government repeatedly tried to erode, under the pressure of public debt, the funds allocated to public housing. It was sometimes successful, sometimes not, in 1988 being prevented by a sentence of the Constitutional Court.

In spite of the fact that in the early 1990s a – still timid – debate on the necessity for a new formulation of housing policies and of a reform of the public-housing sector had started, in 1993 a law to privatise 50 per cent of public housing was abruptly approved. The justification was that this stock was deteriorating, that its management was ineffective and very costly, and that there was little control on occupancy after the first letting, which implies that very few re-lets were available. Part of this evaluation is true, but the solution does not lie in merely selling. Recent research demonstrates that reform of the sector and an improvement in the management of the stock could produce results which would be far better on both the economic and social fronts (ANIACAP, 1994; Coppo, 1994b).

HOUSEBUILDING

Among the most interesting elements of change in new housing construction and in the size of housing stock during the 1980s and 1990s, we may quote:

- the sizeable decline, for the first time since the end of the war, in the number of new houses built;
- the drop in the number of withdrawals from the housing stock;
- and, lastly, the fall-back in the increase of the number of second homes.

The outlook would appear to be a shift away from the 'opulence' of the previous decade (Figure 13.1).

As regards housebuilding, the 1980s were marked by a sizeable fall. Between 1981 and 1991, the number of new dwellings built was 250,000–300,000 per year (the variation depending on the sources utilised),[3] as against the 500,000 of the two previous decades. Looking more carefully at the period from 1981 to 1994, we find that housebuilding, which was still proceeding at a good rate at the start of the period, dropped down to a minimum in 1988. A recovery between 1988 and 1993 did not match the output seen at the beginning of the period. Housebuilding estimated at 302,000–366,000 dwellings per year in the first half of the 1980s went down to 200,000–250,000 in the second half of the 1980s (Tables 13.6 and 13.7).

As far as future prospects are concerned, current indications are that there is unlikely to be a major recovery in the housebuilding sector in the

Table 13.6 Housebuilding by geographical regions, Italy, 1961–90

	1961–70	1971–80	1981–90	1981–85	1986–90
No. of dwellings, average per year (in thousands)					
North-west	158	127	55	66	45
North-east	97	104	45	53	36
Centre	99	91	44	54	33
South	148	182	109	129	89
Italy, total	501	504	252	302	203
%					
North-west	31.6	25.3	21.9	21.8	22.1
North-east	19.3	20.5	17.7	17.6	17.8
Centre	19.7	18.1	17.3	18.0	16.4
South	29.4	36.1	43.1	42.6	43.7

Source: ISTAT, CENSIS, No. of dwellings built in the previous decade

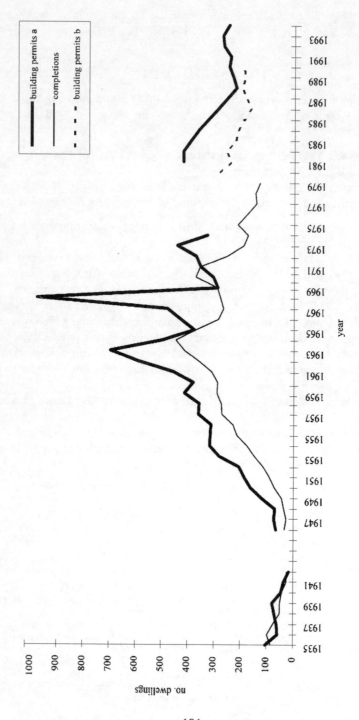

Figure 13.1 Housebuilding, Italy, 1935–94
Sources: a – CRESMO
b – Census 1991

Table 13.7   Housebuilding, Italy, 1981–94

Year	No. of dwellings
1981	428,000
1982	428,000
1983	400,000
1984	370,000
1985	335,444
1986	298,514
1987	263,516
1988	229,742
1989	241,631
1990	256,815
1991	251,394
1992	277,627
1993	282,087
1994[a]	259,252

Sources: CRESME, 1981–1984 Cresme-Notizie, 1985–94 Cresme-Progetto e gestione
Note: [a] Estimated value

next few years. It is useful to recall that, while the rate of housebuilding in Italy had fallen back considerably, it was still some way ahead of that in other European countries. In 1986 in Italy housing completions per one thousand inhabitants were 5.3, the same as in France, but higher than in Germany (4.1), or the United Kingdom (3.8) (Boelhouwer and van der Heijden 1992).

As regards withdrawals from the housing stock, a comparison between housing production and increase of the housing stock shows that, during the 1980s, some 2.5–3 million new dwellings were constructed and the stock increased by 3.1 million units. These data imply a sharp reduction in the number of withdrawals from the housing stock, a phenomenon which had been quite significant in Italy in previous decades. Withdrawals of 1,730,000 dwellings in the 1960s were reduced to 538,000 in the 1970s, and in the 1980s there were net additions to stock (Table 13.8).

Table 13.8   Withdrawals from the housing stock by decades and geographical regions, Italy, 1961–90

	1961–70	1971–80	1981–90	1961–70	1971–80	1981–90
	No. of dwellings			% of total stock		
North-west	−477,286	−219,246	−49,276	−11.2	−4.1	−0.8
North-east	−360,760	−169,389	58,145	−14.0	−5.3	1.4
Centre	−221,103	−58,575	170,620	−8.7	−1.8	4.1
South	−734,011	−91,073	388,692	−15.2	−1.6	5.3
Italy	1,793,161	−538,283	568,181	−12.6	−3.1	2.6

Source: ISTAT, CENSIS

Table 13.9   Housing stock at Census years, Italy, 1951–91

	1951	1961	1971	1981	1991
Italy: No. of dwellings (in thousands)					
Occupied	10,756	13,032	15,301	17,542	19,736
Non-occupied	655	1,182	2,133	4,395	5,293
Total	11,411	14,214	17,434	21,937	25,029
% of non-occupied dwellings, by geographic regions					
North-west	6.0	7.8	11.3	17.3	17.8
North-east	4.2	7.4	10.8	18.1	18.0
Centre	5.6	8.9	13.6	19.8	19.8
South	6.4	8.9	13.1	23.7	26.1
Italy, total	5.7	8.3	12.2	20.0	21.1

Housing stock increase from 1951 to 1990, occupied and non-occupied

	1951–60	1961–70	1971–80	1981–90
Italy: No. of dwellings (in thousands)				
Occupied	2,275	2,270	2,240	2,194
Non-occupied	527	950	2,263	897
Total	2,803	3,220	4,503	3,091
%				
Occupied	81.2	70.5	49.7	71.0
Non-occupied	18.8	29.5	50.3	29.0

Source: ISTAT, CENCIS

The combined effect of new construction, demolitions, and processes of re-use and rehabilitation between 1982 and 1991 produced an increase of 2.2 million occupied dwellings. When compared with the increase of 2.3 million of the two previous decades, this figure seems to show a relatively stable rate of growth of the occupied housing stock since the end of the war (Table 13.9). Where the reduction occurred was in the non-occupied sector, the growth being 900,000 units, as against 2.3 million in the 1970s, years of massive growth in the second-home sector. As a consequence, the non-occupied sector, which grew from 5.7 per cent in 1951 to 20 per cent in 1981, stabilised during the 1980s, at around 21 per cent of the total stock.

These data would seem to indicate a change of trend in the development of the Italian housing stock. The three decades from 1951 to 1981 were characterised by what we might call a low level of efficiency on the part of housing production to increase the stock of housing used as primary residences. The ratio between new buildings and the increase in primary stock remained very low: for every 100 new dwellings produced, the real increment in permanently occupied dwellings was only 44. Compared to this situation, the 1980s marked a turning point, and the increase in the number of dwellings used as main residences varied (according to data used) between 87 and 71 for every 100 new dwellings produced.

Housing rehabilitation

The question of the rehabilitation of the existing housing stock came to the fore later in Italy than it did elsewhere in Europe. It was only in the 1970s that this issue entered the debate in both the academic and political spheres and started to gain increasing importance in the building sector. While, at the beginning of the 1960s, investments in rehabilitation accounted for less than 15 per cent of total housing investments, by the end of the 1970s this figure had risen to 42 per cent, settling at around 47 per cent in the first half of the 1990s (Table 13.10). Over this period some significant changes may be identified in the orientation and character of rehabilitation policies (Padovani 1990, 1991).

A first phase, covering the years from 1974–75 to the early 1980s, was characterised by a strong social connotation and by the close attention paid to housing needs. Public housing and public investment had an important role to play. Implementing housing policies that focused directly on rehabilitating older housing stock – much of it situated in historic centres and inhabited by disadvantaged families – was seen as a change to responding to housing needs while guaranteeing a more balanced use of the urban built environment and of the existing housing resources.

Thus, from 1975 on, a number of Italian local governments, including some of the largest ones, put forward plans for housing rehabilitation. The Ten Year Plan for Public Housing Act of 1978, with a whole chapter on rehabilitation, provided them with financial and planning tools (among them the Urban Renewal Plan, *Piano di Recupero Urbano*).

Table 13.10 Housing investment, Italy, 1981–94 (in billion lire at 1985 value)

Year	Housing new	Housing rehab.	Housing total	Housing % rehab.	Non-housing public works	Total
1981	29,530	21,560	51,090	42.2	43,229	94,319
1983	28,411	22,597	51,008	44.3	39,643	90,651
1985	27,025	22,291	49,316	45.2	40,141	89,457
1987	24,566	22,585	47,151	47.9	43,368	90,519
1989	26,159	22,737	48,896	46.5	47,107	96,003
1991	27,555	24,240	51,795	46.8	48,948	100,743
1992	27,671	24,263	51,934	46.7	46,709	98,643
1993	27,492	24,020	51,512	46.6	41,017	92,529
1994[a]	n/a	n/a	45,246	n/a	34,283	79,529
1994[b]	26,000	25,000	51,000	49.0	n/a	n/a

Source: CENSIS, CRESME
Notes: [a] Provisional data, Census
[b] Provisional data, CRESME

The picture changed during the implementation stage and particularly during the course of the 1980s. The reasons are many and beyond the scope of this chapter, but some of them seem pertinent. The first is connected with the unforeseen result of the first years of experimental public action in rehabilitation, which prompted the revival of private rehabilitation and increased the interest of the market in the historical centres as well as in the nineteenth- and early twentieth-century parts of towns. The second is the gradual withdrawal of public-sector involvement in housing schemes, which eventually also affected housing rehabilitation. The third is the emergence of the problems of urban and economic revitalisation which took away attention and resources from housing rehabilitation.

Thus from the initial, strongly socially oriented approach, which was alive to the needs to restore decaying historical centres and which would have involved large-scale public-sector intervention in housing, the emphasis gradually shifted to rehabilitation of buildings. A wide range of smaller-scale upgrading projects took off, often linked to a single dwelling, in many cases proposed and undertaken by the households themselves. Normally these were outside the areas identified by Renewal Plans and did not involve making recourse to public funding set aside for such projects (subsidised housing). Most of these projects were aimed at improving the dwelling's functional features, while the problems of the communal parts or the building's actual structure were ignored, along with issues regarding urban quality and housing need, which had been important issues of the Ten Year Plan.

This period of intense, small-scale, and essentially private rehabilitation, was associated, according to local conditions, either with housing-price increase, or with changes in tenure with large-scale shifts of dwellings from the private-rented sector to home-ownership. This led to an increasing incidence of eviction of households and traditional businesses from the stock that was being upgraded. Building quality of the housing stock was improved, but social and urban problems were neglected.

The current situation, more concerned with problems of urban restructuring, is one in which the public and private sectors are being encouraged to work together on area-oriented upgrading schemes, aimed at improving urban quality across a wide variety of different situations throughout Italy. Housing issues have also recently entered this debate, and in 1992 a new planning tool was introduced by a public housing law adopted in 1992: the Integrated Action Programmes (*Programmi Integrati di Intervento*). The aim of this scheme is to organise and co-ordinate initiatives and investments, both public and private, in housing as well as in other urban rehabilitation projects. They can be promoted either by the public sector or by private concerns. The idea is to transform renewal projects from rehabilitation of individual dwellings towards urban upgrading schemes. The definition of the procedures to implement these schemes is still on the way. Some experimental projects in this direction were promoted and implemented in the late 1980s by some regional governments such as Liguria and Lombardy (Padovani, 1990; Secchi, 1993).

HOUSING INVESTMENT, FINANCE AND SUBSIDIES

Before analysing the processes of change in housing and in housing policies in the 1980s and 1990s, it seems useful to recall some fundamental aspects of the basic structure of public-sector involvement in housing in Italy as well as of the different features it assumed during two distinct phases into which the post-war period can be divided.

The structure of the public-housing sector

Public-sector involvement in housing in Italy involves three main areas of intervention:

- provision of a fully subsidised form of housing (*edilizia sovvenzionata*). Funding is granted mainly by central government. Public authorities (IACP and municipalities) are responsible for the production and management of this stock, which is rented to low-income citizens. The institutional frame that organises the financing, planning, implementation and management of the public-housing sector, as well as the institutions and the procedures involved, is defined by national housing legislation. Funds are provided by state budget, – for the largest part – by a special fund made up by contributions of employees and employers;
- the development of owner-occupied housing (*edilizia agevolata*). Various forms of grants and loans (particularly low-interest loans and mortgage assistance) are provided to public authorities, builders and households (singly or, more frequently, organised in co-operatives). Revenue limits dictate access to these loans. In addition to direct financial means, other indirect incentives are provided to support owner-occupation. Tax relief is one of the most important;
- provision of low-cost land for social housing, either for public-sector subsidised housing programmes or a special programme (*edilizia convenzionata*). Local authorities are empowered to acquire, by compulsory purchase, land for housing programmes.

The first phase: the 1950s and 1960s

If, as said before, some structural features of public action in housing persisted from the immediate post-war era through to the 1980s, nonetheless within this period two phases may be identified, characterised by

201

important differences in the way housing issues and policies were defined. These two phases are closely linked to the processes of economic and social development the country experienced during that period.

The first phase extended from the 1950s through to the mid-1960s. These were the years of the boom in the construction industry, but also of the so-called 'Italian economic miracle' of industrial development, and of intense processes of urbanisation. Most of this was concentrated in a few, limited parts of the country: the urban areas of the north-west of Italy (mainly within the triangle defined by Milan, Turin and Genoa). One consequence of this unbalanced model of growth was migration on a massive scale from rural areas and from the South towards the northern regions. A second consequence was a dramatically high pressure of demand for housing and land in the regions of growth.

In this period housing policies, as well as urban policies, were marked by what could be defined as a *laissez-faire* orientation. In the area of housing, the focus was on indirect public intervention to support private house-building and to encourage home-ownership, which at the time was restricted to a limited proportion of households (40 per cent in 1951).

Little effort was made to create a public-housing sector and to extend its size even though the state's total investment in public housing was considerable. It is worth noting that in the 1950s public investments were more than 17 per cent of total investments in housing and that in the twenty years from 1951 to 1970 some 800,000 public housing dwellings were completed. This did not increase the size of the public housing stock because from 1951 to 1971 different forms of privatisation removed some 850,000 dwellings from the public stock (Mortara, 1975; Padovani, 1984). In this period public housing was also conceived of as a means towards home-ownership for middle- and low-income groups.

It is well to note that, at this time, the private-rental sector was quite important (60 per cent of total occupied dwellings) and that, since the end of the war, various rent control laws had put on the private rented stock the burden of offering housing at very low rents, sometimes for long periods.

The model adopted at this time in Italy was markedly different from the 'comprehensive state-involvement models tied to the welfare state that had taken root in northern European countries' (Tosi, 1990: 199).

The outcome of this first phase, which marshalled massive investment in house construction, was an improvement in housing conditions, albeit at a high cost. On the down side was the persistence of an important area of housing stress, unable to benefit from the growth in the housing stock, as well as the creation of a system of incentives to support new construction that was heavily penalising housing maintenance and rehabilitation. However, the greatest negative impact was felt on the landscape. In the areas of growth, the negative consequences were congestion, increased housing costs, waste of land, and destruction of the natural landscape, as well as of historical and cultural resources. Meanwhile, in the regions from which people were migrating in large numbers, the consequences were abandonment and decay of the existing stock.

I have taken some space to describe this phase because this was the period in which the basic and somehow persistent features of the Italian model of housing policies were designed and developed.

The cycle came to a halt with the recession that occurred in the mid-1960s.

The second phase: the 1970s

The second phase lasted through to the end of the 1970s. The political and social situation had changed. The broad-based coalition that in the previous phase had called upon the government to provide undifferentiated support to new house building had fragmented and lost power. New issues emerged, such as the problems of redressing the dualistic aspects of growth, rationalising urban development, encouraging a more efficient use of the resources available, housing stock included, and promoting rehabilitation. As regards the housing sector, demands came, specifically from the political parties of the left,[4] for government to make a more concerted and targeted effort towards the areas where specific housing problems had been identified, rather than using resources for expanding the country's overall housing stock. There followed a laborious and often contradictory process of reform. This included innovative legislation in the area of town planning, as well as in housing. The implementation of these laws was seriously hampered by the unresolved conflicts of interest between the coalitions formed in the previous phase, less powerful but still active, and the supporters of the new housing policies (Padovani, 1984).

This difficult path towards housing reforms ended at the end of the 1970s with the adoption of three important legislative measures which substantially changed the institutional framework for intervention in housing. This legislation included the Land Regulation Act of 1977, which provided local authorities with innovative tools to plan the development of areas allocated for construction, renewal or rehabilitation; the 'Fair Rent' Act (*Equo Canone*) of 1978, which introduced new regulation of the rental market; and the Ten Year Plan for Public Housebuilding (*Piano Decennale*), of the same year, which redefined the public housing system. The Ten Year Plan for Public Housebuilding was meant to cover more than just the public-housing sector. The aim was to realise some 100,000 dwellings per year, either wholly subsidised or state-aided, laying considerable stress on providing the means for housing to be made available in the 'social' private-ownership sector. A strong emphasis was put on rehabilitation.

These three acts widened the scope of the instruments available to the public authorities for regulation, intervention and administration in both

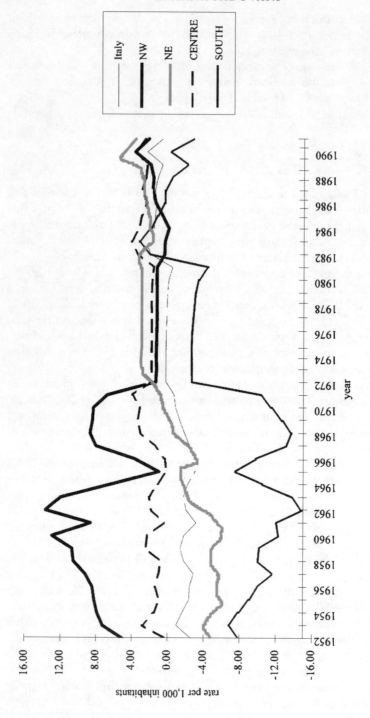

Figure 13.2 Migratory balance, by geographical regions, Italy, 1952–91

new housing development and rehabilitation. The housing-policy model in Italy was thus moving closer to the concept of 'comprehensive state involvement'.

The 1970s were also characterised by a marked change in the geographical distribution of the population and the economy. Reverse migratory movements to the ones occurring in the first post-war period were taking place (Figure 13.2). Population moved away from the traditional areas of urban concentration in the North-west and the other major urban centres, towards the South of Italy and smaller municipal areas. This halt in urban growth and the development of previously marginal or peripheral areas was not peculiar to Italy but was happening in many other countries. What seems more specific to Italy is the extent of the migratory movements. We get an idea of the dimensions of this shift if we consider that:

- in the metropolitan areas, where 72 per cent of total population growth and 40 per cent of housing stock increase were concentrated in the period 1961–71, the corresponding increases in the following decade were 27 and 17 per cent respectively;
- at the other extreme, the smaller municipalities, outside the metropolitan areas, which had suffered a major deceleration in population growth during the 1960s, attracted a 63 per cent demographic growth and 72 per cent of the increase in housing stock in the 1970s.

It took some time for researchers and policy-makers to understand the quality and the intensity of these new processes, which were changing the overall pattern of housing needs.

DEVELOPMENTS IN HOUSING POLICY

At the start of the 1980s, a more complex scenario was taking shape. On the one hand, the long period of debate and demand for reform, along with the search for suitable instruments with which to bring in more effective housing and urban policies, had achieved the desired results with the adoption of the laws discussed earlier. This would allow the promotion of more socially oriented housing policies than in the past.

On the other hand, though, between the end of the 1970s and the early 1980s, the very terms of the issue were being changed by the processes of economic and urban change, as well as by the strength and persistence of the recession and by the emerging problems of cuts in government spending.

Thirdly, the ongoing processes of geographical redistribution of activities and population contributed to further accentuate historical and structural differences between Italian regions. North-western regions were experiencing problems typical of post-industrial societies, while in the North-east a new phase of innovative industrial development was taking place and the

South was facing problems of unemployment, growing poverty and a widening of the gap with the other regions of Italy.

These last events brought to the attention of the public policies new elements in the definition of issues and priorities, sometimes conflicting with the social instances expressed in the first point. In a certain way, the shift in housing policies agreed upon at the end of the 1970s and characterised by the new institutional frame arrived too late. It was more the final act of the long cycle of debates developed through the previous years, than an effective tool able to cope with the new dimensions assumed by housing problems.

In this context, in contrast with the need to develop a process both of interpretation of the new aspects of housing issues and of redefinition of housing policies, the choices made during the 1980s – particularly by central government – were for the most part based on the implicit assumption that housing was no longer a priority item on the agenda of the public sector. Widespread availability and the good average quality of the housing stock led housing specialists to believe that the time when government had to address issues either of net housing shortage or of severe lack of maintenance was now at a close. The areas of housing stress still existing were considered to be of a residual character. The public sector, or specific local-government initiatives, would take care of them.

It is probably correct to state that in Italy during the 1980s, housing problems were far from central in research, political debate, or government action, at either central or local level:

1 Very few resources were allocated for innovative research which might have helped identify the new dimensions of housing needs and the new issues of housing policies. Institutional research in Italy remained outside the debate on the rethinking of social housing policies that occurred in this period in other European countries.

2 The ambitious programme of the Ten Year Plan, never openly discussed, was eroded by a number of pieces of legislation enacted in the 1980s, which, while not necessarily being explicitly new housing policy, made considerable inroads into the approach to housing developed at the end of the 1970s. The apparent paradox of a 'boost in public intervention' in Italy in the early 1980s (Tosi, 1990: 208), when other countries were adopting measures of 'state withdrawal', is to some degree mitigated. The Italian way seems a softer way to withdraw from public intervention in housing.

3 New public intervention in housing was very poor and, in large part, was motivated by questions other than housing problems. Action taken was mainly concerned either with sectorial issues and approached in what we might call a 'conjunctural' fashion, or with constraints in public expenditure.

The result was that privatisation (or the targeting of funds at emergency needs) was seen as the best solution in the face of important matters such as the recasting of public-sector housing (which was considered too costly and unsuited to emerging needs) or the dramatic reduction of the private-rented sector, or the problem of illegal housing. Furthermore, the decision to privatise half of the public housing stock, introduced by a law adopted in 1993, was conceived more as a way of reducing the budget deficit of the IACP and of cutting public expenditure on housing, than as a tool to increase the effectiveness of this sector. In the case of evictions in the private-rental sector, which have been a major problem in the larger metropolitan areas, the solutions adopted have tended towards stop–go attempts to stagger evictions, or to making special transfers of public housing funds to families under eviction orders. Another example is that of special 'amnesties' (condono edilizio) which enabled owners to legitimise existing illegal building by paying a one-off 'fine'. These were introduced solely as part of the debate on reducing the country's fiscal deficit, with little or no account taken of the urban implications of such a move.

Other important issues, such as the problem of the acquisition of land for residential development[5] or the preparation of a new plan for public housing (the Ten Year Plan of 1978 having expired in 1987), were not tackled by the government during the 1980s.

It is only very recently that the question of housing in general and the problem of the effectiveness of the model of state intervention in housing in particular, have come back to the fore in political debate, and have resulted in proposals and some pieces of legislation being enacted.

CONCLUSIONS

The above confirms the necessity to continue to enlarge the scope of the debate on housing and housing policies. Some important problems need to be faced if the perspective is the redefinition of the concept of public intervention in housing.

A first problem is that of the difficulties and limits of housing policies, and not only in Italy, to cope with the more deprived components of housing demand. This subject has regained a topical interest with the emergence of the new areas of severe housing stress.

A second problem is related to the crisis of 'comprehensive' models of public policies, as well as to the crisis of welfare policy, on which the Italian public-sector decision-making processes are still predicated.

Another problem to be tackled is the need to overcome the 'sectorial' character of housing policies with its strong emphasis on construction. This has meant that policies, to date, have tended to consider only the financing and production sides, ignoring other crucial dimensions, such as the land-use and social aspects. The former is important not only to allow a more accurate interpretation and evaluation of the geographical distribution of housing needs but, above all, with the aim of better knowledge and understanding of the ways housing needs are formed 'in' and 'on' the land. The second dimension is conceived not only as a tool to get a more accurate image of the social profile of the people to whom the policies are addressed, but also to introduce such considerations into the planning stages and implementation of housing policies. This would enlarge the network of the actors involved, with new actors entering alongside the institutional ones. Included among these new actors are the households, no longer conceived as passive recipients of an institutionalised process of definition of their needs, but actively involved, by means of their own housing practices, in the process of housing provision (Tosi, 1994).

The aim of overcoming the limits of the 'comprehensive', hierarchical, top-down approach to housing policy, along with the need to give greater priority to the territorial and local dimensions, poses important problems of redefinition of responsibilities and tasks between the various levels and sectors of government. This implies more open and innovative frameworks of co-operation and collaboration between central and local government as well as with the other actors and institutions involved (Cremaschi, 1994).

In this framework it seems misleading to try to organise housing demand in a limited set of categories which rigidly address defined public actions, based on the criteria that have traditionally formed housing policies. The issue is to articulate policies in a much more broad-based and better targeted range of typologies of public action than in the past.

In Italy in recent times there has emerged the view that there has occurred an end to a long cycle of a sort of 'social pact' (De Rita, 1994) between state and population in which housing, understood in highly simplified institutionalised terms of an average number of square metres to be occupied by each family, was seen as a fundamental stage in the development of Italian society. In order for this objective to be achieved, many compromises were made: from speculation, through the waste of land, to low urban quality of life, to illegal construction.

The recent increase in taxation on housing property would appear to be the first sign that such a change has taken place. The challenge now is to define a new relationship which would mark the shift from the home being seen as a 'single product' to a 'multiple product' which would better respond to the needs of a changing population, such as students, the elderly, third world immigrants, and people moving from one area to another in search of work, so that there might be a more flexible supply of types of housing covering the spectrum between the disadvantaged who need help and those who can afford market-driven solutions. The first interesting steps in this direction may be found in the already quoted scheme of Integrated Action Programmes, or in some local public 'multi-purpose' initiatives like the one recently launched by the Venice local government which is trying to develop a housing scheme addressed to both the student population (to meet the serious problem of finding accommodation in Venice) and older home-owners (with problems of money and loneliness).

NOTES

1 The 1951 Census reported 10.8 million dwellings for 11.8 million households – a net deficit of 1 million dwellings – at a time when more than half of the dwellings were overcrowded (22 per cent severely overcrowded, with an average of 3.3 inhabitants per room), lacked basic amenities (only 40 per cent of dwellings were equipped with internal lavatories and only 10 per cent had bathrooms), and disrepair was prevalent.
2 IACPs (*Istituti Autonomi per le Case Popolari*) are special authorities, operating on a sub-regional level, responsible for the production and management of public housing.
3 Sources: Census 1991, dwellings built after 1981, for the first figure; and CRESME (Centro di Ricerche Economiche e Sociali sul Mercato Edilizio, Rome) estimates on new housing for the second.
4 In 1963 a centre–left government coalition succeeded to the previous centre–right governments.
5 Some articles of the law on compulsory land purchase for residential development were declared unconstitutional, the consequence being that at present local governments meet serious difficulties in programming housing development plans.

UNIVERSITY OF GLAMORGAN
PRIFYSGOL MORGANNWG
Learning Resources
Centre

14

THE UNITED KINGDOM

Paul Balchin

In very crude terms, housing need in the United Kingdom has largely been satisfied. In the three decades after the Second World War there was a substantial growth in owner-occupation and local-authority housing and a decline in private-rented accommodation, while the condition of most of the housing stock was greatly improved. By the early 1970s a crude surplus of dwellings over households was achieved (for the first time since 1938) and by 1980 reached 1,026,000, only to diminish subsequently to 822,000 in 1991 (Table 14.1).

The crude surplus in 1991 (albeit at a lower level than in the 1980s) did not, however, indicate the true relationship between supply and need. Of the 23.6 million dwellings in 1991, there were well over a million unfit dwellings or homes lacking basic amenities, dwellings lying empty[1] or undergoing extensive conversion or improvement, and second homes, while there were about half a million concealed households (such as couples sharing with their parents/in-laws) among the 22.8 million recorded households. Taking these concealments into account, in the United Kingdom there was currently a substantial shortage of housing, approaching 3 million.

With regard to England and Wales alone, Niner (1989) reported that there was a current need for an additional 2 million dwellings, and that (depending upon whether there would be a relatively small or large increase in the number of households) a further 1 to 2 million dwellings would be required by the year 2001. Holmans (1995) similarly suggested that in order to meet

Table 14.1   The number of dwellings and households,
United Kingdom, 1980–91

	1980 (000s)	1985 (000s)	1991 (000s)	% change 1980–91
Dwellings	21,426	22,350	23,622	+10.3
Households	20,400	21,400	22,800	+11.8
Surplus	1,026	950	822	−19.9

Sources: Central Statistical Office; Department of the Environment

housing demand and need in England over the period 1991–2011, about 240,000 new homes a year will be required with approximately 40 per cent being in the social sector.

Housing need will clearly remain far from satisfied while housebuilding is at a low level. Whereas in the peak year of 1967 a total of 447,100 dwellings were started in Great Britain, of which 213,900 were in the local-authority sector, the total number of starts had plummeted to 156,000 in the slump of 1992 with only 36,400 in the social sector.

Housebuilding was clearly a victim of tight macroeconomic policy. As a means of reducing the rate of inflation and of applying policies of privatisation, cutbacks in public expenditure were planned throughout each of the Conservative periods of government well into the 1990s. Total public expenditure on housing decreased in real terms by 60 per cent over the period 1979/80 to 1993/94. Moreover, whereas social security and health and personal social services were absorbing an increasingly large share of public expenditure, respectively 30 and 15.5 per cent in 1993/94 (compared to 25.9 and 14.3 per cent in 1979/80), housing's share plummeted to 2.0 per cent in the same year (compared to 7.3 per cent in 1979/80).

A major outcome of this reduction was that the number of homeless households soared. In 1976 there were 26,083 homeless households accepted by local authorities in England, with nearly a third of this number (8,036) officially homeless in London. Because of the increased severity of the housing problem, the number of homeless households accepted (in England) increased to 145,800 by 1991. In addition, the number of homeless single people increased dramatically to about 80,000 by 1991.

It was a tragic irony that at the same time as council waiting lists and the number of homeless reached record levels in the early 1990s, 450,000 building workers had been made redundant in the worst peacetime slump in the construction industry since before 1914.

TENURE

Although a large number of factors determine household tenurial preferences, government policy has undoubtedly had a major impact on tenure choice over the past three-quarters of a century and particularly since 1979. Conservative policies have aimed specifically to increase the proportion of the housing stock under owner-occupation and to reduce the attractiveness of local-authority housing. At the same time, there was an attempt to resurrect private renting and to expand the housing-association sector – as if to compensate for the decline in local-authority housing. Owner-occupation thus accounted for nearly 67 per cent of the housing stock of Great Britain by 1994, whereas in the same year the proportion of local-authority dwellings had fallen to only 19.5 per cent of the total stock – with the

211

Table 14.2   Tenure, Great Britain, 1950–94

	Owner-occupied %	Housing associations %	Local-authority %	Private-rented %
1950	29.0	–	18.0	53.0
1961	42.3	–	25.8	31.9
1971	50.6	–	30.6	18.9
1981	56.6	2.2	30.3	10.9
1991	66.0	3.2	21.2	9.6
1994	66.9	4.0	19.5	9.7

Source: Department of the Environment

private-rented and housing-association sectors accounting for a negligible 9.7 and 4.0 per cent of the total stock respectively (Table 14.2).

Given either free-market conditions or an equitable distribution of public expenditure and/or tax allowances, a very different pattern of tenure would have emerged, a pattern reflecting a genuine household choice between owning and renting, and a choice between renting in either the private or social sector.

Clearly, from 1979 until the 1990s, Conservative governments presided over the greatest onslaught on the direct provision of local-authority housing since its inception, and simultaneously promoted the growth and dominance of owner-occupation. Apart from making considerable reductions in capital expenditure on local-authority housebuilding, the size of the social sector was systematically reduced through the application of a mandatory 'Right to Buy' policy and other processes of privatisation. Meanwhile, aided by a range of tax benefits and a favourable economic climate for investment, the owner-occupation sector became, by far, the largest tenure.

HOUSEBUILDING

Since the late 1970s, it has been increasingly recognised that the rate of housebuilding has been failing to satisfy needs. The Green Paper, *Housing Policy: A Consultative Document* (Department of the Environment, 1977), recommended that in order to keep pace with the 'baby boom' of the 1960s, replace unfit housing and facilitate household mobility, there was a need for 310,000 housing starts per annum until the end of the century.

The housing shortage of the 1990s prompted the *Inquiry into British Housing: Second Report* (Joseph Rowntree Foundation, 1991) to emphasise the need to build 228,000 to 290,000 houses per annum up to the year 2001 and to stress that, within this total, 100,000 should be built for rent.

The housebuilding cycle

The cyclical nature of housebuilding has been evident over the last century but has been particularly pronounced since the 1960s (Table 14.3). How-

Table 14.3   Houses started, Great Britain, 1965–94

Year	Starts (000s) Social sector (local-authority and housing-association)	Private-sector	Total	
1965	181.4	211.1	392.5	
1966	185.9	193.4	379.3	
1967	213.9	233.6	447.6	Boom
1968	194.3	200.1	394.4	
1969	176.6	166.8	343.5	
1970	154.1	165.1	319.2	Slump
1971	136.6	207.3	343.9	
1972	123.0	227.4	350.4	Boom
1973	112.8	214.9	327.7	
1974	146.7	105.3	252.1	
1975	173.8	149.1	322.9	Slump
1976	170.8	154.7	325.4	
1977	132.1	134.8	266.9	
1978	107.4	157.3	264.7	Boom
1979	81.2	144.0	225.1	
1980	56.4	98.9	155.2	
1981	37.2	116.7	153.9	Slump
1982	53.0	140.5	193.4	
1983	48.0	169.8	217.7	
1984	40.2	153.7	193.9	
1985	34.1	163.1	197.2	
1986	32.9	180.1	213.6	
1987	32.8	196.8	229.6	
1988	30.9	221.4	252.2	Boom
1989	31.1	169.9	201.1	
1990	27.2	135.4	162.6	
1991	26.5	135.0	161.5	
1992	36.4	120.1	156.5	Slump
1993	43.3	143.1	186.4	
1994	42.9	158.9	201.8	

Sources: Ministry of Housing and Local Government; Scottish Development Department;
Department of the Environment, Housing and Construction Statistics

ever, the overall trend has been downward since 1967, reaching a nadir of
153,900 starts in 1981 (less than in any year since the late 1940s), with
output in the social sector plummeting to a mere 26,500 in 1991 – lower
than in any peacetime year since the First World War.

In the private sector, the housebuilding industry is mainly speculative.
Houses are mainly built in expectation of being sold during or shortly after
construction.

The private housebuilding industry – particularly because of the preponderance of small firms – is very sensitive to fluctuations in cost. It is quite clear that at times of high interest rates and a tight monetary policy, the number of housing starts fall to a relatively low level (for example in 1974, 1980–81 and 1990–92) and at times of low interest rates and relaxed monetary policy the number of housing starts rises to a high level (for example in 1972, 1978 and 1986–88).

Although housebuilding is cyclical in a macroeconomic sense with booms and slumps normally following each other in response to changing rates of interest, it became apparent in the 1980s that major fluctuations in housebuilding also coincided with the 'political cycle'. In the election year of 1983, base rate had been brought down to a comparatively low 9.83 per cent and a mini-housebuilding boom occurred, and in 1987 (another election year) a base rate of 9.74 per cent stimulated a boom the following year. In the early 1990s, however, the link between political and housebuilding cycles appeared to have been severed. A base rate of 9.42 per cent in the election year of 1992 failed to produce a notable boom in 1993.

Housing renewal policy

The number of houses demolished or closed in Britain declined from an annual rate of 61,785 to a mere 4,187 over the period 1975/76 to 1991/92. This was not only a reflection of cuts in central-government expenditure but a major shift of emphasis away from demolition and redevelopment to improvement and repairs.

The Housing Act of 1969 had been a major factor in restoring many properties which otherwise would have remained in poor condition or have been demolished. The annual number of grant- and subsidy-aided renovations increased dramatically to 398,000 by 1973, and in the same year housebuilding in Great Britain had fallen to 327,700 starts from 343,500 in 1959 – rehabilitation now being on a greater scale.

The government had clearly committed itself to rehabilitation as one of the major planks in its housing policy – new housebuilding falling to abysmal levels in 1980 (Table 14.4).

By 1983 the Conservative government seemed eager to give rehabilitation a boost. Support for private-sector renovation therefore increased from £573 million in 1982/83 to £1,064 million in 1983/84 – the number of grants to private owners and tenants rising from 135,000 in 1982 to 320,000 in 1984 – a record and considerably higher than the previous peak in 1974 (Table 14.4). In both the private and public sectors renovation greatly exceeded the volume of new housebuilding.

The Treasury became sceptical in the 1980s about the cost-effectiveness of improvement-grant expenditure. Local authorities had awarded almost £6 billion of improvement grants over the period 1969 to 1990 (Leather and

Table 14.4   Renovations with the aid of grant or subsidy, Great Britain, 1972–92

	Grants paid to private owners and tenants	Housing starts, private sector	Dwellings (000's) Work completed for local authorities and housing associations	Housing starts, public sector	Total renovations	Total housing starts
1972	153	227	141	123	294	350
1974	242	105	125	147	367	252
1976	83	155	89	171	172	325
1978	72	157	121	107	193	265
1980	95	99	118	56	213	155
1982	135	141	128	53	263	194
1984	320	158	144	40	464	198
1986	163	180	221	34	384	214
1988	157	221	263	31	420	252
1990	137[a]	135	333	28	470	163
1992	30[a]	120	260[b]	36	290[b]	156

Source: Department of the Environment, *Housing and Construction Statistics*
Notes: a Excluding grants paid under the Local Government and Housing Act 1989
b Excluding work completed for housing associations

Mackintosh, 1993), while annual expenditure on grants had increased from £200 million in 1979/80 to £650 million in 1983/84. The government therefore considered that much greater selectivity in the award of grants was essential. The Local Government and Housing Act 1989 thus targeted grants towards the worst housing, and to households in greatest need.

The Act replaced the grant system of earlier legislation with a new and largely mandatory regime of grants, and reformed the system of area improvement. Renovation grants, housing in multiple occupation (HMO) grants, common parts grants, minor works grants and disabled facilities grants were introduced and all were means-tested. Owner-occupiers and tenants would be unlikely to qualify for grant assistance unless their incomes and savings were no higher than the level which would render them eligible for income support or housing benefit; and landlords would not qualify unless (without a grant) their outlay on improvement and repairs failed to exceed their rental income.

The impact of this Act on the volume of rehabilitation in the private sector was undoubtedly very mixed. While the number of grants targeted at low-income households and the elderly increased notably in the early 1990s, the total number of grants awarded in 1992 (96,557) was considerably less than the number awarded during the 1980s peak (320,000 in 1984). Most private owners, moreover, were ineligible for grant assistance regardless of the condition of their housing, and the overall pace of grant-aided rehabilitation in the private sector remained depressed.

The overall level of grant expenditure in England (at 1990/91 prices) plummeted from £1.5 billion to less than £0.5 billion over the period 1983/84 to 1990/91, falling to just under £0.4 billion in 1994/95 (Leather *et al.*, 1994). Thus at 1994/95 levels of grant expenditure, and taking into account the estimated cost of comprehensive renovation (£69 billion) and the assumption that the condition of housing does not get any worse, it would take over 170 years to fully rehabilitate England's private-sector housing stock.

HOUSING INVESTMENT, FINANCE AND SUBSIDIES

The *Inquiry into British Housing, Second Report* (Joseph Rowntree Foundation, 1991) attributed the net housing shortage, deficiencies in the quality of the housing stock, and the denial of tenurial choice to two underlying factors: inadequate investment in rented housing and a heavy financial bias towards owner-occupation.

Since the 1930s, there has been a virtual absence of investment into the building of private-rented housing, and very little private investment in the modernisation of existing property; and from the late 1970s expenditure on new building by local authorities and new towns plummeted, for example from £4.7 billion in 1976/77 to only £0.6 billion in 1989/90 (at 1988/89 prices) (Hills, 1991). The housing associations similarly invested far less on new building and rehabilitation, their capital expenditure falling from £1.3 billion to £0.6 billion from 1976/77 to 1988/89 (op. cit.). Overall, there was clearly a marked shift of emphasis from 'bricks and mortar' expenditure (mainly object subsidies) to assisting home-owners and tenants to meet their housing costs (largely subject subsidies). In 1979/80, for example, 68 per cent of government expenditure on housing was capital expenditure in the social sector, whereas only 22 per cent of 'spending' was in the form of mortgage-interest tax relief and 10 per cent was targeted at rent rebates and allowances. But in 1994/95 only 27 per cent was bricks and mortar expenditure (notably Housing Association Grant (HAG), local-authority capital spending and Housing Revenue Account Subsidy), whereas 24 per cent was in the form of mortgage-interest tax relief and income support for mortgage interest and 49 per cent was on housing benefit to help social-sector tenants pay their rents (Department of the Environment, 1995a).

Transfer and resource expenditure

For most private house buyers, mortgage finance is a necessity. During the boom and slump years of the late 1980s and early 1990s, net advances secured on dwellings increased from £19.7 billion in 1985/86 to £39.9 billion in 1988/89, and then decreased to £15.8 billion in 1993/94 (Table 14.5). But less than 10 per cent of advances each year were used to purchase

Table 14.5 Housing investment, United Kingdom 1985/86 to 1993/94

	Private-sector net advances	Capital expenditure (£ million cash)		Housing-sector borrowing requirement	Public-sector borrowing requirement
		Local authorities	Housing Corporation		
1985/86	19,658	1,721	737	22,116	8,445
1986/87	27,183	1,613	677	29,473	2,485
1987/88	29,554	1,588	740	31,883	−1,434
1988/89	39,967	1,374	738	42,079	−11,598
1989/90	33,920	1,232	907	36,059	−9,131
1990/91	33,164	1,695	1,153	36,012	2,116
1991/92	26,785	1,799	1,638	30,222	7,704
1992/93	18,494	1,622	2,304	22,420	28,892
1993/94	15,836	1,433	1,786	19,055	43,042

Sources: Council of Mortgage Lenders, *Housing Finance*; Department of the Environment (1993) *Annual Report*; Central Statistical Office, *Economic Trends*

newly built houses, the rest was spent on the purchase of 'second-hand' dwellings. Thus only a small proportion of mortgage finance facilitated resource expenditure (on housebuilding), the larger proportion funding transfer expenditure – the owner-occupied market acting very largely as a 'capital guzzler' with very little increase in its size at the end of the day except through the privatisation of council stock.

Capital expenditure by local authorities and the Housing Corporation (via housing associations) is, however, very largely resource expenditure – spending on housebuilding and renovation. Whereas, in cash terms, investment by local authorities remained fairly constant over the period 1985/86 to 1993/94 (although falling in real terms), Housing Corporation investment increased dramatically, overtaking the local authorities by 1992/93 (Table 14.5).

In the late 1970s, however, far more resources had been attracted into increasing the size of the total housing stock than into renewal. There was thus a need to divert investment funds to repairs and rehabilitation. The *English House Condition Survey, 1981* (Department of the Environment, 1983), showed that although local-authority housing (which accounted for about 30 per cent of the total housing stock) contained only 6 per cent of the total number of unit dwellings and 5 per cent of those in serious disrepair (private-rented housing being in a considerably worse condition) these low proportions masked very substantial renewal problems. The Association of Metropolitan Authorities (1983, 1984, 1985) estimated that the repair and renovation of council houses built mainly in the 1920s and 1930s (using traditional methods of construction) would cost £9,000 million to implement, that the repair costs to council houses built in the 1950s and 1960s (using non-traditional techniques such as steel frames and reinforced concrete) would amount to £5,000 million, and the repair costs to council houses constructed in the 1960s and 1970s (using systems building techniques) would cost a further £5,000 million to put right. In total, the £19,000 million needed to be spent on repairs can be contrasted with the government's expenditure plans for public-sector renovation in 1985/86, which was £1,312 million – a severely inadequate figure. In addition, in the private sector the cost of repairs amounted to £25,000 million in 1985 (Carvel, 1985) – a sum which dwarfed the amount of government support for private-sector renovation and clearance in that year, £608 million.

Capital expenditure and rents: local authority housing

Following the publication of the Housing Green Paper (Department of the Environment, 1977), the Labour government began to limit the number of houses which local authorities could build – a stance compatible with the party's adoption of monetarism in 1976/77. Although loan sanctions were no longer required, each local authority had to formulate a Local Housing Strategy showing the need for new council building, improvement grants, lending to housing associations, lending for home-ownership, and the need for private building for sale. From this, they would then submit Housing Investment Programmes (HIPs) and the Department of the Environment would then allocate spending permission. Allocations were decided on the basis of total demand and the total amount which the Treasury thought the country could afford.

The Local Government and Planning Act 1980 retained this procedure, but replaced block allocations of borrowing permission with block allocations of permitted capital expenditure. Expenditure could be supplemented by a proportion of the receipts from the sale of council houses (50 per cent of receipts in 1980, 40 per cent in 1984 and 20 per cent in 1985) – with the remaining amounts 'cascading' over from one year to the next).

Under the Local Government and Housing Act of 1989, however, there was a reversion to the control of borrowing, HIP allocations were to be controlled by means of a *basic credit approval* (BCA) for the following year – the Secretary of State taking into account usable capital receipts.

In addition to borrowing, local authorities could continue to use receipts from the sale of housing and land for capital expenditure. Under the 1989 Act 25 per cent could be used for capital purposes, but 75 per cent now had to be used to repay debt – despite local authorities having enormous problems of repair and modernisation, and despite the housing debt representing only a tiny fraction of the current value of the local-authority housing stock (Aughton and Malpass, 1991).

The Housing Act 1980 introduced a new rent subsidy consisting of a 'base amount' (equal to the total subsidy paid in the previous year) *plus* a 'housing cost differential' (representing the increase in the total reckonable housing costs over those for the previous year) *less* the 'local contribution differential' (the amount a government expects the local authority to pay towards housing through increased rents or local tax contributions). In principle, this gave the local authority the choice between increasing rents and increasing the local tax contribution, but since it was the government's intention to reduce exchequer subsidies the Department of the Environment had powers to specify the target rate of annual rent increase. This resulted in rent increases considerably in excess of the amount that could be fully met from rate contributions.

Over the period 1979/80 to 1987/88, change in rent policy enabled the government to cut subsidies from £1,667 to £464 million – a decrease of 72 per cent; reduce the number of local authorities receiving subsidies from 367 to only 88; raise the level of rents from an average of £6.48 to £17.20 per week – an increase of 165 per cent (whereas gross earnings increased by only 125 per cent); and eliminate any shortfall by requiring local authorities to raise local tax contributions from £306 to £499 million (an increase of 63 per cent).

Soaring rents were reflected in the number of council tenants receiving rent rebates/supplementary benefits/housing benefits. The number increased from 2.1 to 2.6 million (1979/80 to 1987/88), or from 40 to 66 per cent of the total – hardly a trend indicative of the success of public-sector housing policy.

Under the Local Government and Housing Act of 1989, therefore, rent determination was reformed. The Department of the Environment would henceforth assess the total value of all the local-authority housing stock in the country, with the value of each authority's stock expressed as a percentage of the total national value. The government would then decide what the total national increase in rents should be for the following year and would calculate how much each local authority's share should be. In effect, the government would use the capital value of each authority's stock, not to determine rents – as was proposed in the *Inquiry into British Housing* (National Federation of Housing Associations, 1985), but to determine 'guideline' increases, regardless of current rent levels. Local-authority rents consequently rose from an average of £20.70 in 1989/90 to a guideline £29.81 in 1992/93, a 50 per cent increase, greatly in excess of the rate of inflation.

A new *housing revenue account subsidy* (HRA) had been introduced under the 1989 Act to replace the former housing subsidy and to include rent rebates (allowances previously paid as a form of income support by the Department of Social Security). But although the increase in rents towards market levels brought about a reduction in the housing element of the HRA subsidy from £1,356 million to £1,027 million between 1990/91 and 1992/93, it simultaneously led to an increase in the *rent rebate element* from £2,304 million to £2,982 million over the same period.

Capital expenditure and rents: housing-association dwellings

Rather than complementing local-authority housing, housing associations have now become the principal providers of new social housing in Great Britain, as prescribed by the White Paper, *Housing: The Government's Proposals* (Department of the Environment, 1987). Housing-association net capital expenditure thus immediately doubled from £1,157 million to £2,308 million between 1990/91 and 1992/93.

Over the three years 1992/93–1994/95, the government (in 1992) aimed to spend £2 billion per annum on housing-association investment, and to produce a total of 153,000 homes – each association setting out its investment plans in an *approved development programme* agreed annually by the Secretary of State. But with cuts in the size of the HAG for each completed dwelling, more had to be borrowed from the private sector. Although public funding continued (involving HAGs and loans from the Housing Corporation and local authorities), mixed funding schemes were increasingly undertaken – private finance enabling public funds to be stretched over a much greater volume of housing than hitherto. Whereas in 1989/90, HAGs covered 75 per cent of housing-association capital expenditure, the proportion decreased to 67 per cent in 1993/94, 62 per cent in 1994/95 and 58 per cent in 1995/96 – with a reciprocal increase in risk incurred by the financial institutions. With cuts in public funding, housing associations would have to depend more and more on private finance if they were going to satisfactorily perform their role as providers of social housing.

The cutback in the size of HAGs and increased reliance on private finance (which, by necessity, requires a competitive rate of return) have not only had an unfavourable impact on new development and housing standards but an inflationary effect on rents (Whitehead, 1995). The consequences of mixed funding schemes could become even more marked if public funding is reduced to 50 per cent as was suggested in the 1987 White Paper.

Whereas existing lettings would be at fair rents as determined by rent officers, and be subject to rent increases every two years, under the Housing Act of 1988 all new lettings would be at assured or assured shorthold tenure with housing associations setting their own 'affordable rents'. Although affordability is not defined by the Department of the Environment, it was interpreted by the National Federation of Housing Associations as a rent approximately equal to 20 per cent of the tenant's average net income. However, in order to ensure that private capital is attracted into housing investment in this sector, average rents for new housing rose to £48 per week in 1990/91 (significantly in excess of average local-authority rents), and by a further 21 per cent in 1991/92, compared to an increase of only 1.8 per cent in the retail price index and a 5.8 per cent increase in the average income of new tenants. As a consequence, by 1991/92, rents consumed 29 per cent of the income of new tenants. Clearly the underlying reason for these hikes in rent was the government's intention to reduce its share of total investment in this sector. In 1994/95, it was reduced from 67 to 62 per cent and in 1995/96 cut to 58 per cent, leaving the associations to fund the rest primarily from private loans. Rents inevitably rose further, by at least a third over the period 1994/95 to 1996/97, according to the National Federation of Housing Associations.

Subsidies

The demand for housing in both the owner-occupied and rented sectors has been substantially facilitated by subsidies – mainly local tax transfers to local-authority housing (until they were discontinued in 1989/90), housing subsidies, housing benefits in both the public- and private-rented sectors and mortgage-interest relief in occupation.

Throughout the 1980s local tax transfers remained fairly constant, in cash terms, until they were abolished under the Local Government and Housing Act of 1989. But housing subsidies (allocated by the Department of the Environment) declined considerably as a result of attempts by Thatcher administrations to curb public expenditure. By 1987, 80 per cent of local authorities received no housing subsidy at all. However, with the introduction of the housing revenue account subsidy under the 1989 Act (which replaced not only housing subsidies but also local tax transfers), the subsidisation of local-authority housing increased in cash terms, albeit marginally. It should be noted, however, that while housing revenue account subsidies to local authorities are not paid directly to tenants, tenants nevertheless benefit because the subsidies cover a proportion of the building costs and interest charges incurred in the provisions of local-authority housing and thus help to keep rents below market levels.

Housing benefits, by contrast, are an important form of income support (made available by the Department of Social Security) and, as such, are not officially regarded as an item of housing expenditure. However, housing benefits soared throughout the 1980s and early 1990s, in part because of rising unemployment but largely because of soaring rents in both the private-rented and social-housing sectors. If the system of payment is not reformed, benefits (amounting to £8,600 million by 1994/95) are projected to rise to £11.9 million by the end of the decade.

By far the largest subsidy, until recently, was mortgage-interest relief which escalated from £1,450 million in 1979/80 to a peak of £7,700 million in 1990/91 (falling to £3,000 million in 1994/95 because of the house-price slump, lower interest rates and relief being limited to the basic rate of tax). Owner-occupiers also benefit from not having to pay the full cost of acquiring an asset – in effect receiving a further subsidy. Since 1962, they have not had to pay Schedule A tax on 'imputed rent income' (the free use of accommodation equivalent in value to rent) and are also normally exempt from capital gains tax – exemptions respectively worth £7,000 and £3,000 million in the early 1980s (Shelter, 1982), rising to at least £10,000 and £4,500 million by the end of the decade. In addition, local-authority tenants buying their own homes received discounts of over £1,000 million per annum throughout much of the 1980s.

Subsidies had become increasingly inequitable and had exacerbated inflation. They were inefficient and wasteful; adversely affected productive industry, the tax base and impeded household mobility; distorted tenure preferences and rural values; and did not provide effective protection for the poor and did little to facilitate repairs and maintenance.

Because of the distorting effects of housing subsidies, the *Inquiry into British Housing* (National Federation of Housing Associations, 1985), recommended *either* a phased reduction in the eligibility ceiling for relief from £30,000 to zero, *or* the withdrawal of mortgage-interest relief above the basic rate of tax followed by a progressive annual reduction over 10 years to complete abolition. The *Inquiry into British Housing: Second Report* (Joseph Rowntree Foundation, 1991) was more forthright in recommending a phased withdrawal of mortgage interest relief (which it believed to be both inefficient and inequitable), rather than allowing relief to 'wither on the vine' with its real value diminishing with inflation. Between 1991 and 1994, therefore, relief was reduced from an upper rate of 40 per cent to 25 or 20 per cent, and in 1995 to 15 per cent.

With regard to rented dwellings, the *Inquiry into British Housing* (National Federation of Housing Associations, 1985) recommended that tenants in the public and private sectors should pay capital-value rents – comprising a basic element assessed to give a real return of about 4 per cent, plus an element for management and maintenance, plus an element for service charges where appropriate. If the inquiry's proposal to abolish mortgage-interest relief were to be implemented, then the adoption of capital-value rents would put tenants and owner-occupiers on an equal footing since both groups of households would pay – in effect – market rents/prices. The inquiry also recommended that needs-related housing allowances should replace both housing benefits and mortgage-interest relief. Allowances would be 'tenure-neutral' and inversely related to income but positively related to rents or mortgage interest.

The *Inquiry into British Housing: Second Report* (Joseph Rowntree Foundation 1991), however, recommended the introduction of a nationwide rent-setting system. Private landlords would be permitted to charge a rent equivalent to 4 per cent of the capital value of their property immediately, while local authorities and housing associations would be allowed to charge 2.5 per cent of capital value within five years and 4 per cent within ten years. Private landlords would be offered a choice of either tax exemption or 100 per cent capital allowances to improve the yield from capital-value rents, and the phasing-in of capital-value rents in the social sector is intended to ease 'poverty trap' problems resulting from more tenants requiring housing benefit. Reiterating the first report, the *Second Report* also proposed the introduction of needs-related housing allowances to concentrate public funds on the most needy households.

It is clear that the above proposals would be costly to implement: £2.5 billion according to current estimates. But this sum could be more than offset by the phasing out of mortgage-interest relief which totalled £6.1 billion in 1991/92.

DEVELOPMENTS IN HOUSING POLICY

Apart from attempting to cut public expenditure on social housing and presiding over consequential increases in rents up to market levels, government housing policy in the 1980s and early 1990s involved the introduction of major programmes of privatisation and the deregulation of the private-rented sector.

Privatisation

The privatisation of housing by Conservative governments in recent years has taken broadly two forms: first, the selling off of council houses to their tenants; and second, and in part to facilitate rehabilitation, the disposal of parts of the council stock to housing associations, trusts and private companies either for renting or for resale. To some observers, the whole political and ideological purpose of privatisation has been to 'dismantle public rented housing and to promote private ownership and management by all available means' (Daniel, 1987).

Under the Housing Act of 1980 and the Tenants' Rights Etc. (Scotland) Act 1980, local-authority tenants were given the statutory right to buy the freehold of their house or a 125-year lease on their flat. The number of local-authority dwellings initially sold off to their former tenants increased from only 568 in 1980 to 196,430 in 1982 (Table 14. 6). In the early 1980s sales began to exceed local-authority housing completions, and for the first time the local-authority sector began to contract in absolute terms (Table 14.6). The Housing and Building Control Act of 1984 subsequently increased the maximum discount by 1 per cent per year for tenants of between 20 and 30 years' standing, up to a maximum of 60 per cent; and reduced the eligibility period of tenure from three to two years.

By 1986, the pace of council house sales was flagging (there were only 89,250 sales in that year compared to 196,430 in 1982). Overall, seven-eighths of council tenants in England and Wales had failed to exercise their Right to Buy (RTB) and more than 25 per cent of all households (in

223

PAUL BALCHIN

Table 14.6 Local-authority dwellings sold under Right to Buy legislation, local-authority housing completions and the local-authority housing stock, Great Britain, 1980–94

	Sales completed	Local-authority completions	Local-authority housing stock (000s)
1980	568	86,200	6,499
1981	79,430	54,867	6,387
1982	196,430	33,244	6,196
1983	138,511	32,833	6,060
1984	100,149	31,699	5,959
1985	92,230	26,115	5,864
1986	89,250	21,587	5,779
1987	103,309	18,823	5,661
1988	160,568	19,030	5,483
1989	238,286	16,465	5,270
1990	140,680	15,780	5,105
1991	120,664	9,457	5,031
1992	63,986	2,172	4,811
1993	60,274	1,715	4,710
1994	64,235	1,331	4,605
Total	1,648,574	285,118	−1,894

Source: Department of the Environment, *Housing and Construction Statistics*

England and Wales) preferred to remain as council tenants. The Department of the Environment was particularly concerned about the small number of flats which had been sold off (little more than 4 per cent over 1980–85). The Housing and Planning Act of 1986 therefore introduced discounts of 44 per cent on the purchase price of flats (for tenants of two years' standing).

Despite the introduction of more generous RTB discounts in 1986, the government again raised the level of discounts to stimulate sales. In 1988, maximum discounts increased to 60 per cent for council-house tenants of 15 years (starting at 32 per cent after two years), and discounts rose to 70 per cent for flat dwellers after 15 years (starting at 44 per cent on qualification). In Greater London, flats had accounted for only 36 per cent of total RTB sales in 1986, but, with larger discounts, flat sales made up 69 per cent of total RTB sales by 1991. Maximum discounts were raised from £35,000 to £40,000 in 1988 as a further inducement to buy, to take account of the house-price boom and to facilitate purchase in the more expensive areas. With the subsequent house-price slump, however, and rising unemployment, council-house sales plummeted – tenants probably being more adversely affected by the recession than most other households.

Estate privatisation, in its many forms, was motivated by the perceived fiscal need to transfer the responsibility of repairs from government to housing associations and the private sector, and by an awareness that there was a limit to the number of council dwellings that could be sold off under RTB policy. From a peak of 196,430 RTB sales in 1982, sales had decreased to only 89,250 in 1986. Undoubtedly, the problem of repairs and difficulties of resale had deterred many tenants from exercising their RTB. Estate privatisation would thus speed up the process of tenure shift.

Under the Housing Act of 1988 – and subject to local ballot – the government therefore planned to established a number of Housing Action Trusts to repair or rehabilitate housing estates, to improve management and to subsequently sell off the estates to housing associations or private landlords.

The second way in which the 1988 Act facilitated the disposal of council estates was by means of large-scale voluntary transfers (LSVTs) to housing associations and to private landlords. Transfers – requiring consent from the relevant Secretary of State – were already taking place under both the Housing Act of 1985 and the Housing and Planning Act of 1986, but the 1988 Act eased the process of transfer. By 1994, 32 local authorities had transferred the whole of their stock – amounting to a total of 149,478 dwellings, and nearly 250 other local authorities were considering transfer. Clearly, with a total of 94,277 dwellings transferred out of the local-authority sector, LSVTs were more than compensating for the reduction in RTB sales after 1989.

Third, under the 1988 Act, local-authority tenants are able to exercise 'Tenants' Choice' – whereby tenants of council houses individually were able to exercise their right to transfer to another landlord, but tenants of flats were required to decide collectively. Often, the poor condition of the relevant housing deterred potential new landlords from taking over estates. Local authorities were therefore sometimes obliged to pay 'dowries' to the new landlords to cover the cost of essential repairs.

By the early 1990s (if not before), it was possible to see the Conservatives' privatisation policy in perspective. In net terms, the number of council houses had decreased from 6.6 million in 1979 to 4.9 million in 1992, but local authorities still owned 21.3 per cent of the total stock of housing in Britain in 1992, while private landlords owned only 7.5 per cent and housing associations 3.3 per cent. Clearly, the majority of council tenants were satisfied with their landlord and had no wish to either exercise their RTB or transfer to an alternative landlord.

The deregulation of the private-rented sector

The supply of private-rented accommodation in Great Britain has declined from approximately 90 per cent of dwellings in 1914 to only 10 per cent in 1995, compared to 17 per cent in France, 36 per cent in Germany and over 60 per cent in Switzerland (1990); and in Britain, housebuilding in this sector has been virtually non-existent since the Second World War in contrast to most other European countries.

Notwithstanding the fact that rent control or regulation had been in force for most of the period since 1915 and undoubtedly acted as a deterrent to investors, the diminishing role of the sector was investigated by the House of Commons Environment Committee (House of Commons, 1982) which reported that decline was mainly due to slum clearance and transfer to owner-occupation. Private-rented housing fell short on all of the general criteria of investment: the level of risk; liquidity; expected return on capital; and management involvement – the last being very burdensome when compared to large-scale investment in offices, shops and factories or small-scale investment in insurance policies, unit trusts or building societies.

Under the Housing Act of 1980, the government had introduced short-hold tenancies, believing that many properties stood empty because the Rent Acts of 1974 and 1977 had got in the way of landlords and tenants wishing to agree to a lease for a short fixed period. Shortholds were only applicable to new lettings, and at the end of fixed-term agreements of 1–5 years landlords had the right to regain possession. Shorthold was in effect a form of decontrol.

After 1981, if landlords and tenants agreed, market rents on new short-hold tenancies were negotiable outside Greater London – regulation remaining in the capital since some 21 per cent of housing was private-rented (and 30–40 per cent in some boroughs) compared to 13 per cent nationally in 1981.

The 1980 Act also introduced 'assured' tenancies whereby approved landlords were permitted to let their new dwellings outside of the Rent Acts. Building societies, banks, other finance houses and construction firms could be licensed by the government to build homes for rent.

The Housing Act of 1988 and its Scottish equivalent, the Housing (Scotland) Act 1988, further aimed to revive the private-rented sector by reducing the minimum period of shorthold (renamed 'assured shorthold') to only six months and extending assured tenancies to the remainder of new lettings. Assured shorthold lettings are at market rents which take account of the limited period of contractual security of the tenant – and the tenant can apply during the initial period of the tenancy to a rent assessment committee for the rent to be determined. Assured tenancies, on the other hand, although being relatively secure, are at rents freely negotiated between landlord and tenant and therefore at market levels. Existing regulated tenants would (ostensibly) continue to be protected by the Rent Acts.

Critics of the 1988 Acts argue that since rents would rise dramatically (particularly in areas of housing shortage), so too would the need for housing benefits. In 1987, as many as 60 per cent of private tenants received housing benefit but, with rent increases inevitable, the proportion eligible for benefit would escalate – producing thereby a poverty trap. It would be more advantageous for many tenants to be out of work and in receipt of housing benefits than to have a job and be ineligible for assistance (McKechnie, 1987). Rather than rents being permitted to rise to market levels, the private-rented sector might be more usefully assisted by the provision of tax breaks[2] or government grants (of up to, say, £12,000 per property) to enable landlords to secure an adequate return on their investment – estimated at 3.4 per cent higher than the prevailing (1995) gross rental yield of 7.6 per cent. This could revive the private sector, generating an additional 23,000 houses to rent (Crook *et al.*, 1995).

CONCLUSIONS

The concept of the welfare state was arguably one of the most important influences over public policy in the United Kingdom in the twentieth century. Consolidating and extending many of the ideas of the Liberal governments of 1906–14, the Beveridge Report (1942) highlighted the need to protect 'from the cradle to the grave' all individuals and the family from want, disease, ignorance, squalor and idleness. An improved system of national insurance, the national health service, educational reform, a comprehensive town-planning system and Keynesian economic policy were all consequently put in place in the immediate post-war period, and each remained broadly intact throughout the 1950s, 1960s and 1970s irrespective of which party was in power. Housing policy, however, was by comparison erratic and much influenced by the prejudices and nostrums of consecutive governments. However, compared to later divergences in policy, a consensual approach often prevailed.

But in the 1980s and early 1990s, the inefficiencies, inequalities and lack of choice characteristic of United Kingdom housing markets became major causes of concern to most housing specialists. Under the ideology of Thatcherism, housing was singled out to bear the biggest cut in public expenditure at a time when the housing shortage was getting increasingly severe, housebuilding had plummeted and unemployment had soared, the housing stock remained in a poor condition, and local-authority and housing-association rents had risen faster than inflation. Meanwhile, owner-occupation (at least until 1991) had been promoted by more and more regressive mortgage-interest relief and tax exemption, whereas the supply of accommodation at reasonable rent had been depleted by the privatisation of local-authority housing and the deregulation of the private-rented stock – a situation undoubtedly promoting the development of a 'dualist' rather than 'unitary' system of renting in the United Kingdom (Kemeny, 1995).

Based on the White Paper, *Our Future Homes* (Department of the Environment, 1995b), the Housing Act 1996 extended the right to buy to housing-association tenants; replaced the housing association grant (HAG) with a social housing grant (SHG); gave local authorities and housing action trusts the power to offer one-year introductory tenancies to deter 'anti-social behaviour' among tenants; and limited the duty of local authorities to offering only short-stay accommodation to homeless households in the first instance instead of priority consideration on the housing register (waiting list). In effect, local authorities will need to rely increasingly on the private-rented sector where tenants will face higher rents and restricted housing benefit. The legislation failed to implement the White Paper's proposal that local authorities should be encouraged to transfer their housing stock to new local housing companies, housing associations and Housing Investment Trusts – a process that would have hastened the privatisation of public-sector housing.

The 1996 legislation clearly did little to increase the supply of affordable rented housing to those most in need, or to reduce the magnitude of homelessness. Neither did it conflict with what were allegedly the Conservatives' long-term policy objectives. These included: an end to housebuilding in the social sector; the elimination of all housing subsidy (including housing association grants, housing revenue subsidies, housing benefits and mortgage interest relief); the introduction of a new personal subsidy determined by income; and a switch in the role of housing associations from development to management. Home-ownership, meanwhile, would be expected to rise continually, possibly to 80 per cent by 2000, but only as part of natural choice in an unsubsidised free market (*Roof*, 1994). But clearly without a dramatic increase in public investment, the total supply of housing would not increase sufficiently to meet demand or needs in the twenty-first century, rehabilitation would continue to be on a scale insufficient to produce a marked improvement in the condition of housing, and unemployment in the construction industry would remain high, while more and more would find it difficult or impossible to satisfy their housing requirements at a price they could afford.

NOTES

1 There were 804,000 empty houses in England alone in 1995, 690,000 of which were in the private sector – a high proportion being in a poor condition.
2 From 1 April 1996, institutional investors and housing associations were able to set up Housing Investment Trusts to fund and manage private rented housing. Investors were eligible for capital gains tax exemption and a reduced rate of corporation tax (at 24 instead of the standard 33 per cent). Rents, under assured tenancy, would nevertheless, be at market levels.

Part IV

HOUSING IN TRANSITION

15

INTRODUCTION TO HOUSING IN TRANSITION

Paul Balchin

Because of political and economic pressures, market economies emerged in the countries of central Europe in the 1980s and early 1990s and rapidly began to replace the system of centrally planned command economies which had been set up in the Eastern bloc after the Second World War. Under pressures from the International Monetary Fund and western governments, 'shock therapy' began to bring about the liberalisation, stabilisation and privatisation of the economies of central Europe (see Gowan, 1995). Although in Hungary a limited market was developed under communism in the 1980s, it rapidly expanded after the centre–right won the free elections of March 1990, and there was little indication that the further development of the market economy would be adversely affected by the former communists gaining a parliamentary majority in the elections of May 1994. In Czechoslovakia, the birth (or re-birth) of the market economy was signalled by the resignation of the communist leadership in November 1989 and by the subsequent victory of the Civic Forum movement in the elections of June 1990 – the development of the market transcending the transformation of Czechoslovakia into the separate states of the Czech Republic and Slovakia. In Poland, the apparent downfall of communist government was marked by the triumph of Solidarity in the elections of June 1989, yet the return of former communists to power in September 1993 as in Hungary was unlikely to reverse or retard the development of a market economy.

Thus, within the context of rapid political change, emerging housing policies in Hungary, the Czech Republic and Poland (like other social and economic policies in transition) needed to take account of the legacy of previous policies before attempting to introduce new solutions to often old problems. Within central Europe there had been three common features of housing provision during the communist period of government (Turner, 1992). First, there was almost a complete absence of the private-rental sector – private investment property being nationalised after the Second World War. Second, because of wartime destruction and the lack of building during the war, there was a substantial shortage of dwellings in the 1950s

231

and 1960s which necessitated the large-scale construction of state or state-sponsored housing in high-rise estates – the precise form of promotion varying from one country to another. Third, although there were large home-ownership sectors throughout most of central Europe (in large part because of the comparatively large rural population), when communist governments took over in the 1940s it was not considered appropriate to nationalise this sector of housing – for both political and management reasons. Nevertheless, for about thirty years, owner-occupation at best received little assistance from the state (problems of mortgage finance and capital gain remained unresolved), and at worst home-ownership was severely constrained by planning restrictions and starved of resources. By the 1970s, however, the sector began to expand and a market developed – in part, a result of the 'push factor' (the anonymity, sterility and unpopularity of high-rise estates), and partly because of the 'pull factor' of an increasingly overt demand for ownership among privileged groups in society – 'in combination with the failure of the state to provide satisfactory social housing for these groups, or to reward them in any other way' (Turner, 1992)

TENURE

Constructed almost entirely after the Second World War (or made available through the nationalisation of private-rental stock), social-rented housing in the late 1980s and early 1990s represented a large proportion of total dwellings in the future Czech Republic and in Poland – being owned either by the state or by co-operatives, municipalities, or industrial enterprises or

Table 15.1 Housing tenure in Hungary, the Czech Republic and Poland

		Social-rented %	Owner-occupied %		Other tenure (including private renting) %
Hungary	(1990)	22.6	59.8[a] 11.9[b] 5.6[c]	} 77.3	0.1
Czech Republic	(1991)	59.4		40.5	0.1
Poland	(1988)	24.3[d] 19.4[e] 12.6[f] }	56.3	41.1	2.6

Sources: Hungarian Central Statistical Office, *Housing Statistical Yearbook*; Polish Central Statistical Office, *Statistical Yearbook*; SNTL, *Statistical Yearbook of Czechoslovakia*
Notes: [a] Family homes
[b] Condominiums
[c] Co-operative ownership
[d] Co-operatives
[e] Municipal housing
[f] Owned by industrial enterprises and institutions

institutions. The state-owned rented sector in Hungary, in contrast, was relatively small (Table 15.1).

Largely because a high proportion of households still resided in rural areas, owner-occupation as a share of total tenure was fairly high in both the Czech Republic and Poland (higher even than in the Netherlands or Germany), while in Hungary the owner-occupied stock was proportionately one of the largest in Europe – not only with its pre-eminence in the countryside but because of a rapid increase in home-ownership among the newly affluent sectors of the urban population. Whereas in the Czech Republic and Poland, owner-occupied dwellings were very largely family homes, in Hungary owner-occupation – particularly in urban areas – also took the form of condominiums and ownership co-operatives.

Private-rental housing throughout central Europe is very largely a residual sector – the remnant of mostly old tenancies in existence before the nationalisation of the housing stock in the immediate post-war period. Private landlords had very restricted rights; for example, in Hungary rent was limited, there was no right of repossession and landlords only had control over the exchange of tenants. In recent years, however, a new but small private-rental sector has emerged, mainly for the use of foreign tenants at uncontrolled rents. In the Czech Republic, rents in a very restricted supply of private-rental housing were subsidised out of local budgets – municipalities paying 60 per cent of a tenant's management and maintenance costs. As in Hungary, the demand for private-rented accommodation by foreigners is inflating rents – to the detriment of middle-income households who do not qualify for social housing but are unable to buy (Long, 1993).

HOUSEBUILDING

After a slow start in the post-war years, the level of housebuilding increased markedly in the 1970s. Large-scale public-sector housebuilding (involving the use of prefabricated technology and the development of high-rise estates) contributed substantially to the reduction in housing shortages in much of central and eastern Europe, despite marked increases in the number of households. Shortages decreased, for example, from the equivalent of 13.4 per cent and 40.5 per cent of the respective housing stocks of Czechoslovakia and Poland in 1970 to about 11 and 26 per cent of their stocks a decade or so later (Table 15.2).

Although housing deficits continued to decline in the 1980s, they still remained acute, particularly in Poland (Table 15.2). There is thus a very great need to undertake large-scale housebuilding programmes well into the twenty-first century – for example, in Poland, an extra 350,000 houses will be required each year until 2007 to satisfy the needs of 1.6 million households and to replace housing in poor condition (Hajduk, 1992). It is very

Table 15.2   Housing shortages in Hungary, the Czech Republic and Poland

		Number of dwellings	Number of households	Surplus or deficit	% total dwellings
Hungary	1980	3,146,565	3,719,349	−302,789	8.9
	1990	3,687,996	3,889,532	−201,538	5.5
Czech Republic	1970	3,088,841	3,502,718	−413,877	13.7
	1980	3,494,846	3,875,681	−380,835	10.9
	1991	3,705,681	4,051,583	−345,902	9.3
Poland	1970	6,675,200	9,376,000	−2,700,800	40.5
	1978	8,715,300	10,948,000	−2,232,700	25.6
	1988	10,716,400	11,970,000	−1,253,800	11.7

Sources: Hungarian Central Statistical Office, Housing Statistical Yearbook; Polish Central Statistical Office, Statistical Yearbook; SNTL, Statistical Yearbook of Czechoslovakia

Table 15.3   Housebuilding in Hungary, the Czech Republic and Poland, 1980–92

	Hungary		Czech Republic		Poland	
	No. of completions (000s)	Per 1000 population	No. of completions (000s)	Per 1000 population	No. of completions (000s)	Per 1000 population
1980	80.7	7.8	89.1	8.3	217.1[a]	6.1[a]
1989	55.1	5.3	51.5	4.9	189.7	4.0
1992	41.8	4.0	25.8	2.6	135.6	3.6

Source: Hungarian Central Statistical Office, Housing Statistical Yearbook; Polish Central Statistical Office, Statistical Yearbook, SNTL, Statistical Yearbook of Czechoslovakia
Note: [a] 1981

regrettable, therefore, that the number of housing completions plummeted in the period 1980–92 – by 71 per cent in Hungary, 48.2 per cent in the Czech Republic and 38 per cent in Poland. The number of dwellings built per thousand of the population similarly decreased (Table 15.3).

Reduced output was a direct result of a considerable reduction of state expenditure and a restructuring of the construction industry (post-1989), whereby kombinants (large state-owned companies) were replaced by comparatively small private firms. The withdrawal of substantial government support for new housebuilding, on which the construction industry had been heavily reliant (Baross and Struyk, 1993), was reflected in the decreased volume of social-housing production as a proportion of total output, over 1980–89 – particularly in Hungary, but also in Czechoslovakia and Poland (Table 15.4).

Within the context of housing shortages and diminishing housebuilding, it is of very great concern that 'housing standards and quality are low and investment in the maintenance of the existing stock has been negligible over the last decades' (Czerny, 1992). Since the housing stock contains a dis-

Table 15.4 Social housing production as a percentage of total housing construction in Hungary, Czechoslovakia and Poland

	Hungary	Czechoslovakia	Poland
1980	38.8	75.9	74.3
1987	17.1	70.8	68.4
1989	11.5	70.6	63.4

Source: United Nations 1991

Table 15.5 Age of dwellings in Hungary, the Czech Republic and Poland

	Approximate building period of dwellings (% at end of 1980s to early 1990s)			
	Pre–1920	1920–44	1945–60	Post–1960
Hungary	18.0	14.8	11.8	55.4
Czech Republic	21.6	20.3	10.1	48.0
Poland	13.8	16.6	14.1	55.5

Sources: Hungarian Central Statistical Office, Housing Statistical Yearbook; Polish Central Statistical Office, Statistical Yearbook; SNTL, Statistical Yearbook of Czechoslovakia

proportionately large proportion of post-1960 dwellings (Table 15.5), it is comparatively modern housing (particularly on estates developed in the 1960s and 1970s) that is, generally, the most seriously affected by inadequate maintenance. Incentives for good management have been weak as state maintenance firms (with budgets administratively set) enjoyed a monopoly in the district in which they operated 'with little concern for the true cost of good maintenance' (Baross and Struyk, 1993).

HOUSING INVESTMENT, FINANCE AND SUBSIDIES

In the Eastern bloc countries after the Second World War, 'investment in the construction of dwellings was given a low priority as compared with investment in heavy industry' (Czerny, 1992). Consequently, in Hungary, Czechoslovakia and Poland, housing standards were poor and investment in the maintenance of the existing stock was negligible. Nevertheless, in the 1970s and 1980s housing was a recipient of an array of subsidies designed to facilitate investment, but in the early 1990s investment plummeted to a very low level.

The proportion of housebuilding in Hungary receiving interest-rate subsidies from the state decreased from 40 per cent in the 1970s to less than 5 per cent in 1991 – state-financed housebuilding diminishing from a maximum of 39,000 dwellings to only 1,600 over the same period. New dwellings were financed mainly by OTP (National Savings Bank) loans – builders being subsidised by low (3 per cent) interest rates, with the OTP receiving the difference between market rates and the subsidised rate. Because of the withdrawal of this subsidy in 1989 practically no public housing was built in Hungary during the early 1990s (Elter and Baross, 1993).

In Czechoslovakia, investment funding seemed to be in jeopardy in the early 1990s because of the prospect of a similar dismantling of general building subsidies (Kahout and Zajicova, 1992). At that time, the state fully subsidised interest rates relating to the construction of public-sector dwellings and subsidised half the cost of interest in the development of co-operative housing. The state also provided cash subsidies for the construction of owner-occupied housing depending on the marital status and age of the buyer.

Municipal housing in Poland was also fully financed from public funds; housing owned by industrial enterprises and institutions was financed from internal funds and subsidised low-interest credit; co-operative housing was financed very largely by subsidies covering 40 per cent of the cost of interest, but also from the funds of members and unsubsidised bank credit. Owner-occupied housing was mainly privately financed although the state met a quarter of the cost of interest payments. In 1989, however, the state terminated its low-interest housing loans policy – the main instrument for financing new social housing projects through the state budget (Bogdanowicz, 1993). Apart from housebuilding investment diminishing by 14 per cent over 1989–91, funds were increasingly targeted at the co-operative sector which increased its share from 45 to 61 per cent of the total level of investment, in contrast to all other sectors which correspondingly received less funding (Hajduk, 1992).

There were also common features relating to rents in the social sector: the landlord was the state (normally represented by a local state-owned maintenance company), low regulated rents were set by the central government and utility costs were subsidised. Means-tested demand or 'subject' subsidies were therefore unnecessary, and tenants enjoyed security of tenure (Hegedüs et al., 1992). But there were serious flaws in this system of renting (Baross and Struyk, 1993). First, the relatively high cost of providing housing and associated services was not reflected in the low rents charged – in Hungary, for example, rents were equivalent to only 3 per cent of the average household's income in 1992 – compared to 15–25 per cent in EC countries (Csomos, 1993); in Czechoslovakia in the 1980s they accounted for only 2.7–5 per cent of income in the case of municipal housing and 5–10 per cent in respect of co-operative housing (Kahout and Zajicova, 1992); in Poland rents covered no more than 25 per cent of the total cost of rent and services in municipal and company housing in the late 1980s (Kozlowski, 1992). Second, there were large supply or 'object' subsidies – badly targeted (with the largest subsidies going to households in bigger and better units). Third, low rents deterred household mobility and encouraged under-occupancy. Households were unwilling to move to more appropriate accommodation when family size decreased. Fourth, tenants – in effect – enjoyed ownership rights, notably the right to exchange and the right to inherit public-sector rental accommodation. Fifth, rents normally failed to cover repair and maintenance costs, with adverse effects on the condition of the housing stock.

Although housing subsidies in central Europe might loosely be referred to as 'subject' subsidies or 'object' subsidies, the distinction between the two is often blurred since the construction industry can capitalise on a subject subsidy and build it into price or rent where it would become transformed into an object subsidy. Nevertheless, in value, most subsidies start off as subject subsidies and are clearly intended to keep rents or house prices affordable. In Hungary, for example, of the 11 housing subsidies existing in 1992 (a product of an earlier regime and totalling in value 127.4 Ft billion), as many as seven were subject subsidies (amounting in value to 87.7 Ft million). This system of subsidies therefore

> contributes very little to the efficient operation of the housing system. A great proportion of the subsidies...does not influence new construction or new acquisitions. These subsidies...are received by households which built or bought their housing in past years ... [and] ...the state must continue to carry these burdens.
>
> (Hegedüs *et al.*, 1993)

Among the present subsidies in central Europe, moreover, very few are carefully targeted and these are often regressively allocated. The introduction or extension of housing allowances, however, would undoubtedly allocate a subsidy more progressively through the process of means-testing, and the need for other subsidies could be considerably reduced if the allowance covered the difference between cost-rents and the percentage of income households could reasonably be expected to pay.

Of the three countries, Hungary, the Czech Republic and Poland, only the last has implemented a complete housing allowance programme. Dating from the 1980s, Polish allowances (which cover rent and utility charges, and are payable from local-authority budgets) are related to income and floorspace (Millard, 1992). It was estimated in 1990 that, in the municipal sector, about 20 per cent of households with earned income and 60 per cent of retired persons would be eligible for allowances, and even higher proportions could qualify in the co-operative sector (Kulesza, 1990). In Hungary, under the Social Law of December 1992, all local authorities have been mandated to introduce allowance programmes, while in Czechoslovakia a housing allowance programme was incorporated in draft social-security legislation in 1993 (Kingsley *et al.*, 1993). In each country, programmes envisage a step-by-step movement to cost-rents and then to market rents over several years (Baross and Struyk, 1993). It is possible, however, that none of the schemes proposed or implemented to date is as well targeted as those in western Europe; nevertheless, when introduced, allowances will initially cover the difference between cost-rents and the proportion of income a tenant can reasonably be expected to pay, and will subsequently eliminate the need for most other housing subsidies.

DEVELOPMENTS IN HOUSING POLICY

With the emergence of a market economy in the early 1990s, it was increasingly acknowledged that investment in state housing and particularly in renovation would only be feasible if rents were raised towards market levels, subsidies were targeted at lower-income households by means of allowances, and housing management and maintenance were transferred to private companies or local authorities (Baross and Stryk, 1993) (see Table 15.6). But, because of political sensitivity, these processes – to some extent – had to proceed slowly, for example rents on municipal housing were frozen in Hungary from summer 1992 to summer 1994, while in Poland rents still covered only a small proportion (30 per cent) of the cost of maintenance in 1993 (Long, 1993). It was not surprising that from 1994 apartment rents in Poland were set to rise annually by 3 per cent of the dwelling's capital value to help facilitate escalating maintenance costs – prompting increases in housing allowances.

With the extension of the market economy, private-sector investment has been expected to compensate (at least in part) for the reduction of state assistance to social housing. In the Czech Republic, the introduction of savings and mortgage banks in 1994 to supply credit (at subsidised rates of

Table 15.6 Rents, housing allowances and management in state housing, Hungary, Czechoslovakia and Poland, 1990–95

	Rents	*Housing*	*Management allowances*
Hungary	January 1990, 30% increase, small relative to inflation Moratorium on rents 1992–94.	No real allowances introduced until 1993.	Transferred from state maintenance companies (IKVs) to public service companies, 1993–95.
Czechoslovakia	July 1992, 100% increase. Rents still remain small compared with costs.	Plans to introduce allowances in 1993–94.	In some districts in Prague, controlled by local government.
Poland	Several increases but at less than the rate of inflation.	Allowances for deserving households.	No transfer to private management.

Source: Based on Baross and Struyk (1993)

interest) for owner-occupation was but one outcome of the development of a free-market economy and the liberalisation of the financial sector; while, in Hungary, very high house-price/earnings ratios of up to 16:1 (three or four times as high as in western Europe, but only a quarter higher than in the Czech Republic and Poland) resulted in OTP (the National Savings Bank) providing long-term roll-over credit (at rates of interest adjusted to inflation) – assisted by state 15-year repayment subsidies. Even so, loans to dwelling value rarely exceeded 30 per cent (to the disadvantage of average families wishing to build or buy) and the mortgage debt amounted to no more than 6 per cent of the gross domestic product (in 1989) – compared to 100 and 58 per cent in the United Kingdom and 80 and 22 per cent in Germany respectively (Csomos, 1993).

Housing investment in the owner-occupied sector in Poland (as in the Czech Republic and Hungary) is financed privately with the assistance of low-interest bank credit – subsidised by the state (in Poland at 25 per cent in the early 1990s). In 1992, the PKO BP Bank (the largest in Poland) became the first to offer mortgages, but only to borrowers who could prove that they would set aside 25 per cent of their monthly incomes to meet repayments (Long, 1993) – a requirement that would probably exclude all but the most affluent households. It was the view, however, of the government that, in the long term, the prosperous would be left to provide their own housing, the middle-income group would be eligible for preferential mortgage credit, subsidised or guaranteed by the state, and the poorest households would have access to only a low standard of housing, allocated by local authorities and financed by a National Housing Fund (Millard, 1992).

Clearly, in each of the three countries discussed, the development of a building society system to convert savings into mortgages, the introduction of deferred mortgage repayment loans (at initially low rates of interest – giving way to higher rates of interest in later years), and the provision of mortgage-interest tax relief would all facilitate the extension of owner-occupation, and enable housebuilding in the private sector to offset the large fall in numbers of state-constructed housing (UNICEF, 1995).

To further both the development of a market economy and a reduction in state expenditure, three different forms of privatisation have become major issues in housing throughout most of central and eastern Europe in recent years. First, in the Czech Republic and Poland, properties have been transferred from state ownership to their former private owners (or their heirs) through the process of restitution – in Czechoslovakia up to 30 per cent of the national stock of housing was eligible for restitutional transfer (Kingsley et al., 1993). Second, state-promoted co-operatives have been converted into condominiums, for example in Czechoslovakia in 1992 and in Poland in 1993. Third, and most importantly from an economic standpoint, state housing has been sold to existing tenants – normally at a discount (rather than at historic cost or a market price), and with the additional incentive that rents might otherwise rise quite substantially in the future, reflecting escalating maintenance and repair costs (Clapham, 1993).

The economic benefit of 'selling cheap' is that this would quickly raise a large amount of revenue for the state, even if in the long term more revenue could be collected by means of a high-price policy, and the state would no longer have to incur the high costs of repairs and maintenance. The economic costs of selling cheap include the loss incurred in the long term by not selling at prices close to or at the market level, the loss of accessible and affordable rented housing (an important social asset), and the difficulties faced by low-income buyers in funding the costs of repairing and maintaining their properties (Clapham, 1993). Clearly it would be the best units in the best locations which would be privatised first, leaving a residue of welfare housing for the socially and economically disadvantaged – a situation similar to that found in the 'dualist rental systems' of the liberal-welfare regimes of western Europe and the United States. Although a low-price strategy has been widely adopted in most countries in central and eastern Europe, notably in Hungary since 1969 (particularly following the transfer of state housing to the local authorities in 1991), by 1993 a national policy had not yet emerged in Poland, while in the Czech Republic the impetus after 1991 was left with individual municipalities.

Despite the very substantial arguments both for and against privatisation, 'it has been taken for granted that it is a "good thing"' (Clapham, 1993). It is assumed that a free market in housing is a more efficient way of determining production and exchange than the previous system of central planning, that the market will encourage owners to keep their properties in good order, and that privatisation and the extension of the market is a step towards western-style democracy and a palliative to the problems of unemployment and a lowering of living standards. In Hungary, therefore, about 38 per cent of the 1990 stock of public housing had been sold off to former tenants by 1993 (Csomos, 1993), while in Poland 25 per cent of the 1989 stock had been privatised or re-privatised by the end of 1992 (Bogdanowicz, 1993) (Table 15.7).

Hegedüs et al. (1992), however, attempted to show that under certain circumstances the privatisation of the public-rented stock was compatible with the development of a social market in housing. Housing in the public-rented sector after 1991 had been sold off by local authorities both at high and low discounts, and (for political reasons) with rents remaining low. With high discounts and low rents there was clearly every prospect of municipal insolvency and the disappearance of the rental stock. With low discounts and low rents, however, there was little incentive to buy. But with the introduction of housing allowances, rents could be increased with far less political opposition than hitherto, although local authorities would still

240

Table 15.7   The privatisation and re-privatisation of social housing in Hungary and Poland, 1989–93

		Number of dwellings privatised	% 1990 public rental stock
Hungary			
	1990	54,023	7.7
	1991	82,118	11.7
	1992	74,133	10.5
	1993	58,133	8.3
	1990–93	268,665	38.2

		Number of dwellings privatised or re-privatised			
		Municipal	Co-operative	Company	Total
Poland					
	To 1989	180,831	n/a	22,051	n/a
	1990	8,189	n/a	714	n/a
	1991	11,339	n/a	1,497	n/a
	1992	10,744	n/a	11,298	n/a
	To 1992	211,103	1,341,161	35,821	1,612,458
	% 1989 stock	10.41	45.60	2. 42	24.88

Sources: Csomos (1993); Bogdanowicz (1993)

need to decide whether to sell at a large or small discount. Hegedüs *et al.* (1993) suggested that selling at a low minimum discount (that is, at a high rather than low price) would be the economically preferred course of action, since both higher prices and higher rents would make the sector self-financing and (with housing allowances) would not be detrimental to low-income tenants. High rents, moreover, would stimulate the development of a private-rental sector (where housing allowances would also be available), and, through the process of competition, create an integrated rental system similar to that found in corporatist or social-democratic countries of western Europe. Inevitably, local government would be opposed to introducing a high-price strategy after most of the best houses had been sold off at relatively low prices, but at the same time they would welcome an increase in both sales revenue and rent receipts to finance the rehabilitation of the remaining public-sector stock. It is probable that the political power of those who favour the 'give-away' strategy will be greater than those who wish to sell at near-market prices, and it is notable that in Budapest many district authorities continued to adhere to a low-price policy (Hegedüs *et al.*, 1993).

Even at a low price of purchase, however, many low-income households would find the cost of repairs and services a considerable burden. 'Policy should therefore focus not only on the transfer of state-owned dwellings at low cost to tenants, but also provide support for maintenance costs and utilities for low-income households' (UNICEF, 1995).

CONCLUSIONS: HUNGARY, THE CZECH REPUBLIC AND POLAND

With substantial reductions in state-funded housing investment, with rents rising (albeit slowly in some cases) to market levels, with housing management being transferred from central government organisations to private agencies, with an increased reliance on private finance to expand owner-occupation, and with massive programmes of privatisation depleting the public-rental stock and relegating it to a residual tenure, it is very evident that (at least as far as housing is concerned) a liberal-welfare regime is emerging and there is little possibility of an integrated rental system being developed.

It might be considered that Hungary, the Czech Republic and Poland are wasting a major opportunity since a great asset available to these countries

> is that they have inherited a large public rental stock after half a century of communism which could form the basis for a cost-rental sector. One alternative to selling the dwellings into individual owner-occupation would be to sell or transfer the stock to cost rental organisations...owned and run by tenants themselves in the form of tenant rental co-operatives.
>
> (Kemeny, 1995)

The stock could clearly be broken up into a variety of cost and profit rental tenures as well as owner-occupied housing – competing with each other on equal terms and facilitated, in part, by tenure-neutral subsidies. But, because of the haste of the transition process, a neo-liberal housing regime 'has been promoted as the true representative of genuine free market housing policy' (Kemeny, 1995), rather than the social market. It is remarkable that, despite central Europe's historic cultural, social and economic ties with Austria and Germany, the collapse of the Russian hegemony in 1989 failed – within the housing arena – to signal the adoption of social democracy or corporatism, but a regime somewhat characteristic of that of the United States or the United Kingdom in the 1980s (Kemeny, 1995). The adoption of a social market economy and a unitary rental system in Hungary, the Czech Republic or Poland cannot, however, be ruled out in the long term, particularly if neo-liberal policies appear to fail in the years to come.

FOOTNOTE ON FORMER YUGOSLAVIA

Prior to a subsequent consideration of housing policy in Slovenia and Croatia, it is important to take an overview of policy development in the former Yugoslavia. From the 1940s, Yugoslavia's brand of communism was characterised by self-management and decentralisation (Nord, 1992). Soviet-style state socialism was eschewed in favour of a system of 'socialist

242

democracy' based on the 'diffusion of power to enterprises and local units'. Housing was no exception to this pattern of decentralisation (Mandic, 1991). During the first two stages of reform, the responsibility for housing provision was transferred from a central to a republic and municipal level, and subsequently to self-management enterprises. The third stage involved the formation of self-managing housing communities. Housing consumption was, to an extent, assisted by means-tested rent subsidies in the social-rental sector, and loans for house purchase were at sub-inflation rates of interest.

By the 1980s, however, Yugoslavia suffered from economic problems at least as severe as those being experienced elsewhere in central and eastern Europe; inflation, for example, exceeded on average 50 per cent per annum over 1980–86, while unemployment soared to unacceptable levels. Housing, moreover, was in a state of crisis, as housing shortages continued to be a major problem – especially disadvantaging younger households, the less skilled, and employees in low-productivity industries (Mandic, 1991). The comparatively small social-rented sector – amounting to less than one-fifth of the total stock – was of little help to the less economically productive sections of the population since the provision of social housing was normally related to employment. In general, therefore, peasants and unskilled workers were expected to build their own houses, while the middle classes queued for social housing (Seferagić, 1985).

In the years leading up to the fragmentation of Yugoslavia in 1991–92, there was an apparent shift towards a market economy in housing. The privatisation of the social-rented sector and the further promotion of owner-occupation (already approaching 70 per cent of the total housing stock) were considered desirable goals. Although the economic and housing history of Yugoslavia had been markedly different from that of Hungary, Czechoslovakia and Poland throughout the forty years after the Second World War, the outcome – in terms of the aims and objectives of housing policy – was remarkably similar. This became particularly evident in Slovenia and Croatia following their secession from the federation.

16

HUNGARY

József Hegedüs and Iván Tosics

This chapter describes and explains the transition from a state-controlled housing system towards a market-oriented housing system in Hungary. The emphasis of the analysis is on the processes of decentralisation and privatisation which are the most important elements of the restructuring of the institutional framework of the housing system.[1]

After the Second World War, the east European countries – including Hungary – were drawn into the Soviet Union's zone of influence and were inevitably led to adopt the Soviet economic model in which the living standards of the population were subordinated to the goals of economic growth defined by the communist party.

In the first phase (1949–56), housing production and consumption dropped practically to the minimum level. The chief task of the housing policy was to establish a network of organisations and legal institutions dominated by the state. State control over the housing system affected both demand and supply. On the demand side, strict control over incomes, the strictly controlled housing finance and subsidy system, and the political and legal property right limitations (nationalisation, the 'one-family–one-flat' principle, and restrictions on allocating rental flats and loans to families moving to towns) were the main elements of regulation. On the supply side the monopoly of state enterprises, severe restrictions on private building (small-scale industry), complete abolition of speculative housebuilding (building for sale) and limited supplies of land and building materials, along with the complex licensing procedures, were typical measures of state control. Together these factors enabled the state to redistribute a portion of the gross domestic product (GDP) from infrastructure development (housing, education, health, etc.) to industrial development. In this transformation housing development was completely subjected to the requirements of economic policy. There evolved a network of controlling state institutions: the councils, the industrial enterprises controlled by the ministries, and the National Savings Bank, which were incorporated into the central planning system.[2]

The considerable extent of state redistribution did not exclude the construction of private houses or private transactions from the system. Even in

244

urban housing construction, not all loopholes could be closed, such as the rental black market and unlicensed building – even though the volume of these was restricted. Housebuilding in rural areas continued to be based on the efforts of the families themselves.

The socialist housing model described above has been modified several times in Hungary since its introduction in the late 1940s. For instance, the period between 1956 and 1968 is characterised by concessions which were based partly on political considerations (the end of the Stalinist era) and partly on economic ones (such as allowing private housing to release tensions caused by the unproductivity of state investment in housing) in order to release tensions. The third period of the east European housing model was marked by the emergence of housing factory technology in the early 1970s. Centralised decision-making and industrialised technology, as well as a significant increase in state investment in housing constituted the basis for the housing policy to which other sectors of the housing system like financing, land provision, building material supply, housing standards and building regulations, were subordinated. In our more detailed analyses (Hegedüs, 1987, 1988; Tosics, 1987, 1988) we showed the huge problems (inefficiencies and inequalities) of the system in respect of allocation of subsidies, the use of investment into the building industry, the urban consequences, and so on. Thus, even in the period of relatively high budget expenditures on housing in the 1970s, the socialist housing model was a quite ineffective way of allocating this money. And being almost totally dependent on the state budget, housing policy had to be changed with the first signs of budget difficulties.

TENURE AND HOUSEBUILDING

The public-rental sector

The public-rental sector was an important part of the housing sector, but the role of the state (state housing provision) was much wider. In Hungary the most important step towards the creation of an extensive public-rental sector was the nationalisation of the urban private-rental stock in 1952. Unlike the Soviet Union or Yugoslavia the urban housing stock in Hungary was not totally nationalised: only flats in houses with more than six rooms were transferred to state ownership, though this process was carried out without compensation to the owners. More than 200,000 dwellings were nationalised in Budapest where, as a result, the public sector grew from 30 per cent to 75 per cent of the stock. However, taking the whole country into account, the public-rental sector never dominated the housing stock: its 25 per cent share of the housing stock remained quite stable in the socialist period. Only in the bigger cities, including Budapest, did the share

Table 16.1   The percentage of inhabited dwellings in Hungary and Budapest by tenure, 1980 and 1990

	Public-rental %	Owner-occupation and co-operative %
1980		
Budapest	57	43
Hungary	25	75
1990		
Budapest	53	47
Hungary	22	77

Source: Central Statistical Office (CSO)

of the public-rental sector exceed 50 per cent of the local housing stock (Table 16.1).

There were two interrelated key elements in the Hungarian rent system, which explain a series of strange elements in the socialist rental sector. The first was the change in the meaning of leasing: as a redefinition of property rights, tenants were entitled to exchange the rented dwelling for another rented or owner-occupied dwelling. According to practice, the reallocation of vacated homes, mostly in cases when tenants had died without offspring, was in the hands of state agencies. But the tenants had the right to make private transactions (such as swapping with other tenants inside the state sector, and even with owners of privately owned homes) which is practically identical with selling the right of tenancy. Thus the state passed over the 'right of transfer' to the tenants. These homes could be almost freely swapped and inherited (if the heirs were living in the same rental home). In this process, the state did not receive any payment (until 1971 when 'key money' was introduced for state allocation of rental units). However, during the market transition, public-rental units had their market value determined usually at about half the value of an owner-occupied flat of similar type. The local authority (that is, the owner and landlord) had very limited possibility of controlling these processes, and basically no possibility of evicting families from the rental sector or forcing them to exchange their homes.

There were two basic ways of getting into the public-rental sector: one was through the distributional system of the councils (the principles of which could be according to position or social status), the other way was through the market of tenancy rights. According to a representative survey (Household Survey, 1983), one-third of households moved to their present rented homes through private transaction in the period 1957–82.

Another critical feature of the public rental sector was the low level of rent. Rents were set on the basis of square-metre floor area with some variation according to quality and only some minimal adjustment for location within a district and a very limited adjustment between cities and towns of different sizes. Rents were only a fraction of the estimated market levels and have even been consistently less than the amount required to cover operating costs.

The management of 90 per cent of the state-owned housing stock was provided by state-owned management companies (IKVs). These companies were not owners of the stock but only managers with some property rights. Decisions on new construction and the allocation of public-rental housing to new tenants was always the responsibility of the government and the local councils. IKVs were allowed to collect and use the residential and commercial rent revenue and with the help of extra budget subsidies IKVs could decide about priorities of maintenance. These companies, however, were quite bureaucratic and inefficient and were blamed for the deficiencies in the maintenance of the stock (Hegedüs et al., 1996).

New investment in the state-rental sector (that is, new construction) was financed by the budget. According to the planning procedure, the central agencies (National Planning Office and ministries) defined how many new units should be built in the current planning period in different counties and cities, and the cost of the investment was channelled from the central budget. Within the budget, this was the most expensive form of housing investment. There was always an attempt to minimise the number of social units each year and the share of this type of housing (Table 16.2).

Table 16.2 Newly built state-rental units, Hungary, 1971–90

Period	Number/year	Share of state rental unit in total building %
1971–75	29,735	33
1976–80	32,449	35
1981–85	16,297	22
1986–90	5,789	11

The state-rental stock (in 1990) had two very different types of units. The first is the stock which was originally nationalised in 1952. Most of this is located in the inner parts of the (larger) cities. The second type is the new units built mainly after the 1960s and usually situated in the outer ring of the cities. These represent a part of the so-called 'new housing estates'. Even if the regulation of the state-rental stock is handled similarly in respect to the two types, the social and financial problems related to each of them are radically different.

The owner-occupied sector and new construction

At the beginning of the socialist housing system the owner-occupied sector was very much disadvantaged. From the late 1950s, however, a gradual process of change started in which the disadvantageous elements of the regulation have gradually been withdrawn and some incentives have been given to the private sector. In the first decades the emphasis was put on 'semi-private' forms such as co-operative housing or new units built by the National Savings Bank. These forms were built with huge state subsidies and could therefore be considered more state than private forms (despite the fact that families moving into the completed flats became virtual owners). The 1970s and the first part of the 1980s were the 'best years' of this kind of housing policy in Hungary, showing a relatively high level of new construction (almost 10 new units per 1000 population) with the dominance of prefabricated technology in towns.

In these decades housing finance was assigned to a secondary role; it followed the decisions on new construction: its volume, its distribution among settlements, its composition as regards technologies used, as well as by flat size and amenities. The financing considerations were raised primarily as budgetary factors.

Housing finance actually has three sources: state budget, credits and the resources of the population. The source of the budgetary appropriation for housing is derived from enterprise and state institutions outside the sphere of wages and income distributions.

There was a wide variety of central subsidies. Council-rental flats were created entirely from budgetary subsidies (no loans were involved). Homes built for sale (National Saving Bank (NSB) – flats and state co-operative dwellings) were also given subsidies. Price allowances for the products of the state building industry were applied, which reduced the selling price. The social policy allowance introduced in 1971, and still existing in an altered form, was granted according to the numbers of dependants of families involved in organised housebuilding in order to reduce the down-payment. Another form of subsidy was the subsidised land use for which the council received only nominal amounts. Until 1983, these subsidies could only be used for state-provided housing.

Housing loans were given for the long term (35 years) with low interest rates (1–3 per cent) only for new units. The credit policy systematically discouraged owner-building (in which builders themselves organised the construction process, and in most cases used their own or their family's work), with lower loan ceiling and higher interest rates. (For example, in this period, employer loans were not given at all for self-building.) Credit served as a sort of subsidy born of the difference between the low official interest rate and the 'quasi-market interest rate' (or inflation).

Privately acquired resources include personal incomes, direct savings, and in the case of self-building, informal loans and help from friends and relatives.

The population pieced together the financial resources needed for building by adding their own work and the help of friends and relatives to their own savings. The volume of self-building indicates its significance in the Hungarian housing system: even in the 1970s, when state housebuilding was as its height, self-building amounted to 40 per cent of new building. This process, known as *kalaka*, can be defined as an informal co-operation between families, lending work to each other for long terms. This served as a sort of mass response to the defects of the loan system (Hegedüs, 1988).

From the beginning of the 1960s, the system of housing finance can be described as 'deeply subsidised'. The overwhelming bulk of budgetary subsidies went to state housing, while private provisions (especially owner-building) were based on self-help and self-credit using a relatively small amount of NSB loan. The housing reform of 1971 introduced a multi-channel system that was designed to increase the population's contribution to state housing investment and to introduce the use of income brackets for the most heavily subsidised forms of housing. In spite of these reforms, the financial system remained essentially unchanged. After 1971, the dividing line between different channels of access to housing continued to be between state and private housebuilding, illustrated by the data in Table 16.3.

Table 16.3 Financial sources of newly built homes in the state and private sectors, Hungary, 1973[3]

	state housing %	Private housing %	Total housing %
Budgetary subsidies	53	1	32
Loan	35	24	31
'Cash'	12	75	37
Share	52	48	100

(*N* = 80123)
Source: Calculated from NSB reports. Data do not include the interest-rate subsidy. At the same time, the budget components do contain the cost of new rental flats. The 'cash' includes the estimated amount of self-help work.

This table clearly shows that housing finance is based on deep subsidies. In the state sector – with half of the newly built homes – the 'cash' contribution by the population is an average 11–12 per cent of the cost of the home as opposed to 75 per cent in 1973. In the private sector, however, 'cash' contributions have always been proportionately high. The credit sector is somewhat more balanced, as was mentioned above, yet there are significant loan differences to the detriment of self-building.

HOUSING INVESTMENT, FINANCE AND SUBSIDIES

Towards the end of the 1970s, signs of the budget crisis could already be noticed, and it left its mark on the housing policy. From the mid-1970s to the late 1980s, inflation affected the building industry and the real-estate market more heavily than other sectors of the economy. This is partly because of internal actions (ineffective use of foreign loans) and unfavour-able circumstances in the international economic sphere (deterioration of the terms of trade because of the rise of oil and raw material prices).

By the end of the 1970s, the position was taken by housing policy-makers that the financing system must be transformed to make the housing system more effective. The way was shown by private and self-building which had produced good-quality dwellings with minimal state subsidies. It was sug-gested that state house construction could be replaced by increased support for private housebuilding. Some measures were taken in this direction at the end of the 1970s (land regulation, improving of borrowing terms for private building) but the real breakthrough came with the Housing Act of 1983.

The new housing policy practically abolished the system of deep sub-sidies based on the gap between the state and private sectors. The new housing policy gradually revoked the large budgetary subsidies granted to state-provided housing and the discriminative treatment of private house-building.

Changes in the regulation of the public-rental sector

The Housing Act of 1983 tried to solve the increasing problems of the public-rental sector. The rents had been increased in 1982 (after a decade of rent freeze), but because this was a 'one-time action' and the conditions to adjust rents to inflation were not established, the increase remained insig-nificant and did not compensate for inflation. In the 1980s, rents typically covered less than half of the maintenance costs and so cross-subsidisation through commercial rents was crucial, besides the direct central budget subsidies. Until 1991, when the local governments attained the ownership of the public management companies, the management of 90 per cent of the state-owned housing stock was provided by the state-owned IKVs. Until 1990, when the central state subsidy was discontinued, IKVs got significant subsidies (45 per cent of total IKV budget on average).

Under these circumstances the crisis of the public-rental sector (dete-rioration, dominance of private transactions, huge financial deficit) was steadily growing in the 1980s.

At that time, state-initiated and controlled rehabilitation activity started, which, because of the financing schemes, reached only a very small portion of the stock and served more 'political prestige' purposes than radical changes in the physical condition of the stock (e.g. these activities were mostly limited to the tourist and central business areas of the cities).

After 1971 a mandatory sum of 'key money' had to be paid to the local council on taking over a rented dwelling. Tenants giving up their tenancy were entitled to the reimbursement of their 'key money' payment. But the official 'key money' was only a fraction of the 'market key money' (the sum of money actually paid in the course of transfers to tenants' rights). In 1982 a new system was introduced offering three to ten times the face value of the key money for vacated dwellings, creating incentives for tenants to give back their public-rental unit to the housing authority instead of exchanging it on the market for another rental or an owner-occupied unit. However, in bigger cities, where the majority of rental units are concentrated, market transactions remained the main form of exchange, because the market 'price' for a public-rented dwelling was even higher than the increased level of repayments.

These efforts, albeit with limited results, were clear signs that the state wanted to improve the functioning of the rental sector. The main reason for this was the onset of state budget difficulties which resulted in the output of state-rental units (and of co-operative housing, which was the other main form of deeply subsidised housing) declining as a proportion of total housing production.

Table 16.4   The share of the main housing provision forms within new housing, Hungary, 1950–90

Period	state provision (%)			Private provision (%)	Total new construction
	state rental housing	state co-op housing	state NSB housing	Single family housing and private co-operative	New units per year
1950–60	27.7	–	–	72.3	43,400
1961–70	23.7	9.4	3.7	63.2	56,700
1971–80	22.9	11.9	13.9	51.4	87,900
1981–90	13.6	3.6	22.9	59.9	64,200

Source: Hegedüs and Tosics (1992: 170)

These changes show the clear tendency of a slowly shrinking state-rental sector which takes over more and more social functions instead of being a tenure category, similar to owner-occupation, open to all strata of the population. Thus as a consequence of the changes in state housing policy in the 1980s the role of state-rental housing started to change from a comprehensive type into a residual type.

Changes in the regulation and subsidisation of the owner-occupied sector

The Housing Act of 1983 expanded the eligibility conditions of the state subsidies to the private sector. As for budgetary subsidies, social-policy allowances were also given to self-builders. The disadvantageous credit conditions of self-building were erased and – in theory – builders in the private sector were also eligible for employer loans. Bank credit (though with smaller allowances) was obtainable for private builders and real-estate transactions. Local-council grants were introduced, which meant an interest-free loan or lump-sum grant could be given by local councils. It also became possible to use council or state resources for preparation of parcels of land for owner-built housing.

Several elements of the newly introduced system of subsidising depended on the size of family (number of dependants), thereby moving towards a subsidy system with a more welfare-conscious character. Yet families in the highest income group, who were most likely to build luxury homes, were also eligible for support.

One of the central features of the Housing Act of 1983 was that it passed the growing burdens of the housing economy not directly onto the population but largely to the subsidised credit sector. Therefore the surplus costs caused by rising inflation would have to be paid by a future generation. Between 1980 and 1985 the share of credit in the total costs of housing rose to 46 per cent from 27 per cent in 1970–75. The share of direct budget cost decreased while those of the population remained unchanged.

Low and fixed interest rate credit is a special type of subsidy whose cost does not appear immediately but takes the form of interest subsidy over the years after the issue of the credit. Thus the increase in the share of credit meant an increase in the future claim on the budget. Some signs of this were already discernible in the early 1980s; while in 1970–71 the interest-rate subsidy amounted to 9 per cent of budgetary spending on housing, it rose to 25 per cent in 1980–85 and to 64 per cent by the end of the 1980s.

Restructuring the system of subsidisation entailed serious consequences for the construction industry. The state building industry, organised to mass-produce housing such as rental flats in housing estates and NSB housing lost its monopoly position within the finance system when the support obtainable for the preferred one-family houses substantially increased, gradually equalling subsidies for state housing. Though certain institutional factors delayed the crisis of the state building industry, it was eventually unavoidable. The housing policy, which was influenced strongly by the finance ministry, in short 'sacrificed' the state building industry by cutting the indirect subsidies from the state budget.

The budget and loan sector tried to keep abreast of inflation, increasing subsidies, yet the lag continued to grow. In 1983–86 these measures were sufficient for the growth of private house construction to make up for the decline in new state housing.[4] After 1987, private building came to a halt while at the same time state house construction underwent an accelerating decrease.

Yet the radical transformation of the subsidy system failed to meet expectations. One explanation may be that the institutional framework of production did not react to changes in demand, and supply was inflexible to respond to the increase in demand. The data bear out this assumption in

Table 16.5   Housing expenditures in the budget, Hungary, 1981–94
(in billion forints)

	1981–86	1987	1988	1989	1990	1991	1992	1993	1994
Total	27.5	36.2	57.6	103.9	92.5	63.2	76.5	69.7	74.8
GDP	907	1,226	1,440	1,723	2,089	2,308	2,805	3,335	4,309
Housing expenditures related to GDP (%)	3.0	3.0	4.0	6.0	4.4	2.7	2.7	2.1	1.7
Inflation	6.9	8.6	15.5	17.0	28.9	35.0	23.0	22.5	18.8
New housing construction per 1000 inhabitants	7.1	5.4	4.8	4.9	4.2	3.2	2.5	2.1	2.0

Source: Ministry of Finance, yearly budget reports; Kopint-Datorg, 1995; Hungarian National Bank reports

that after the temporary growth in housing construction, the rise of prices became continuous (Table 16.5).

Despite the reforms there had remained systematic supply-side obstacles hindering the building of one-family houses. First, there were shortages in the building materials market prevailing between 1983 and 1986. Also, the supply of land did not increase dynamically; newly designated building parcels were the exception, not the rule. Without adequate resources, the parcels offered for building increasingly lacked basic infrastructure and these costs were pushed onto the builders. Easing bans on building was protracted when political prejudices and interests were involved, further slowing the spread of self-building. However, the basic hindrance was the peculiar social structure of self-building. House-building in this sector depends not only on money inputs but also on non-monetary relations necessary to create kalaka or informal network financing. These social relations cannot be established quickly and cannot be expanded flexibly. In economic terms, this means that the demand elasticity is less than one. A market-type building appears in the owner-building sector when the kalaka-type working relations are replaced by relations expressed in monetary terms. However, because of the limits to entrepreneurship, this sphere failed to expand efficiently. The increase in demand was met by increasing the prices but not production.

State building enterprises could not respond to the changes in demand, but their decline was gradual, since, as was seen above, the private sphere was also unable to react flexibly to the transformation of the system of subsidies. State building firms tried to build one-family houses out of prefabricated slabs, but their prices were not competitive with those of self-building. The decline (liquidation) of state building companies was slowed by the fact that the local interest mechanisms favoured the large enterprises, but, even so, the decline of state building companies was much faster than in the USSR or Czechoslovakia.

DEVELOPMENTS IN HOUSING POLICY

As we have analysed it, in Hungary significant changes occurred years before the 1989–90 political changes. These changes are called with the typical language of this period 'reforms', pointing out the fact that in these changes the state tried to improve the regulation without giving up too much of its control over the processes. Even if the state-rental sector remained strictly controlled (rents and allocation principles did not change), the creation of the possibility of subsidised transfer from the state sector to the private-housing sector (privatisation) and the changes in the regulation of the private sector (the modification of the deep subsidy system) were remarkable compared to other central and east European countries where the real changes (leading to the collapse of the socialist housing system) started later, only following the political changes, and were more sudden.

The Hungarian attempts at reform, however, were not very successful. One possible explanation of the failure of the 1983 Housing Act was the inflexible supply facing the restructured financing system and changed demand. This, in turn, increased the prices and inflation rather than boosting housing production.

The typical approach of the housing system of centrally planned economies is such that state control (price and wage control combined with subsidies) leads to under-investment in housing. The 1970s and the 1980s in Hungary represent two periods in which the subsidy system led to an over-investment in housing: in the 1970s as a consequence of deep subsidies directly connected to state (and NSB) housing forms, and in the 1980s as a consequence of deep subsidies indirectly connected to private investments.

Another, additional, explanation of the failure of the housing reform in this period could be found in the unchanged and rigid institutional set-up and in the property relations. The wider use of market forces helped the collapse of the system, as the ruling social groups could trade off their power position into a market position. Housing was an important part of the process. But it did not lead to a more efficient production and allocation system. The behaviour of the institutions (enterprises, banks, developers, real-estate brokers) were motivated by the subsidies transferred to households and not by the 'profit motive'. This period can be characterised as a 'perverse' interaction of market and state.

The transition in process (1990 onwards)

The political changes of 1989–90 were the outcome of the economic failure of a social system, and led to radical changes of the institutional set-up of the country. The definition of the state was changed by law immediately, and the role of the state gradually, as markets could not take over the task the state played in the centrally planned economy.

After the new parliamentary elections (spring 1990) and local elections (autumn 1990) housing was not considered as one of the priority areas. The political responsibility for housing was not clarified, the idea of creating a 'Housing Office' was rejected and there were six ministries involved in this area with none of them taking overall responsibility. The segmentation of housing was sharply criticised but the structure has not been modified.

Given the very weak and uncoordinated central regulation of housing policy in the first three years of the new political system, the most important development from the point of view of housing in that period was the emergence of autonomous local governments. The Local Government Act of 1990 gave a wide range of responsibilities to local government, but at the same time it led to a fragmented system of administration.[5] Also the resource allocation system has been changed substantially: subsidies for public housing maintenance were terminated in 1991 and replaced by a new form of intergovernmental transfer to local governments (6.4 billion Ft in 1991). This was labelled as a normative grant for housing-related expenditures, but they were not obliged to spend this money on housing; there was not even a systematic monitoring of how the local governments were spending this money (the typical use was as grants or interest-free loans for households building or buying homes, based on the general practice in the 1980s).

The Property Transfer Act of 1990 transferred the ownership of the state-rental housing stock to the local governments (in the case of Budapest to the district governments). Partly because of the very slow clarification of responsibilities at the ministerial level, the new central regulations, the Rental Housing Act, the Housing Concept, and also the Social Law (containing the subsidy system proposal) were all only passed as late as 1993. Thus the creation of the new framework for local-government housing policy was very slow and local governments had a three-year period in which they had responsibilities but their rights and the main intentions were unclear.

The shrinking public-rental sector

Privatisation (theoretically possible since 1969) started in the mid-1980s, some years before the political changes of 1989–90, when the strict constraints on privatisation of bigger buildings were lifted and a system of huge financial discounts was introduced for sitting tenants.

Most rented dwellings were sold for 15 per cent of their market value, this being the selling price of any public dwelling which had not been extensively modernised during the previous 35 years.[6] Moreover, tenants had to pay only 60 per cent of the discounted sale price if they paid for the property in cash. The other option was to pay by instalment: in this case 10 per cent of the sale price had to be paid in cash and the remainder in monthly instalments over 15 years at a low fixed interest rate; the interest rate was set at 3 per cent for the whole repayment period even though inflation was between 20 and 30 per cent by the end of the 1980s. Compensation vouchers could also be used for paying, and the local authorities were obliged to accept these at full face value despite their low (20–30 per cent) market value.

According to the regulations, privatised apartments could be resold or rented out by the owner immediately after the purchase without any restrictions. Moreover, there was no restriction on turning the apartments into offices or shops, and these changes did not have to be reported to the local authority.

This very liberal regulation of privatisation, which favoured sitting tenants by giving them a huge 'gift' in the form of a substantial and marketable discount, has been sharply criticised by many housing experts. The main push for this 'give-away privatisation' came on the one hand from the local governments (as a consequence of their inability to increase rents in the public-rental sector, this could easily be considered as a negative-value asset), and on the other hand from the main beneficiaries, the families living in the best public-rental flats.

Privatisation started from the central level, with government decree having equal effect on all districts (similar to the decree about rent increase). Local authorities could influence the volume of sales only through the speed of the administrative process and their willingness to sell or not, as it was necessary to obtain a decision from their respective Executive Committee for each dwelling. However, from the end of the 1980s not only the volume of sales but also the financial terms began to differ between local authorities.

Less than 2 per cent of the public-rented stock in Budapest was purchased by tenants in 1988 and 1989. In contrast, 20 per cent of the stock was sold to sitting tenants between January 1990 and January 1992. Over the period 1990–94, 56 per cent of the public-rented stock in Budapest had been sold, compared to 51 per cent in the country as a whole (Table 16.6).

Housing again became a political issue at central-government level in June 1993 when the government started to recruit political supporters for the next election and changed in the last stage of the parliamentary process the draft of the Rental Law. Instead of giving local governments the decision-making right over their rental sector the government introduced the right to buy for sitting tenants, forcing local governments to sell at low

Table 16.6   The sale of public rental housing, Hungary, 1990–94

		1990	1991	1992	1993	1994	Total
Budapest	Number	22,156	46,991	47,282	40,133	60,966	217,528
	Share[a]	5.7	12.1	12.2	10.3	15.7	56.0
Other districts	Number	31,867	35,127	26,851	18,258	30,993	143,096
	Share[a]	10.1	11.2	8.6	5.8	9.9	45.6
Country total	Number	54,023	82,118	74,133	58,391	91,959	360,624
	Share[a]	7.7	11.7	10.6	8.3	13.1	51.4

Note: [a] Percentage of the public rental stock at the end of 1990

Table 16.7   The share of public housing within the housing stock, Hungary, 1990–95

	Public-rental housing stock 1 January 1990		Number of sales 1990–94	Public-rental housing stock 1 January 1995	
	Number	Share		Number	Share
Budapest	387,878	48.9%	217,528	170,350	21.0%
Country total	701,447	18.2%	360,604	340,843	8.6%

Sources: Housing Statistics Yearbooks 1990, 1991, 1992, 1993, 1994

prices all houses in which at least one of the tenants wanted to buy.[7] The rule of forced 'give-away' privatisation was even meant to be extended to the case of commercial leases, which were the main revenue sources for many of the local governments (commercial rents having been much higher than residential rents).

The Court of Constitution modified at the end of 1993 some parts of the Rental Housing Law (eliminating the rule of forced privatisation of commercial leases and reducing the time period of right to buy from five years to one year). Even so, the new legislation is leading to substantial changes in the housing sector. Table 16.7 shows that the share of the public-rental sector decreased by January 1995 from about 18 per cent nationwide in 1990 to about 9 per cent, and from about 49 per cent in Budapest in 1990 to 21 per cent. When the right to buy terminated in December 1995 the proportions were probably as low as 5–7 per cent in the country as a whole and 12–15 per cent in Budapest. (This means that Hungary will be among the countries with the smallest rental sector in the whole of Europe.) This change leads to the secularisation of the public-rental sector (the remaining public-rental stock is dominated by one-room flats without basic amenities). Another serious consequence of mass 'give-away' privatisation is the drop in the volume of social allocation of vacated flats. Also affordability problems are increasing: many of the new owners are unable to cover all the costs associated with their new property (the increasing utility prices and the maintenance of the building especially cause hardship).

In January 1992 the Metropolitan Research Institute and the Urban Institute carried out an extensive empirical research programme (BRHPS, 1992) to explore the impact of privatisation measures. The equity consequences of 'give-away' privatisation were evaluated by examining the magnitude and beneficiaries of two different types of housing subsidies, the rental subsidy and the value gap,[8] how big the subsidy is to be and whom it will benefit.

According to the survey data the upper income group with an income above the 75 per centile receive 31.7 per cent of the rent subsidy, while the low-income group, those in the lowest 25 per cent income range, receive 21.6 per cent of the rent subsidy – less than their proportional distribution in the population. The investment value was distributed even more unevenly than the rent subsidy, which follows from the fact that the dwellings in the best condition were bought first. Forty per cent of the total investment value went to the upper income group, while the lowest income group obtained 17 per cent.

To sum up, 'give-away' privatisation increases inequalities, favouring the higher-income tenants. This, of course, gives political strength to local officials and representatives. Also, privatisation at a highly discounted price works against future housing policy. Besides the emerging social problems, deferred maintenance is likely to remain a political problem in privatised dwellings, as only some of the new owners will be able to shoulder maintenance costs and many of the new owners will most likely be unable to meet the heavy financial burdens of covering renovation costs.

Privatisation of management

In 1990 central state subsidy for the maintenance of the public-rental stock was discontinued. The new source, the revenue from privatisation of the rental units, was considerably less than the former central subsidy, because of the huge discounts on the sale price and the instalment payment option.

Because of the financial problems and the well-known inefficiency of IKVs, new local politicians made a lot of efforts to reorganise these companies. Many of the companies, however, had begun their internal reforms even earlier, before the local election.

The main directions of the IKV-initiated reforms are that they form separate limited companies from those divisions that can work independently. The majority of stocks are held by the IKV headquarters, but typically the local management and employees have some shares. Each company contracts with the IKV, but it has the right to get other contracts on the market. In this way, IKVs have gradually cut off from their main organisation their technical departments and branch offices. The IKVs are forced to move to the market partly because of the privatisation process.

In another view the owner's role should be clearly and completely taken over by the city administration, specifically its housing department. The housing department would be responsible for the district's housing policy and planning strategy as well as its directives and methods. They would have to gather all information concerning the rental stock, to form the new management system, and to control it. For the services of building management independent companies would be contracted. Operation and maintenance, relationships with tenants, and collection of the rents would be the tasks of the managing companies, either completely newly established or formed from the remains of the IKV.

Experience shows that in Budapest the district local governments tend to delay the reforms of the IKV and are only taking small steps towards change. They are unaware of the number of flats they are going to keep, therefore decisions on the system to adopt would very likely be premature. The reliability of the maintenance firms is unknown as well. In most of the districts the number of rentals and the amount of work is rapidly decreasing in the IKVs. Most of the IKVs have been temporarily turned into local-government companies and ownership duties are taken on by the housing department.

Evaluation: local options versus compulsory central regulation

Practically all the tasks and responsibilities for public-rental housing were transferred in 1990–91 to the local level. Thus from early 1992 onwards local authorities could decide on rent levels, and regarding privatisation, on the volume of sales and the level of discount on the market price (i.e. whether to decrease the very high discounts). In this situation there were basically two options for the local governments regarding their local rental housing policy (Hegedüs *et al.*, 1993). One of the options was to preserve most of the public-rental stock and try to establish new mechanisms through which prices could express real values (sale price discounts are minimalised and rents are increased to cover at least maintenance costs) and subsidies be targeted to poorer families. The other option was to reduce the volume of public-rental stock as much as possible. The main difference between these two options was in the choice of rent levels and in the regulation of privatisation.

Local authorities who chose the first option had to increase rents, and increase the sale price of rented dwellings by reducing the level of discount on the market value of the property, and/or find other ways to slow down privatisation. Rent increases, however, were almost impossible at the local level because of the absence of a comprehensive housing allowance system. To increase sale prices by reducing discounts was also difficult politically because the practice of selling public stock at very low prices started some years earlier and some of the best flats had already been sold at a low price,

259

mainly to families with good positions in the previous political and eco-nomic hierarchy.[9] Some local authorities reduced the discount level but more of them chose indirect methods to slow down the privatisation process:

• they maintained a high entry threshold for sales within apartment blocks by requiring that a minimum proportion of tenants in each building make application to buy before any one sale was permitted;
• they established extensive prohibition lists of buildings which were not available for privatisation;
• some local authorities even introduced a moratorium on sales for a period of time before the Rental Housing Act was passed in 1993.

Local authorities who chose the second option took the opposite course of action and increased their rate of sales at a low price. They applied the previous central regulations on the terms of sale, which were very advanta-geous to sitting tenants. Furthermore they applied a low threshold or none at all, reduced prohibition lists to a minimum, and tried to speed up the bureaucratic processes of privatisation.

Choice between the two options varied according to different local circumstances. In the case of many local governments where the rent gap between current and potential rent was large and therefore the push from tenants to buy was substantial, the majority of the public-rented stock was sold within three years, between 1990 and 1993. In contrast, some other localities with a lower rent gap and a very mixed stock, had more opportunities to resist 'give-away' privatisation for a long time and introduced a moratorium on sales. This picture with different levels of residential privatisation changed dramatically at the beginning of 1994 when the new Rental Housing Act gave the tenants the right to buy. This meant both the elimination of the first option, and a reduction in the decision-making power of local authorities with regard to the public-rental sector.

Owner-occupied housing

In 1989 extremely low-level fixed-interest rates for new construction loans were increased to a near market level and the loan instrument became an adjustable rate mortgage. The withdrawal of budget subsidies from state construction as well as from heavily supporting private building activities resulted in a dramatic drop in new construction and in a shift from multi-family to single-family housing. These tendencies are illustrated in Table 16.8.

The decrease in new construction (which occurred in the other central and east European countries as well) is related to the economic and institu-tional changes and cannot be explained by any one reason. A proper

Table 16.8   New housing construction, Hungary, 1988–94

	Number of flats	Average floorspace m^2	Percentage of single-family houses	Percentage of construction in the private housing sector
1988	50,566	85	64.2	–
1989	51,487	88	67.6	–
1990	43,771	90	72.3	69.8
1991	33,164	90	63.6	73.4
1992	25,807	93	66.9	79.4
1993	20,925	95	69.3	86.3
1994	20,947	97	70.0	86.5

Source: Housing Statistics Yearbook

Table 16.9   Financial resources of new housing investment, Hungary, 1989–94[11]

	1989	1990	1991	1992	1993	1994
Loan (%)	31	38	26	21	19	14
Subsidy (%)	18	19	16	11	11	11
Own resources (savings, self-help) (%)	51	43	58	68	70	75
The share of housing investments within total investments (%)	25.7	24.7	20.2	15.9	15.9	14.9
The share of housing investments within the GDP (%)	5.1	4.2	3.3	3.1	3.1	3.1

Sources: Hungarian Central Statistical Office, NSB and budget reports

explanation of this relationship is necessary to enable government to create a housing policy more compatible with change.[10] In any case, one of the main reasons for this drop in new construction can be found in the un-favourable changes to the finance and subsidy systems. As a result, the share of required 'own resources' (i.e. savings and/or self-help) increased drama-tically compared to the other possible sources for new construction (Table 16.9).

Since 1989 the housing finance and subsidy system has undergone a series of changes. The main aim was to reduce the burdens on the central budget. Another important aim was to distinguish between two types of subsidies: those with social aims (to target subsidies on families who could not improve their housing situation with market interest-rate loans), and those with efficiency considerations (to create incentives for new construction in a high-inflation environment with decreasing real wages and decreasing state expenditures on housing subsidies). As a first step towards these aims the finance and subsidisation of housing had to be separated. The finance element had been based on the adjusted-rate mortgage (later, from 1994 a new instrument, deferred-payment mortgage had also been intro-duced). The social-type element of subsidy was the social policy allowance, targeting subsidies to households with children who build or buy new flats. The efficiency-type of subsidy was a repayment subsidy, covering a certain percentage of the repayment, an amount stepping down every five years.[12] The repayment subsidy was a significant achievement compared to the earlier version of this subsidy in the form of a low interest rate: it meant a change in the form of the loan subsidy to one which contributes towards a market-led loan rather than one that obscures the cost of the money.

Many of the problems of the housing finance and subsidisation systems became clear in the 1990s, for example:

- None of the existing subsidies were targeted to lower-income families.
- Most subsidies did not correlate with social status.
- The number of children, showing only a loose relationship with the household income, became the key indicator for targeting subsidies in the housing subsidy system; this targeting measure was, however, inappropriate since the households buying or building new housing were obviously not from the lower end of the income scale.

The year 1994 was an important one in the restructuring of housing finance and subsidisation, with significant changes to both loan and down-payment subsidies. The repayment subsidy was replaced with a reduced form of mortgage assistance available to new home buyers only.[13]

In November 1994, the social policy allowance was increased for families with more than one child (and its name was changed to 'Housing Construction and Purchase Grant'). This was a very costly way (for the budget) to give an incentive to new housing construction. Five months later (in April 1995) policy on this allowance changed again with a drastic restriction of eligibility, sharply reducing ceilings for unit price and number of rooms. This change represents the first real effort at social targeting for home-ownership subsidies.

The linkage of subsidies to new construction remains very strong, and outweighs social goals. New housing is clearly more likely to be purchased by better-off households: in 1993 buyers of new homes had substantially higher incomes and assets worth more than twice those of buyers of existing homes, and they were far more likely to own a second home. In addition, the new units were not only more expensive (by more than a third), but also substantially larger than existing units, and three times as likely to be single-family homes.

Referring back to Table 16.9 it is clear that the housing finance system, despite the innovations of recent years, has serious troubles as a result of which loans only cover a very small (14 per cent) proportion of building costs as opposed to the 70–90 per cent share of loans in building costs in western countries. There are two main reasons for this small share of loans:

- crowding out: resources available in the economy are used for other purposes (the government wants to finance its deficit from the financial market and offers very high return on government bonds and this is a much better and lower-risk operation for banks than lending for housing and other purposes);
- affordability: the great majority of households are unable to pay the price with their available financial resources, and also government bonds are a better investment for the population than other types of saving.

It is easy to argue that further restructuring of the housing finance system is necessary in order to achieve a higher share of loans in new investment and through this give incentives for new construction. However, both constraints given above, the high level of government deficit and the low level of household incomes, are related to the current achievement of the economy, and so the problem cannot be handled just with changes in the housing finance system.

Public housing investment has decreased close to nil, and until the public-rental housing sector is stabilised (when rent fees cover expenses and privatisation is completed), new public-rental housing construction programmes of the municipalities cannot be funded and therefore cannot be anticipated.

Also, the investments of the private sector in building houses for sale have fallen back (more than 85 per cent of the new houses were built by the residents themselves). In spite of these facts this market is under development and instead of recession we can speak about a restructuring process.[14]

One difficulty in the development of the market is the presence of the black economy (20–50,000 people causing an additional budget deficit of 5–7 billion Ft through not paying taxes or social security). Housing policy has to help building for sale because it is a goal of the economic policy that the professional developers (who pay taxes) should control a growing share of the market. The advantages and disadvantages of the alternative decisions must, however, be analysed carefully: the increase of legal and financial control over private construction (pushing back the black economy) will better fit the new economic system but will result at the same time in a smaller amount of new houses.

Pressures resulting from the structure of the housing stock

The housing situation in Hungary is not, in general, bad, as the descriptive indicators do not show a worse housing situation compared to other countries of similar economic development (Table 16.10). The overall housing shortage is over (the number of housing units and households is in equilibrium at a national level). But Hungary compares unfavourably with western countries regarding the size of flats and in respect of the infrastructural services (even compared to other eastern countries).

263

Table 16.10   Comparative housing data: Hungary and other countries with similar
economic attributes (1993 data)

	Poland	Slovak Republic	Czech Republic	Hungary	Portugal	Ireland	Spain
GNP/capita ($US)	1.910	1.930	2.450	2.970	7.450	12.210	13.970
Dwelling stock per 1000 inhabitants	296	306	395	385	386[a]	299	440[a]
Percentage of dwellings with 4 or more rooms	n/a	22.8	18.6	40.5	74.2	86.8	85.9
Average useful floorspace (m^2)	n/a	n/a	n/a	n/a	84[a]	120[a]	93[a]
Percentage of dwellings with:							
Piped water	88.3	n/a	n/a	84.0	n/a	97.0	n/a
Bath or shower	76.3	89.1	90.0	79.6	82.2	94.1	94.0[a]
Inside WC	76.2	81.8	88.5	75.3	88.8	96.5	n/a
Central heating	66.7	n/a	n/a	40.1	n/a	58.7	n/a
Built since 1945	71.6	82.8	n/a	68.4	55.0[a]	55.0	73.0[a]
New construction per 1000 inhabitants	2.5	n/a	3.0	2.0	5.6	6.0	5.3

Source: *Annual Bulletin of Housing and Building Statistics for Europe and North America*,
UN, 1995
Note: [a] Statistics on Housing in the European Community, 1993

To evaluate the amount of housing demand we have to consider some structural problems of the current situation. A significant number of the households are overcrowded or in a substandard dwelling. At the same time the majority of these households do not have adequate effective demand. In the country there are about 300–800,000 households which should be supported. It is, however, clear that there is no possibility of a new construction programme of this magnitude. The current housing problems could be eased by other processes as well: by developing the free market (to fill vacancies) or supporting new house constructions which create new vacancies. This requires essentially the improvement of efficiency and targeting in the subsidy system.

Expenses related to housing

According to statistical information in the 1980s the share of housing expenses (rent, overhead expenses, credit instalments) was around 10–15 per cent of the total household expenditures. In the 1990s housing expenditures grew considerably: according to the information of the Hungarian Household Panel households spend 26 per cent of their incomes on expenses related to their house (one-third of households spend more than 30 per cent of their incomes). This growth is significant but in international terms these ratios cannot be considered as too high.

High housing expenses, of course, cause great problems to families having very low incomes. Although families in better income positions pay higher overhead costs compared to families with small incomes, this amounts to only 15 per cent of their total income (including instalment payments as well). Families with the smallest incomes, on the other hand, spend about 42 per cent of their incomes on housing expenses.

Affordability in the rental house market

One part of the expenses related to rental houses is the rent which was previously determined directly by the central housing policy. The ratio of rent to household income still shows the effects of rent control because it is not as much as 10 per cent even for those families with the smallest incomes.[15] It is very important to emphasise that the increase in the rents was less than the increase in overhead expenses.[16] The electricity, water and sewerage prices increased rapidly and the utility expenses of tenants rose well above 20 per cent, which is one of the highest ratios among the former socialist countries of eastern Europe.

Private rents are around twenty times higher than the fixed municipal rents (taking any local public rent increase into account), so from an average monthly income even the cheapest private-rental unit is unaffordable. With such big differences the private-rental sector will not play a very important role in solving the housing problem. However, it is a favourable new tendency that private rents do not grow with inflation because of the limited effective demand.

Outstanding rent and public-utility bills

The amount and value of outstanding rent and public-utility bills have increased at an extremely high rate (the dues registered by the Municipal Fee Collecting company tripled between 1991 and 1994). This has created very serious problems for the municipalities and for the public-utility companies, putting their operation at risk. In response to the problems the billing system of the public-utility companies was restructured, the fee and due collecting methods were changed, and the municipalities changed their housing subsidy systems in order to decrease the amount of outstanding bills and to prevent the further growth of arrears. There is an increasing regional differentiation in the amount and pattern of arrears.

Cost of housing credits

Loan repayments started to become a very important part of the household budget in the last few years, for families who moved into their new house after 1989. According to the logic of the housing finance system, even in the case of the credits taken after 1989 the real value of the instalments decreases rapidly (because of high inflation) and the instalments account for a diminishing part of the household budget. This means that today the instalment of a credit taken after 1989 is a relatively high burden (currently 18.5 per cent of household expenses) but in a few years inflation will depreciate these instalments. In fact, a high proportion of instalments in the household expenses shows the presence of a modern house financing system and a low proportion the lack of housing credits and the importance of cash.[17]

The house price to income ratio

The house price/income ratio is the most widely used indicator of the housing system. This indicator shows how many years the income should be cumulated to reach the price of an average house. The ratio of the medians of house prices and incomes is 6.0. This value is a lot higher than in most developed countries (2.5–4.5) but is relatively moderate compared to the surrounding countries. The house price/income ratio decreased in the last few years (1992–95), from 7.1 to 4.8 in the case of new households that appeared on the market. The reason for this is that the increase of house prices in the last four years was below the inflation rate.

If we compare household incomes with the price of new houses then we get much higher values – approximately 1.5–2 times more. This is one of the factors leading to the low level of new construction.

Housing consumption and household incomes

Our analysis shows that there is no significant relation between household income and house-value (which is one of the indicators of housing consumption). The 1992 data (Hegedüs and Kovács, 1992) show that the incomes of families belonging to the bottom 20 per cent are one-seventh of the incomes of the families belonging to the top 20 per cent and at the same time the value of the houses owned by the lowest group is only half that of those owned by the highest group. One of the reasons for this is the low income elasticity of the housing consumption of households because the market is very inflexible (high transaction costs, etc). The other reason is that households, when their incomes increase, do not think about improving their own living conditions but think about saving money for the housing needs of their children who will at some time in the future enter the housing market as potential purchasers. This is a reaction to the lack of a good housing credit system.

CONCLUSION

Housing policy after transition: what kind of central regulation for the decentralised and privatised housing sector?

Years after the political changes of 1989–90, the housing sector is still in transition. The privatisation and decentralisation moved forward fast, the sector operates more and more under market forces, and the possibilities for central government for direct intervention are less and less. Even in this situation, however, it is crucial to have a clear housing policy to use in a proper way all the remaining indirect tools the government still has. At the moment there are many open questions for the future of the sector, and basically it is not decided which housing model is the target.

The first overall housing policy decree after the political changes was accepted by the government in May 1993. This assigned the most important tasks in order to establish the directives of a National Housing Concept. As one of the steps the government established an intergovernmental committee to co-ordinate this work. With its lack of staff and financial means, however, this committee could not work very effectively. Finally, in 1995 the Ministry of Finance got the main responsibility for housing policy and a National Housing Committee was established. Having clarified the responsibility for housing at the central government level the next step should be the acceptance of a clear direction for housing policy. The main questions of such a new policy are, on the one hand, the fate of the remaining public-rental sector, and, on the other hand, the future of the housing finance and housing subsidy systems.

The fate of the remaining public-rental sector

It is clear that as a result of the transition process the share of the remaining public-rental sector has decreased to a very low level (Table 16.11). As a special outcome of the give-away privatisation process most public-rental units remain in mixed-ownership buildings.

The low share of public-rental units is leading to a residualised (as opposed to a comprehensive) rental sector with strong social tasks. At the same time, the ratio of housing expenditure to income is becoming higher as a result of increasing utility charges.

Under such circumstances it is expected that rents will be increased as far as possible to cost-covering levels to finance repairs to the high proportion of public-sector housing in bad condition (Table 16.11), and means-tested housing allowances will be introduced. Housing management will become privatised. All these systems will be regulated on a local level with very different local incentives.

Table 16.11 Estimate of the distribution of the former public-rental housing stock according to ownership after privatisation, 1995

	100% private condominiums	Private majority in condominiums	Private minority in condominiums	100% public ownership	Total %
1 The public-rental housing stock of 1990 according to the ownership of the building in 1995	10	71	10	9	100 (414,000)
2 Number of flats according to the physical condition of the building:					
(a) bad	13	20	59	71	28
(b) medium	43	57	32	27	50
(c) good	44	23	9	2	22
Total	100	100	100	100	100
3 The remaining public-rental sector (after privatisation) according to the organisational form of the building in 1995	–	40	27	33	100 (113,000)

Source: Estimate is based on February 1995 data from the Budapest Rental Panel Survey (MRI) sponsored by USAID

The basic question for the future of the public-rental sector is the role the central housing policy should play. According to one of the views the aim should be a 'self-financing' central policy leaving all the 'public' tasks (the regulation of rents, subsidies, management) totally to the local level (except for some central subsidies tackling the problems of deferred maintenance of mostly mixed-ownership buildings). The other view emphasises the necessity of more central intervention, in the form of creating more equal chances (i.e. having a central budget subsidy system and clear principles given) for local governments in very different economic positions to determine their own local rental policy.

The other important question is about the role local governments should play regarding the public-rental sector. Here the option of creating semi-public housing associations should be discussed; this would mean a clear split between the tasks of determining the principles of local housing policy and being the landlord of housing units.

There are, of course, other important problems at the local level, as well. One of the central problems of the future will be how to tackle the problems of renewal of owner-occupied housing units (in mixed-ownership buildings) taking the low and differentiated incomes and the decrease in state budget subsidisation into account. The other question is the possible future of private-rental housing: under what circumstances (subsidies, tax regulation, etc.) could this sector play a more significant role on the local housing markets?

Towards a market-based housing finance system with means-tested social subsidies

As discussed earlier, there are serious macroeconomic constraints on lending playing a bigger role in housing investment. Thus the restructuring of the housing finance system itself can only bring limited results. Even so, it is important to continue the building up of market-type financial institutions (especially to introduce all the legal and financial elements of real mortgage lending), and continue the separation of finance and subsidisation.

As a result of transition there is a very substantial differentiation of population incomes. This must be taken into account when restructuring the housing subsidy system. The upper (relatively thin) strata of society do not need any more subsidies to solve their housing problems; thus incomes and wealth must be taken into account when allocating subsidies. The broad middle classes do not have enough income or savings to be able to build or buy new housing and it is a political and macroeconomic question whether their housing consumption should be increased by efficiency-type subsidies (such as interest subsidies on market loans) or whether initiatives should be given to other types of population investments. The lower end of the income scale, households living on or below the poverty line, needs means-tested social-type subsidies, especially in connection with the affordability problem. Here again it is more a political question as to how much housing consumption should be subsidised, because the available subsidies could also be used in subsidising other sectors of the economy, such as education and health care.

NOTES

1 This chapter is based on a series of earlier papers and research reports. The housing finance sections are based on joint work with Katharine Mark published in Hegedüs *et al.*, 1995.
2 The classification of the different forms of housing provision should be based on this political and economic environment. The institutions of the so-called co-operative sector, for example, were also incorporated into this decision-making

system. Therefore also co-operative housing and the other housing types controlled by the formal institutions had to be regarded as state-sector (that is, 'state provision') irrespective of the fact that an enterprise, the NSB, or a co-operative, played the dominant role. The housing types remaining in the private sector (that is, 'private provision') consisted mostly of owner-built housing. Private transactions such as the sale or exchange of flats between individuals evolve in contradiction to the state-controlled system.

3 The state provision includes the new rental flats and the state-built condominiums sold through different 'channels' that vary according to types of waiting list and financing. The private provision included owner-built housing. It is worth noting that the table does not cover private trading of state-owned and privately-owned flats. If such data were available, it would probably further widen the gap between state and private provision.

4 In the 1981–85 period, 288,000 private homes were built compared to the planned 255–270,000 meant to supplement the decreasing state building.

5 The number of local governments has increased from 1,542 to 3,089 (plus 22 district governments in Budapest).

6 The price was set at 30 per cent of the market value if extensive modernisation had been undertaken within the previous 5 to 15 years, and 40 per cent if the modernisation had been undertaken within the previous 5 years.

7 The regulation made it compulsory to offer long-term instalment payment for the sitting tenants on 90 per cent of the sale price. This and the unofficial rule that the sale price should not exceed 50 per cent of the market value ensured the give-away character of the compulsory sales.

8 The value gap is defined as the difference between the value of the unit as property, on the one hand, and the value of the unit as a rental plus the purchase price paid, on the other.

9 This argument (the 'communists' were allowed to buy cheap, so why should we, the simple people, be excluded from it?) was the main constraint in the efforts to rationalise the regulation of privatisation.

10 The regulations (designed to meet socio-political priorities) put in place in November 1994 were not very effectively applied and the budgetary expenses caused by these regulations were too high compared to the results attained.

11 Data in the table are mainly to show tendencies and do not give an accurate picture. Basic information is missing, such as loans issued by other banks than NSB (but NSB controls the great majority, probably 90 per cent of the loan market), loans issued by employers (this is decreasing as many big employers are in increasing financial trouble), and it is unknown how much local-government use from the normative housing grants they get for housing investment purposes (we calculated from the total amount).

12 The highest subsidy was 80 per cent repayment reduction, for families with three children in the first 5 years. Two-child families got 70 per cent and one-child families 40 per cent reduction. In the second 5 years the reduction has been halved. The maximum amount of loan was 500, 400 and 300,000 Ft for the three categories of families, respectively.

13 The government pays 4 per cent of the loan balance in each of the first 5 years; the figure steps down to 3 per cent for the next 5 and then to 1 per cent for 5 more years. The maximum loan amount eligible for this '4–3–1' subsidy is Ft 1.5 million for families with children and Ft 600,000 for households without children. This maximum is the only subsidy feature related to household size.

14 Earlier (in the second half of the 1980s), companies building for sale got substantial governmental support (cheap land, governmental house construction

industry giving support, direct and indirect governmental support). Now this activity is totally market-driven.

15 Our information does not include the most recent increase of rents in different cities around the country.

16 This way there is a completely opposite overhead/rent ratio in Hungary as compared with the west European housing market.

17 The new loan construction of the OTP, the delayed payment credit, introduced in 1994, could modify this relation.

17

THE CZECH REPUBLIC

Luděk Sýkora

The Czech Republic has a population of 10.3 million people (the former Czechoslovakia had 15 million). The largest city and capital is Prague with a population of 1.2 million. The Czech economy is more stable than that of any other former Eastern bloc country. It shows a reasonable rate of growth, and attracts a substantial amount of foreign direct investment to the region. The macroeconomic figures for 1994 were: a growth in the gross domestic product (GDP) of 2.7 per cent (4 per cent growth was expected for 1995), an inflation rate of 10 per cent, and unemployment of 3.5 per cent.

Recent changes in the political and economic system of the Czech Republic, commonly referred to as a transition from a centrally planned to a market-oriented economy and from a totalitarian to a democratic political regime, have had an immense impact on housing. The most important trends within the Czech housing system consist of a withdrawal of the state from financing new housing construction, rent deregulation and the introduction of housing allowances, and the privatisation of state and municipal housing.

The post-1989 developments in the housing system are closely inter-linked with basic economic reforms and deeply rooted in the heritage of the communist housing system. Therefore, this chapter describes both the evolution of the communist housing system between 1948 and 1989 (see also Short, 1990; Michalovic, 1992; Musil, 1987; Anderle, 1991; TERPLAN, 1993; Telgarsky and Struyk, 1991), and the post-1989 changes.

COMMUNIST HOUSING POLICY

The main aims of communist housing policy were a sufficient supply of housing for those in need and a just distribution of housing among the population. Relative social equality was maintained by state control over housing construction, the non-market allocation of using, and constraints put on the exchange and letting of flats. Housing provision was based on estimates of 'objective housing needs' expressed in norms, such as size

272

standards (m^2 of living space per person) and technical parameters of dwelling equipment. Communist housing policy was based on the premise that everybody has a right to live in an affordable dwelling. Consequently, rent was regulated and heavily subsidised and new construction was fully financed or substantially subsidised by the state.

There were four main types of tenure: state, enterprise, co-operative and private (family) housing. The state-owned housing stock consisted of apartment houses built prior to the communist take-over in 1948 (mostly pre-war and nineteenth-century buildings) nationalised during the 1948–89 period, and newly constructed mostly prefabricated blocks of flats. State housing accounted for 45 per cent of all dwellings in 1960 and 39 per cent in 1991.

The state housing stock was managed by Housing Services Companies (HSC) established by and subordinated to local authorities (National Committees). While the housing departments of local authorities were in charge of housing allocation to families in need, the Housing Services Companies were in charge of collecting rent, basic maintenance and repair of buildings. Because of rent regulation, rent revenues of HSCs from both residential and commercial premises located in state properties amounted to less than half of HSC expenditures. A large part of HSC expenditures had to be covered by state subsidies. Financial resources for maintenance and repairs were limited and many old apartment houses fell into disrepair.

State housing was produced within the Complex Housing Construction (CHC) programme. CHC included building state housing as well as the provision of land and technical and service infrastructure (retail, schools, cinemas, etc.) for all forms of housing construction (state, enterprise, co-operative and private). The construction of state housing was predominantly based on industrialised prefabricated technology delivered by large construction companies. The power of construction companies to influence decision-making processes shaped the character of housing schemes. High uniformity of housing design was a reflection of standardised production – promoted by producers. From a town planning perspective, smaller districts of prefabricated housing for a few thousand inhabitants constructed in the early 1960s evolved into 'New Towns' in the case of Prague in the 1970s and the 1980s for up to 100,000 people. Furthermore, construction companies strongly preferred new housebuilding to the rehabilitation of older housing. Consequently, clearance and renewal projects formed the character of urban rebuilding activities and rehabilitation gained importance only in the second half of 1980s. Clearance schemes and transfer of housing declared as uninhabitable to non-residential use resulted in a heavy loss of dwellings, which exceeded half the number of newly constructed apartments during 1960–91.

Enterprise housing was a new form of tenure introduced in 1959. The costs of construction were covered by the state budget and the allocation of apartments was controlled by particular companies. Enterprise housing served as a tool of labour policy with the aim of attracting labour to preferred industries and regions. It played a marginal role in the 1960s, increased through the 1970s, and was abandoned in the 1980s, when enterprises provided housing for their employees through co-operatives.

Co-operative housing had been rapidly increasing its share of new housing construction since its introduction in 1959. Building Housing Co-operatives (BHCs) could be established by citizens under state supervision, approval and control. BHCs became the main developers of housing in communist Czechoslovakia. Co-operative housing was the dominant tenure of newly built dwellings in the Czech Republic from 1965 until 1992. Its annual share of all newly constructed dwellings fluctuated between 28 and 66 per cent. The only exceptions were in the years 1976–77, when more dwellings were built in the private (individual self-build) family housing sector than by co-operatives. Co-operative housing enabled people to have quicker access to housing in exchange for financial participation in its construction costs. Furthermore, some co-operative schemes offered better-quality housing in low-rise apartment houses. However, most co-operative houses were built with prefabricated technology and in the same locations as state housing.

There were three sources of financing co-operative housing construction. First, individual contributions by future tenants were given in terms of cash amounting to about 20 per cent of the total costs. Second, state allowances covered approximately 40 per cent of the construction costs. They included a basic subsidy per apartment equal to 9,400 crowns and a subsidy of 910 crowns for each square metre of living floorspace of the flat under construction (figures are given for the 1980s). Special allowances were given in the case of difficult physical conditions on site or in the case of long distances for transporting construction material. Third, credit from the state savings bank was given at low interest at a rate of 1–3 per cent for a period of 30–40 years. When the co-operative sector took over enterprise housing, companies could pay a part of the membership fee (up to 15,000 crowns) to attract or retain employees. This was given in the form of a loan, of which 90 per cent was not refundable when an employee signed a ten-year contract.

The introduction of co-operative housing challenged the dominant position of state housing provision as early as the first half of the 1960s, when about 50 per cent of newly constructed apartments were built in the co-operative and private sectors. In the second half of the 1960s, co-operative housing's share of total housebuilding increased to 60 per cent. In that period, the two housing tenures in which households participated financially (private and co-operative housing) totalled nearly three-quarters of all new housing output.

The 1970s were characterised by a high share of enterprise housing (around 20 per cent) and a steady increase in private housing's share of the number of newly constructed apartments (from 20 to 30 per cent). The growing role of private housebuilding was influenced by a programme of state subsidies to support family housing construction. In the 1980s, the shares of new construction for co-operative, private, state and enterprise housing were respectively 40, 30, 25 and 5 per cent (Table 17.1).

Private family housing construction (individually self-built detached, semi-detached or terrace houses) was supported by loans from the state savings bank offered for up to 40 years at 2.7 per cent interest and available up to a maximum of 250,000 crowns. Furthermore, labour-stabilisation housing allowances (up to 25,000 crowns per person, plus a further 10,000 crowns for employees in preferred industries, plus 20,000 in declared regions) could be given by the state (when a ten-year contract was signed by the employee). The allowance could reach a maximum of 78,000 crowns per dwelling.

In the 1980s, state involvement in housing finance consisted of investment grants for state housing construction, state allowances for individual (private) housing construction and subsidies for co-operative housing construction, cheap credit for co-operative housing construction, and cheap loans for private family housing construction. The main goals of housing policy for the 7th and 8th Five Year Plans (1981–91) were to stabilise and attract new labour for preferred industries in specific regions, retain a high intensity of housing construction, remove housing shortages, and gradually modernise the housing stock.

Table 17.1 Share of completed new dwellings in a particular tenure of the total new construction, and intensity of housing construction, Czech Republic, 1980–94

Year	state/municipal %	Enterprise %	Co-operative %	Private/self-build %	Completions per 1,000 population
1980	21.45	20.00	34.03	24.52	7.81
1981	19.94	17.57	33.97	28.52	6.12
1982	20.48	10.82	38.31	30.39	5.92
1983	20.44	4.30	42.76	32.50	5.53
1984	20.62	3.67	44.25	31.47	5.55
1985	22.69	5.86	43.88	27.58	6.45
1986	22.14	2.44	43.08	32.35	4.55
1987	24.13	3.01	40.87	32.02	4.73
1988	28.58	3.09	39.28	29.05	4.89
1989	30.83	1.49	38.20	29.48	5.32
1990	20.09	3.16	38.25	38.51	4.30
1991	23.70	4.59	46.71	24.99	4.05
1992	19.73	4.45	41.48	34.34	3.53
1993	19.72	4.36	30.49	45.43	3.05
1994	23.26	5.31	30.84	40.59	1.76

Source: Czech Statistical Office (CSU), own recalculations

Housebuilding had a distinctive spatial pattern, depending on tenure. Private family housing construction prevailed in most regions, especially in rural areas, villages and small towns. State housing was concentrated in large towns and cities and in industrial districts. Enterprise housing was built especially in certain backward frontier areas and districts with new industrial developments. Co-operative housing was characteristic of medium and large towns and cities.

Housing production gradually increased from 1948 to the mid-1970s when 85,000–97,000 dwellings were completed annually. As a consequence of a general recession, annual housing production declined to 50,000 dwellings through the 1980s. While the intensity of housing construction was quite high (7–9 completed dwellings per 1,000 people) through the 1970s it declined to rates around 5 by the end of 1980s (Table 17.1). In the first half of the 1990s, the intensity of housing production dropped further to less than 2 completed dwellings per 1,000 people in 1994.

Housing indicators from the April 1991 census can be used to illustrate the state of housing at the end of central planning. According to this data, 58 per cent of all dwellings were built after the Second World War and 31 per cent were located in prefabricated buildings. The average gross floor area of an apartment was 70.5 m^2 or 25.5 m^2 per capita. The occupancy rate was higher in comparison with many west European countries. The Czech Republic had 359 dwellings per 1,000 people, compared to 466 in Sweden, 424 in Germany, 412 in the UK, 385 in France, 376 in Austria, 335 in Italy, 296 in Spain and 278 in Ireland (ESHOU, 1995). On average, 2.76 people lived in each apartment. Out of the total number of 3.7 million dwellings, 3.4 million were linked to piped water supply and 1.9 million to piped gas; 2.2 million had central heating, 3.5 million a bath and 3.3 million an indoor WC. The most common dwelling units were two-bedroom (1.3 million) and one-bedroom apartments (1.2 million).

TENURE 1970–91

In the Czech Republic, public-sector (state/municipal and enterprise) housing substantially increased its share of the total housing stock up to 39 per cent by 1991. Private-sector housing, in contrast, decreased to 40.5 per cent by 1991. The most radical changes were developments in the co-operative sector, which increased its share of the total housing stock from virtually zero to 20.4 per cent by 1991 (Table 17.2).

There was an important shift in type of housing, influenced by the mass production of prefabricated apartment houses. While in 1970 more than half of all dwellings were still family houses, their share was reduced to 41.2 per cent by 1991. Dwellings located in apartment houses increased their share from 44.6 to 58.0 per cent over the same period (Table 17.2)

Table 17.2   Tenure, ownership and type of housing in the Czech Republic: 1970, 1980 and 1991

Share of dwellings in		Public sector	Co-operative sector	Private sector	Not identified	Total
Family	1991	1.52	0.49	39.15	0.00	41.16
houses	1980	1.45	0.47	43.99	0.01	45.92
	1970	3.00	0.36	50.29	0.03	53.68
Apartment	1991	36.75	19.86	1.36	0.05	58.02
houses	1980	36.35	14.50	1.98	0.04	52.87
	1970	32.92	8.23	3.33	0.09	44.57
Other	1991	0.73	0.06	0.03	0.00	0.82
houses	1980	1.10	0.09	0.01	0.01	1.21
	1970	1.55	0.15	0.05	0.00	1.75
Total	1991	39.00	20.41	40.54	0.05	100.00
	1980	38.90	15.06	45.98	0.06	100.00
	1970	37.47	8.74	53.67	0.12	100.00

Source: Czech Statistical Office (CSU), own recalculations
Notes: Family house can include a maximum of three apartments and can have a maximum of three floors (ground, first and second floors); apartment house has two or more apartments accessible from an internal house corridor with common main entrance, unless it is a single-family house; in both family and apartment houses more than half of total floor area must be used for residential purposes; in 1970 and 1980, family houses could contain a maximum of 120 m^2 of living floorspace, while in 1991, this limit was 150 m^2; larger houses and villas containing more living floorspace are in the category 'apartment house'; other houses include, for instance, municipal and enterprise dormitories, apartments in student housing, apartments in social-care buildings for the elderly or handicapped, apartments in hospitals, hotels and enterprise build-ings; public-sector housing includes municipal housing and housing of state organisations including (state) enterprise housing; co-operative housing contains houses of building co-operatives, People's Housing Co-operatives, agricultural and other co-operatives; private hous-ing includes privately owned housing and housing in ownership of foreign citizens and organ-isations.

Table 17.3   Tenure, ownership and type of housing in Prague: 1970, 1980 and 1991

Share of dwellings in		Public sector	Co-operative sector	Private sector	Not identified	Total
Family	1991	0.31	0.16	11.40	0.00	11.87
houses	1980	0.14	0.14	13.03	0.00	13.31
	1970	0.59	0.16	12.40	0.01	13.16
Apartment	1991	62.38	23.37	1.64	0.08	87.47
houses	1980	63.53	19.18	2.98	0.03	85.72
	1970	66.34	13.60	5.73	0.18	85.85
Other	1991	0.64	0.01	0.01	0.00	0.66
houses	1980	0.91	0.02	0.04	0.00	0.97
	1970	0.97	0.01	0.01	0.00	0.99
Total	1991	63.33	23.54	13.05	0.08	100.00
	1980	64.58	19.34	16.05	0.03	100.00
	1970	67.90	13.77	18.14	0.19	100.00

Source: Czech Statistical Office (CSU), own recalculations
Notes: See Notes to Table 17.2

In urban areas, and particularly in Prague, the state and municipal rental housing was, by far, the dominant tenure, accounting for 63.3 per cent of all dwellings. Co-operative housing increased its share to 23.5 per cent by 1991, while private-sector housing was reduced to 13.1 per cent by 1991. The majority of dwellings in Prague were located in apartment houses (Table 17.3).

HOUSEBUILDING

The 1991 census showed that for each 100 apartments there were 107 households, 276 persons and 266 rooms. The recent need for dwellings in 1991 was estimated by TERPLAN (the State Institute for Territorial Planning) to be 173,003 units (Anderle, 1993). Alternative estimates rise to 278,177 dwellings (ABF, 1992). The future housing need for the year 2000 was estimated by TERPLAN to be 667,450 dwellings (the sum of three components: the current need for 173,003 dwellings, the need for 214,447 dwellings taking into account the growth of households over 1991–2000, and losses of dwellings amounting to 280,000). These figures imply a need for an annual housing construction of 70,000 apartments, which would assure the maintenance of housing standards from the late 1980s.

However, housing construction rapidly declined after 1989 (Table 17.4). In the Czech Republic, only 18,162 apartments were completed in 1994, compared to 31,509 in 1993, 36,397 in 1992, 41,719 in 1991 and an annual construction of 45,000 to 67,000 in 1985–90. The decline is to continue, as only 10,964 housing units were started in 1994, 7,454 in 1993, 8,429 in 1992 and 10,899 in 1991. The annual number of started dwellings dropped in

Table 17.4 Housebuilding in the Czech Republic, 1980–94

Year	Started	Under construction	Completed
1980	69,459	154,271	80,661
1981	53,765	144,954	63,084
1982	48,489	136,388	61,400
1983	54,459	134,304	57,078
1984	60,929	137,763	57,298
1985	47,337	118,844	66,678
1986	51,973	123,946	47,080
1987	57,309	131,325	49,000
1988	61,120	141,291	50,700
1989	55,965	141,721	55,073
1990	61,004	158,840	44,549
1991	10,988	128,228	41,719
1992	8,429	97,768	36,397
1993	7,454	72,356	31,509
1994	10,964	62,117	18,162

Source: Czech Statistical Office (CSU)

1991–93 by 85 per cent from the figure for the 1980s, when construction of 53,000–67,000 new housing units was started annually. The annual stock of dwellings under construction was reduced from 140,000–150,000 in the late 1980s to 62,117 in 1994.

Contemporary levels of housebuilding are very low considering that almost 30,000 dwellings are condemned each year, the generation of the mid-1970s baby boom are reaching maturity, and new social and demographic trends, such as higher separation rates, are already present. Furthermore, a large number of dwellings in the central parts of large towns and cities, particularly Prague, have been converted to office use. In a situation where the initial housing shortage inherited from communism is further deepened, government officials argue that there is no housing shortage and existing distortions will be removed by rent deregulation.

The enormous decrease in housing construction was influenced by a coincidence of several factors. Among the most important was, first, the termination of state housing construction and the withdrawal of state subsidies to co-operative and private housebuilding. Second was the central-government policy of wage regulation, aimed at keeping inflation low and creating a competitive advantage for domestic industries, while constraining purchasing power by keeping real wages below the 1989 level. The 1995 average monthly wage was approximately CZK 8,000, that is, $270 or £200. Third, the rapid liberalisation of prices sharply increased construction costs and raised prices of new housing out of reach of middle-income households. The market could not react in an environment of huge disparities between housing need and demand, and the government was not willing to bridge the gap between the high need (but low purchasing power) of households and the sharply increased costs of housing production. The present housing-policy programmes, that is, housing saving schemes and mortgages, are not likely to change this trend.

HOUSING INVESTMENT, FINANCE AND SUBSIDIES

The withdrawal of the state from financing new housing construction

In the post-1989 period, the state subsidy for housing construction has virtually ceased. The Complex Housing Construction (CHC) programme was terminated at the end of 1990, and there has been no state investment in new housing construction since 1993. There are quite a few projects from the CHC programme that commenced prior to 1990 and are still being paid out of the state budget. These include mostly construction of technical and social-service infrastructure (a typical example is the construction of schools). Housing under construction was transferred to municipalities in 1991. Its completion was first financed through state-issued bonds. Since 1993, it has been paid out of municipal budgets. The state gives only very limited subsidies for the construction of municipal housing, namely for flats intended to house municipal and state employees. An *ad hoc* measure was used in mid-1995 when the state allocated CZK 1,400 million to subsidise new construction of municipal housing. Projects submitted by municipalities had to be financed 50 per cent from local budgets to receive the state subsidy which will cover the second half of project costs. Unfortunately, municipalities were obliged to send an application in one month after the programme's announcement and two weeks after the rules were set out. Using rushed decisions without giving enough time to prepare proper projects cannot be regarded as a well-thought-out approach to housing policy.

The state has also withdrawn from direct subsidies for the construction of individual self-build private family housing and co-operative housing. Furthermore, following the privatisation of the state savings bank, loans are not available on favourable terms (since January 1993). The state does not intend to bridge the gap between the low interest rates desired by the consumer and the interest required by banks. Furthermore, commercial banks have been hesitant to fund housebuilding (credits can be obtained for short-term use but at 13–14 per cent rates of interest).

Government housing policy has introduced only one programme intended to stimulate housing consumption: a housing saving scheme based on Austrian and German experience was launched in 1993. Each citizen can deposit monthly or annually a certain amount of savings to one of the newly established housing savings banks. The interest on savings is 3 per cent. The state gives a contribution equal to 25 per cent of the annually deposited sum of money. However, the contribution is given to a maximum of CZK 4,500. After five to six years, credit equal in value to the savings is available on 6 per cent interest. If one wished to fully use the state contribution, the maximum amount of finance available after five years would be CZK 250,000 (savings plus credit), but this is equal to only a small proportion of the market price of a two-bedroom apartment or a small single-family house.

Up to 1995, housing savings schemes attracted about 650,000 citizens, of whom only one-third are really interested in gaining credit for the purchase or construction of housing. The remainder of housing savings bank clients use the scheme as an alternative way of saving, utilising the advantage of state contributions (when credit is not required, savings can be used for any purpose). An important change in legislation, which took effect in July 1995, enables legal entities, such as housing co-operatives, to use the scheme. Because of the low-interest credit, there is a great demand from institutional investors to enter the scheme.

In 1995, several changes in legislation were made to allow for the provision of mortgages. However, the mortgage legislation will not significantly increase housing consumption, as its design makes mortgages available only to the highest income bracket of the population. Mortgages can be given up to 70 per cent of the value of collateral (property in existing ownership or property being purchased). Forty per cent of the purchased property must be paid prior to using a mortgage scheme. Banks intend to offer mortgages for 10–20 years with 10–11 per cent interest. Under these conditions, a monthly payment for a mortgage (covering half of the price of the smallest single-family house in the Prague area) exceeds the average monthly wage of an individual. The central government will provide a 3 per cent interest subsidy for those mortgages used to finance new housing construction. The person signing a mortgage in the first year of the state programme will get a 4 per cent discount for the entire period of the loan. The 1 per cent bonus is given in order to develop the system and motivate interest. Subsidies are limited to loans of up to CZK 800,000 for an apartment, CZK 1.5 million for a single-family house, and CZK 2 million for a multi-occupied dwelling. Banks will consider household incomes when issuing a mortgage. The monthly payment should not exceed 30 per cent of household monthly income. For instance, a household with an average net income (for example, CZK 12,600 in 1995) could obtain a mortgage for 20 years equal to about CZK 500,000 with a monthly payment of about CZK 4,000. Considering property prices, mortgages will not help middle-income households to acquire new housing. Consequently, they will have only a limited impact on housing production.

There are no housing policy programmes aimed at stimulating housing production. Indirect support can be drawn from the Municipal Infrastructure Finance Programme (MUFIS), established in 1994 and managed by the Municipal Finance Company (MFC), a joint venture between the Ministry of Finance, the Czech and Moravian Guarantee Bank and the Union of Towns and Cities. MUFIS provides long-term capital at fixed interest rates to support construction of new housing infrastructures. Up to 1995, MUFIS received 25 proposed municipal projects worth $20 million. MUFIS is expected to expand to $100 million by 1998 (Reynolds, 1995).

While the share of housing expenditure in the 1989 state budget was about 8 per cent (Kingsley et al., 1993), it declined to 1.5–3 per cent in 1992–95. In 1995 the state budgetary support for housing was equal to 0.6 per cent of the Republic's GDP, which is a substantially lower proportion than that of west European countries. The plan of housing expenditure in the 1995 state budget includes:

- CZK 520 million: CHC developments commenced prior to 1992;
- CZK 350 million: completion of municipal housing;
- CZK 300 million: renovations and reconstruction;
- CZK 200 million: housing allowances;
- CZK 800 million: contribution to housing saving schemes;

- CZK 300 million: support to owner-occupation through mortgage-interest subsidies;
- CZK 1,700 million: houses for the elderly;
- CZK 1,200 million: reserve fund.

Rent regulation/deregulation: towards market rents

At present, rent in the housing sector is regulated in respect of unlimited leases for Czech citizens. The net rent is being deregulated step-by-step for both municipal and privately-owned apartment houses, although the price paid for amenities such as water, gas and electricity supply, and services such as waste collection has been fully deregulated. This part of housing costs has increased by more than 400 per cent since 1990 (Table 17.5). The net rent was first increased in June 1992 by 100 per cent. The second increase of rent followed in January 1994. At that stage, the ceiling for rent increase was lifted by an average of 40 per cent. However, while the increase has been lower than 30 per cent in some locations, it has reached levels of over 100 per cent in others.

Table 17.5   Increase in housing costs, Czech Republic, 1990–95
(January 1990 = 100)

Year	Net rent	Amenities	Housing costs	Inflation
1990	100	100	100	100
1991	100	198	167	152
1992	194	291	261	172
1993	194	345	298	203
1994	272	406	364	222
1995	332	443	407	244

Source: Ministry of Finance of the Czech Republic, Czech Statistical Office (CSU)

Starting in July 1995, the rent ceiling is increasing each year according to three coefficients. The first coefficient is equal to the annual inflation rate, the second depends on the size of municipality (1.19 for Prague, 1.15 for towns of over 100,000 inhabitants, 1.06 for municipalities with less than 10,000), and the third is at the discretion of the central government. For the period of July 1995 to June 1996, the net rent paid for one square metre in first-category apartments in Prague increases by 31 per cent from CZK 6 to 7.85, as the initial rent is multiplied by 1.10 (inflation 10 per cent) and 1.19 (location coefficient). The government coefficient for this period is 1.00.

In 1995, the average rent for a three-room flat in state and private apartment houses was about 15 per cent of the annual income of a typical Czech family with two employed adults, and 25 per cent in the case of a pensioner household. Despite a lower share of household expenditure for housing than in west European countries, the increase in housing costs can mean quite a heavy burden for lower-income families. Real wages are still under the pre-1989 level and expenditures for basic needs of living, such as food and basic services, account for a substantial part of the household budget. The most important change in incomes has been their rapid and radical divergence. Consequently, the situation of many households is substantially different from the average figures.

Since April 1995, an additional rent increase of up to 20 per cent has been allowed for towns with more than 50,000 inhabitants (and up to 10 per cent in smaller municipalities). Rent can also be reduced by 15 and 10 per cent respectively. The decision about the increase of the rent ceiling is at the discretion of each municipal authority. For example, the city of Prague approved a 20 per cent increase for most of its territory, including inner-city as well as some suburban neighbourhoods, a 10 per cent increase for outer-city locations, and a zero increase for a few small rural settlements located within the city's administrative boundary. Consequently, the 1995 state and municipal deregulation of rents allows for up to nearly a 60 per cent increase in rent in Prague's inner city.

The rent of apartment houses completed after June 1993 and not supported by any state subsidy is fully deregulated and can be determined freely by mutual agreement of tenant and landlord. The same applies to rent paid by foreigners, which is also not limited by any regulation. Consequently, there are two housing markets in Prague: first, the domestic and regulated one, and second, the deregulated housing market used by foreigners and wealthy Czechs. This split stimulates a transfer of housing units from the former to the latter segment of the market. Furthermore, the number of transfers of apartments out of the regulated segment of the housing market is even higher because of changes from residential to office use. From July 1995, market rent can be charged for newly signed leases; however, it is difficult to estimate the consequences of this measure.

Housing allowances

The government introduced housing allowances for low-income households to ease the burden of increasing rent. A subsidy is given to a household for a maximum of two years, and the household is expected to find cheaper accommodation and move within this period. Housing allowances are paid in relation to need. Only households that earn a total income of less than 1.3 times the subsistence level are eligible for financial assistance. In 1995, the official subsistence level for an individual was CZK 2,440 ($85, £60) for one adult, CZK 4,360 for two adults, and CZK 7,580 for a family of two adults and two children. The maximum subsidy for rent that a single-individual household can receive is CZK 200; for a two-person household it is CZK 260; and for a three-or-more-person household it is CZK 350. The housing costs (net rent plus amenities) for an average flat occupied by such families will be between CZK 1,500 and 2,000, therefore, on average, the subsidy covers less than 20 per cent of total housing costs.

HOUSING POLICY UNDER TRANSFORMATION: THE 1990s

Developments in housing policy

Contemporary housing policy is conditioned by general changes in society, namely by the liberalisation of the economy pursued by the neo-conservative policy of the government coalition. In January 1991, fourteen months after the political changes of November 1989, economic reform was launched. The major focus of reform was the reintroduction of private ownership and market exchange. Housing as a specific subject was not high on the political agenda. The government believed that the general introduction of a market economy would lead to the establishment of market mechanisms in the housing system. Up to the mid-1990s, major changes in housing were caused by general policies of reform, while explicit housing policy played a rather marginal and passive role.

The quick move towards the market model for housing is a desirable direction for the government. It is believed that the market will allocate and provide housing efficiently. Consequently, state involvement in housing is being quickly withdrawn. The state has completely withdrawn from direct housing production, and subsidies for co-operative and private housing construction have virtually ceased to exist. Privatisation of housing and gradual rent deregulation towards market levels form the cornerstone of the government's approach to housing aimed at the internal transformation of relations within the existing housing stock.

Czech housing policy is institutionally the responsibility of the Department of Housing Policy at the Ministry of Economy. However, some measures are implemented in co-ordination with the Ministry of Labour and Social Affairs and the Ministry of Finance. Responsibilities for the provision and management of public housing have been decentralised to municipal government. Supplementary information on the transition of the Czech housing system can be found in Musil (1992) and Kingsley et al. (1993). A general review of east European housing privatisation is given in Clapham (1995).

Changes in ownership and management of public housing

In 1991, 877,000 apartments (23.5 per cent of total dwelling stock) were transferred from the state to municipal ownership. However, buildings whose floorspace was more than one-third in commercial use were retained in state ownership. Using this measure, responsibilities for public housing were transferred to municipalities. Municipalities also received ideological support for the privatisation of municipal housing management and for sales of municipal housing.

Many municipal governments decided to privatise Housing Services Companies (HSCs) or to abolish them and contract a number of small private real-estate management companies (Kingsley *et al.*, 1993; Sýkora and Šimoníčková, 1994), in the belief that efficiency and maintenance quality would increase substantially. However, there are municipalities which are still using HSCs. Unfortunately, no research has been undertaken to assess the impact of these changes.

Privatisation of municipal housing

Two basic forms of privatisation of state and municipal housing stock were applied: restitution (reprivatisation) and sales of municipal housing (privatisation). In the restitution (reprivatisation) process, properties confiscated by the communist regime between February 1948 and December 1990 have been given back to their original owners or their heirs. Most transfers were accomplished by the end of 1993; at present there are only a small number of cases seeking court resolution.

There are no exact statistical data available for the impact of restitution in the Czech Republic (estimates are around 10 per cent of dwelling stock). However, studies by Daněk (1994), Eskinasi (1994, 1995) and Sýkora and Šimoníčková (1994) give figures for some localities. For example, in central Prague 70–75 per cent of all houses were returned. Figures for Prague inner-city neighbourhoods are lower (30–65 per cent) and are declining to zero for outer-city districts. In České Budějovice, a regional centre of 100,000 inhabitants in South Bohemia, 5.7 per cent of all apartments were returned in restitution. The share of restituted properties was higher in the city centre, where 25 per cent of houses were returned.

In general, restitution has had a clear geographical pattern, as it mostly influenced central parts of towns and cities. A high demand for commercial space in central locations influenced a substantial transfer of residential space in reprivatised buildings to office use. Returned houses could immediately be marketed and therefore the process is seen as the most important impetus for the development of the real-estate market in the Czech Republic (Sýkora and Šimoníčková, 1994).

The housing which was not restituted and remains in municipal ownership can be privatised. The methods of privatisation differ substantially among municipalities as there is no central-government legislation to guide the process. Prior to 1994, only whole houses could be privatised. In Prague, the majority of tenants of a house were offered the property as the first priority. Tenants had to form a co-operative of tenants or another legal entity, such as a limited liability company, to acquire the property. If they were not willing to buy the house it was offered to a minority of tenants. As a last resort, a house could be sold to anybody else, for instance to a real-estate development company (for details see Eskinasi, 1994, 1995).

Since 1994, when the law on ownership of apartments and non-residential premises was approved, municipalities have been able to sell individual apartments to private owners. The first municipal programmes based on the privatisation of apartments were approved by the local governments of larger towns in mid-1995. The privatisation of municipal housing has yet to be advanced. Many municipalities do not privatise at all, others use privatisation as a tool to increase revenues for local budgets. The city of Prague intends to privatise up to 80 per cent of the municipal housing stock. However, only a small fragment of houses and apartments had been sold by 1995. For instance, the largest of Prague's municipal authorities, Praha 4, was quite active in housing privatisation, but by the end of 1994 only 98 houses out of 1,300 which remained in municipal ownership after restitution had been transferred to private owners.

In 1995, 41 per cent of dwellings were in owner-occupation, out of which 39 per cent consisted of single-family housing and 2 per cent were private apartment houses. Out of 59 per cent of the dwellings in the rental sector, 23 per cent were in municipal housing, 4 per cent in state-owned buildings, 21 per cent in the co-operative sector, 9 per cent were located in privately owned apartment houses, and 2 per cent were apartments leased by owners of single-family houses.

Transformations in the co-operative sector

Two Acts were designed to transform the Building Housing Co-operatives. First, the Act on the transformation of co-operatives allows the division of large co-operatives into smaller ones. Second, the Act on the ownership of apartments and non-residential premises enables co-operative members/ tenants to purchase their flats. Applications had to be submitted by the end of June 1995 and flats were to change hands by 31 December 1995. The purchase price was equal to the amount of money needed to repay the outstanding bank loan used by the co-operatives to fund initial construction.

Prices varied considerably, depending on the age of co-operative housing (up to 35 years). Flats in older properties (where loans were already largely repaid) cost a few thousand crowns. Meanwhile, flats in newer prefabricated apartment blocks commanded CZK 50,000 to 100,000. The transfer price of newer co-operative apartments was equal to the level of the average annual income of an individual, and was several times lower than the current market price for co-operative flats in Prague (CZK 700,000–1,000,000 for a two-bedroom 70 m^2 apartment).

Most co-operative housing is located on land in state or municipal ownership (an estimate by the Czech Geodetic and Cadastral Office puts it at 60 per cent). In this case, the land is offered for sale. The price for land will be regulated by state decree (1 m^2 will cost CZK 1,700 in Prague, CZK 25 in small settlements). The co-operative tenant is not obliged to purchase the land but, for a fee, can acquire its use until the building is demolished.

At present, nearly a fifth of the country's population live in 700,000 co-operative-owned apartments. In Prague, about 70 per cent of co-operative members have applied for home-ownership. It is likely that by the end of the 1990s most co-operative houses will be transformed into condominiums, thereby substantially increasing the share of owner-occupation in the Czech Republic.

'Condominium' legislation: transfers from the rental to the owner-occupied sector

The Act on ownership of apartments and non-residential premises (inspired by US condominium legislation) was approved in April 1994. It offers the possibility of selling individual dwellings in an apartment building. The ownership of a dwelling in an apartment building will include shared responsibilities for communally used parts and spaces of the building, such as the roof, stairs or lift. Apartments can be sold to sitting tenants or to any third party.

The new legislation affects municipal, private, and (as examined above) co-operative rental housing. It will have an important impact on transfers of housing stock from the rental to the owner-occupied sector. Municipalities are eager to sell apartments in municipal housing (with or without tenants), thus reducing their responsibilities and expenditures and increasing local budget revenues. Private landlords can benefit from sales because the difference between their current income from regulated rent and potential gains from sales to owner-occupation. However, it is difficult to estimate the consequences of transfers out of the rented stock.

Tenants' rights: landlord–tenant relations

The deregulation of rent on vacant possession and a sharp divide between regulated and deregulated markets led to fears of harassment and speculation, which have been reported in a few cases. However, the legal protection of tenants is relatively strong. The passages of the Civic Code that govern eviction require a landlord to obtain a court order and provide a replacement dwelling of the same standard for tenants. Landlords try to force tenants to move under the pretext of unnecessary building renovations. Tenants, on the other hand, refuse to accept replacement flats. Both sides then seek court resolution. However, there have been several cases where compromise has been achieved. Tenants have moved and houses in prestigious locations have been renovated and re-let as offices and/or luxury apartments. In this way, quite a few buildings were rehabilitated and gentrified in the central and inner-city neighbourhoods of Prague.

The act on ownership of apartments enables sales of flats with their tenants to new owners. Nevertheless, landlords (private owners or municipalities) must give tenants the first option to purchase at the asking price. Tenants can decide during a six-month period whether to buy their apartments. If they choose not to purchase, the landlord can then offer the dwelling to another interested buyer. However, there is an ensuing 12-month period during which the tenant can purchase the apartment for the price offered by the buyer. After this period the apartment can be sold freely. The option of selling tenanted apartments is going to be used by municipalities as well as private landlords.

CONCLUSIONS

The Czech version of the east European housing model (Clapham, 1993) has been quickly dismantled. The contemporary housing system is at the stage of transition; the state-dominated housing provision has ceased to exist and the market has not yet filled the gap. Without a more active housing policy aimed at stimulating housing production and consumption, housing standards may quickly decline.

Contemporary housing policy is using measures that are helpful to higher-income households (wishing to buy) and low-income households who are already housed. Since housing saving schemes have a very limited impact, very little is done for middle- and lower-income households in need of dwellings, especially for newly formed younger households. If these groups do not soon become a target of housing policy, social problems may appear. The worst effect would be a transformation of young people's values, which will draw them to street life, delinquency and crime.

Fortunately, homelessness is not as yet a serious problem as it is in many west European countries. Existing homeless shelters are managed by voluntary NGOs with a marginal involvement by the state. Social segregation was virtually absent during communism. However, it is anticipated that processes of segregation such as gentrification and ghettoisation will transform some parts of large towns and cities. The decline of social and physical factors is likely to appear in the private-rental stock of older housing in the inner cities rather than in more suburban prefabricated housing estates, where middle-income households are largely situated. Municipal housing is not residualised or stigmatised as it is in the United Kingdom.

Future housing policy in the Czech Republic is likely to continue to evolve along free-market lines, assuming the absence of any major change in political representation in parliament.

18

POLAND

Henryk Hajduk

Fundamental changes in the economic system of Poland were introduced by parliament in Warsaw through the adoption of a legislative package in December 1989, and in the first half of 1990 urgent changes that were necessary to ensure the application of an effective policy of economic stability. At the same time changes created the foundation for a new economic system. Here can be mentioned:

- a strict fiscal policy (control of incomes and budget expenditures, restriction of subsidies and the extension of taxes);
- realistic interest rates and reduction of low-rate loans;
- a restrictive income policy;
- internal convertibility of Polish zloty;
- abolition of all forms of administrative distribution of goods and foreign currency;
- extension of free market prices;
- liberalisation of foreign trade;
- demonopolisation of monopolised spheres of trade and production;
- privatisation of state-owned enterprises that will not only lead to changes in ownership but also to an increase in efficiency.

The market-oriented economic reforms started in 1990 and are now in various stages of implementation. Since 1992 various positive trends in the national economy can be observed. The rate of economic growth has been relatively high, and industrial output has constantly risen. The rate of unemployment has slightly decreased since the end of 1994. In 1994 the basic macroeconomic indicators increased as follows:

- gross domestic product (GDP) by 5.0 per cent;
- industrial output by 11.9 per cent;
- construction work by 2 per cent;
- investment outlays by 7.1 per cent;
- income from the export of goods by 25 per cent;
- payments for the import of goods by 12 per cent.

For the year 1995 the further stabilisation of economic conditions was foreseen as well as the stimulation of additional economic growth. The further improvement of living standards and continuation of reforms started in 1990 remain the essential aim. Increased activity aimed at the exploitation of productive possibilities, the acceleration of investment processes and reduction of the inflation rate are assumed. The basic economic indicators are predicted to improve as shown in Table 18.1.

Table 18.1 Forecast percentage increases of economic indicators after 1994

	1995	1996	2000
Gross domestic product (GDP)	6.0	5.5	5.0
Public consumption	2.0	2.5	2.0
Private consumption	3.5	3.5	3.0
Investment outlays	7.0	7.3	7.0
Export of goods and services	23.5	12.0	7.5
Import of goods and services	10.0	7.5	5.0

The performance of the construction sector in Poland, as in other countries, is closely related to general economic trends and is greatly affected by the economic policies adopted by governments. The deep recession in 1990–91 and government expenditure cuts in investments caused a great curtailment of construction. The economic recovery during 1994 and 1995 ended the recession in construction (except for housebuilding). It is temporarily estimated that in 1994 output of the construction industry grew by about 2 per cent with further growth of 4 per cent predicted for 1995 and 5.5 per cent in 1996. It is foreseen that in the year 2000 construction output will be higher than in 1994 by 35 per cent.

TENURE

The existing housing stock belongs to four basic groups of owners, namely municipalities, housing co-operatives (working as non-profit organisations), enterprises and owner-occupiers. There are also, of course, private-rental dwellings. Two forms of use can be distinguished in each of the above groups: the first concerns proprietary dwellings purchased by tenants (occupants), and the second one covers rental dwellings used by virtue of hire contract. In the face of changes which have occurred in the economy, and also in the housing co-operative movement (which is ceasing to be a semi-state organisation and is becoming truly a co-operative organisation within the private sector), this means that jointly with private rented and owner-occupied dwellings 65–70 per cent of the total housing stock will be found in the private sector. The detailed structure of housing stock is given in Table 18.2. The table shows that the structure is relatively stable and changes

Table 18.2 Tenure, Poland, 1990–93 (percentages)

	Municipal	Co-operative	Enterprise	Owner-occupation and private-rented
1990	25.0	17.9	13.4	43.7
1991	25.4	17.6	13.7	43.0
1992	26.4	18.0	13.1	42.5
1993	27.0	18.0	12.4	42.6

evident in the structure are not of great importance. It is appropriate to emphasise that in all groups of dwellings the share of proprietary dwellings is growing, although very slowly. The number of dwellings of this kind is increasing as a result of the sale of dwellings by *gminas* (municipalities) and factories to existing tenants. In co-operatives a part of the housing stock is being converted into proprietary dwellings as well.

HOUSEBUILDING

The annual level of housebuilding has declined since 1978. In that year the number of dwellings completed was the largest of the whole post-war period. After reaching the peak of approximately 280,000 dwellings, production of dwellings began to decline gradually and in 1994 it amounted to

Table 18.3 Number of dwellings completed, Poland, 1975–94

Year	Housing construction	Multi-family		One-family	
		000s	%	000s	%
1975	248.7	190.2	76.7	57.9	23.3
1976	263.5	200.0	75.9	63.5	24.1
1977	266.1	192.5	72.3	73.6	27.7
1978	283.5	209.3	73.8	74.2	26.2
1979	278.0	206.0	74.1	72.0	25.9
1980	217.1	161.4	74.3	55.7	25.7
1981	187.1	140.9	75.3	46.2	24.7
1982	186.1	130.6	70.2	55.5	29.8
1983	195.8	138.1	70.5	57.7	29.5
1984	195.9	141.0	72.0	54.9	28.0
1985	189.6	132.9	70.1	56.7	29.9
1986	185.0	127.6	69.0	57.4	31.0
1987	191.4	131.0	68.4	60.4	31.6
1988	189.7	125.5	66.2	64.2	33.8
1989	149.8	91.4	61.0	58.4	39.0
1990	134.6	86.6	64.3	48.0	35.7
1991	136.8	97.0	70.0	40.0	30.0
1992	133.0	96.0	72.0	37.0	28.0
1993	94.4	61.0	64.0	36.0	36.0
1994	71.6	35.5	49.5	36.1	50.5

only 70,000 dwellings – equivalent to about 26 per cent of output in 1978. A similar decrease in the production of dwellings is characteristic also of multi-family construction: its share diminished gradually from 75 per cent of total output down to 55 per cent in 1994. One-family housing construction was not subject to so significant a reduction as multi-family house-building, and because of that its share in the total housing output in 1994 was assumed to be about 50 per cent (Table 18.3).

It should be emphasised that rural areas contain a high proportion of housing output for single families – one-family dwellings making up over 95 per cent of the rural housing stock. The relative stability of the spatial distribution of housing should be emphasised. Around 80 per cent of newly constructed dwellings are located in urban areas – a share which in general has remained stable, with only slight changes, over the last dozen or so years (Table 18.4).

Table 18.4 Dwellings completed in urban and rural areas as percentages of total housing completions, Poland, 1980–93

Place of housing completion	1980	1985	1990	1991	1992	1993
Urban areas	79	74	74	79	80	76
Rural areas	21	26	26	21	20	24

Although increases in the number of dwellings diminish year after year, there is a gradual increase of the number of dwellings per 1,000 inhabitants (Table 18.5). The indicator has grown from 248 in 1970 to 294 dwellings per 1,000 inhabitants in 1992. If we assume that this index characterises the quantitative level of satisfaction of housing needs, we should consider the level reached in this sphere by west European countries. In comparison with countries of the European Union, Poland has the lowest number of dwellings per 1,000 inhabitants.[1] Table 18.6 shows, moreover, that the number of houses completed every year per 1,000 inhabitants was diminishing during the period 1970–93.

Table 18.5 Number of dwellings per thousand population, Poland, 1970–92

1970	1975	1980	1985	1990	1991	1992	% growth 1970–92
248	259	274	286	286	289	291	12

Table 18.6 Housing completions per thousand population, Poland, 1970–93

1970	1975	1980	1985	1989	1990	1991	1992	1993
6.0	7.3	6.1	5.1	4.0	3.5	3.6	3.4	2.4

A deep recession in the residential construction sector has been taking place since 1980, marked by a sharp decline in the number of dwellings completed. In 1994 only 71,000 dwellings were completed – 24 per cent less than in the previous year. In 1993, the decline reached 29 per cent. In 1993–94, the volume of housing was 50 per cent lower than in 1992.

The average usable area of a new flat had been increasing rapidly and in 1994 it achieved 89.3 m². The average area of a private flat (117.6 m²) was almost twice that of the average area of a co-operative flat (62.3m²). In 1990–94, the average area of a new flat increased by 12.1 m². The average area of a private flat increased by 9.8 m² while that of a co-operative flat increased by only 2.1 m².

For 1995, an output of approximately 70,000 dwellings was foreseen: 30,000 in multi-family construction and 40,000 in individual buildings. In 1996 the establishment of specialised saving-credit systems and financial institutions for funding housing was planned (housing banks and the National Housing Fund and Social Housing Societies) supporting cheap rental construction. According to the growth rate of activities mentioned above and the increase of the population's income it is optimistically foreseen that in the year 2000, the number of dwellings completed will amount to 145,000 (75,000 as individual dwellings and 70,000 as multi-family houses). But in the absence of necessary economic changes the number of dwellings completed might be only around 125,000.

The rehabilitation and maintenance (R and M) sector of construction has stayed relatively stable during the period of overall recession in the house-building industry. A large supply of materials (including imported fittings and materials for finishing) and services has caused an increase in demand for the modernisation and repair of private flats. Allowances in income tax in respect of repair have played an important part in the development of R and M as they strongly stimulated private expenditure in this area. It is estimated that the high level of R and M achieved in former years was preserved in 1994.

The high level of demand for the repair of private dwellings was not accompanied by an increase in outlay on building repairs, especially in the public sector and in private-rental buildings. Housing managers and owners reduced expenditure on repairs because of a shortage of financial resources. The reform of the rental system was introduced only at the end of 1994. It makes the costs of renovation realistic and will encourage the use of income from rents to cover costs of repair. It is estimated that there are about 1 million flats in Poland in need of immediate repair.

Housing stock

Clearly the housebuilding output of the construction industry over the years can be measured in quantitative and qualitative terms.

According to data from the 1988 National Census the housing stock in Poland was as follows:

- 10.8 million dwellings (7.1 million in urban areas and 3.7 million in rural areas);
- 10.8 million dwellings occupied by 37.1 million persons (28.5 million in urban areas and 14.6 million in rural areas).

The quality of dwellings can be characterised as follows:

- dwellings in buildings erected before 1918 represent nearly 14 per cent of the total number of dwellings in urban and rural areas;
- dwellings built after 1944 amount to about 70 per cent of the total housing stock;
- residential buildings with non-flammable walls represent 91 per cent of housing stock;
- basic facilities (running water, toilet, bath, central heating, gas) are installed in 50 per cent of the total housing stock; the share of facilities in urban areas is much higher than in rural areas.

Finally, the size of dwellings and number of occupants per unit may be an important feature illustrating the condition of the existing housing stock. The share of small dwellings – (one or two rooms per unit including kitchen) amounts to 24 per cent of the total housing stock, that of three rooms is 34 per cent, and that of four or more rooms is about 42 per cent. The average floor area per dwelling is nearly 60 m^2 in urban and 80 m^2 in rural areas. On average there are 3.20 persons per dwelling in cities and 3.97 in the country (0.97 person per room in urban and 1.11 in rural areas, 16.5 m^2 per person in urban and 17.4 m^2 in rural areas); 16.5 per cent of the total population lives in overcrowded dwellings with two or more persons per room.

In the post-war period the housing sector in Poland has been characterised by a low level of production and chronic housing shortage. In 1980 there were 14 per cent more households than housing units, a deficit of 1.3–1.8 million dwellings. Through the 1980s, there was little progress in reducing this deficit.

Most of the demographic housing needs occur in urban areas. They are the most difficult to satisfy since the concentration of these needs varies. Nine voivodships (counties) which have the biggest urban complexes give rise to 53 per cent of urban demographic needs. In addition to the demographic needs, there are needs related to the necessity of replacing older housing stock. The huge gap in renovation and modernisation and the accumulated lack of appropriate care for many years have resulted, together with insufficient resources, in the estimated need of stock – replacement being much higher than the actual decrease in housing numbers. The government therefore estimates that the production level required to satisfy the country's housing needs by the end of the decade is about 300,000–400,000 units per year.

Privatisation and the restructuring of the housebuilding industry

The reform of the housebuilding industry has to include both the operation and management of existing housing resources and the development of new housebuilding. The very important changes that have to occur in the building industry involve:

1 The privatisation of construction resources through a speedy process of ownership transformation, which will not only lead to changes in ownership, but also to a noticeable increase in efficiency, dissolution of large inefficient companies, and a change in the organisational structure of companies.
2 The restructuring of the housebuilding industry, which should aim at an increase in the number of small and medium-sized companies, competition between companies, changes in construction technology, and the introduction of new technologies.
3 A reformed land market and process of land-use planning that will facilitate access to land for all potential users, including small and medium investors, and shift land use and development to the responsibility of local government.
4 A modification of financial and taxation policies which would discontinue investment subsidies in respect of newly built houses and replace this facility with credits. Government should also introduce flexible mortgage and equity instruments, make interest rates correspond to savings deposits, and abolish all forms of subsidising the formation of construction and building-material companies.

The chief ways and forms in which the privatisation of the building industry in Poland is being carried out include:

• increasing the number of new private firms, and their production capacity;
• increasing the production capacity and size of existing private building companies;
• privatising the existing state construction enterprises through their conversion (in various forms) into private companies;
• creating new private construction partnerships, joint venture companies and other companies with both domestic and foreign capital;
• purchasing parts or all of the state-owned companies by individuals and companies in the private sector;
• leasing of parts of companies or whole companies to employees or other individuals in the private sector;
• other forms of transferring employment and resources of the construction industry from the public to the private sector.

Along with the increase in the number of business entities in the private sector there was a simultaneous increase in the building output. The

Table 18.7 Output of and persons employed in the private sector of the building industry (percentages of the whole)

	1990	1991	1992	1993	1994
Sale of goods and services	33.8	57.4	78.0	86.3	87.8
Persons employed	36.3	57.1	71.8	73.7	80.0

production volume and the number of persons employed in the private sector of the building industry are presented in Table 18.7. Looking at these figures, it can, by implication, be seen that in quantitative terms the ownership structure of housing production companies in Poland is quite diversified. It includes various types of companies and various forms of ownership. State-owned companies no longer dominate the market. The private sector has an increased share in both construction and repair projects.

The private sector is dominated by very small, small and medium-sized firms. They use traditional technologies, and their organisation and methods of work are very simple. Of building enterprises in Poland, 87 per cent are craft enterprises with 1–2 persons employed, 11.4 per cent are units employing up to 50 persons, 1 per cent are medium-sized units employing up to 200 persons, and larger units make up only 6 per cent of the total (Table 18.8).

In 1994 a further transformation of ownership took place which enlarged the share of the private sector in construction. Establishing new private economic entities, the growth of production and employment levels in existing private enterprises, the decline in the number of enterprises (in

Table 18.8 The organisation of the construction industry, 1980–93

	1980	1985	1990	1991	1992	1993
Number of construction enterprises (state and private) employing:						
less than 5 persons	–	–	–	–	145,212	130,882
6–20 persons	–	–	–	–	14,400	14,720
21–50 persons	61	133	447	–	177	2,441
51–100 persons	48	119	286	147,252	879	879
101–200 persons	157	294	427	643	736	668
201–500 persons	477	554	457	490	501	428
501–1,000 persons	240	258	245	221	265	162
above 1,000 persons	355	294	144	90	110	65
Total number of construction enterprises	1,338	1,652	2,006	148,696	162,280	150,245
Total number of persons employed in construction industry (000s)	1,336	1,282	1,242.7	1,116.3	1,066.2	860.8

production and employment terms) in the public sector, and the privatisation of public enterprises all resulted in the share of the private sector in construction expanding from 78 per cent in 1992 to 86 per cent in 1993 and 87.2 per cent in 1994.

Under these circumstances, the forecast for 1995 confirms the trend towards more substantial recovery of construction activity, except for housebuilding. It is foreseen that construction output will be increasing till the year 2000 by around 5 per cent annually, assuming favourable progress in a range of social and economic reforms as well as the internal and external economic environment.

HOUSING INVESTMENT, FINANCE AND SUBSIDIES

The introduction of market rules into housing requires the creation of a new system of finance, which will realise profit from the capital market and will be adapted to the new pattern of housing tenure. It should only be limited to providing private credit and state assistance during the period of amortisation, but should also define the economically and socially desirable methods of controlling the costs of construction and the cost of acquiring capital for housing purposes.

The very low income of people, high interest rates on loans and the rapid increase of apartment prices have caused the state to discontinue housing subsidies, and will consequently result in a drastic drop in demand for housing in the future, as well as halting housing developments already started.

The continuation of the current system of transforming the financial sector can lead only to a greater shortage of housing. It is necessary to act in many areas. Aside from the continuation of the changes in the financing of apartments for sale and the creation of the mortgage credit system, the system for financing the construction of rental housing must also be created (including apartments with controlled low rent and co-operative tenement apartments).

To make housing a commodity accessible to wider social strata, the financing system – in the short or long run – must rely on institutions and mechanisms which will cause:

- a relative decrease in the interest rate on credits;
- a greater propensity to save for housing purposes;
- more possibilities of feasibly recovering the resources invested in housing.

This means that it is necessary to create many specific savings and credit institutions, both for buyers of apartments and for the developers and construction companies – which should be separate. Many systems for the mobilisation of savings and for issuing loans (different for various income groups) should be accompanied by the development of institutions to guarantee stability and balance within the financial sector. The creation of a new system of finance is not possible without a great deal of government assistance. The assistance should include both the creation of legal conditions for the functioning of these mechanisms and institutions, and also financial assistance.

It is estimated that the housing needs of wealthy families can be satisfied first of all with the purchase of apartments with their own resources and with long-term mortgage loans. The most appropriate form for satisfying the needs of middle-class families will be mortgage loans issued by housing savings-and-loan institutions, under the so-called contractual saving system. The interest rate on these credits will be lower than the market rate. A system of financing social-rental housing will be created for low-income families. These apartments will have a controlled low rent, and credits for this type of housing will come from the resources of the National Housing Fund, from local government and from credits contracted on the market. Loans issued by the National Housing Fund will have a preferential character.

The three proposed forms of satisfying housing needs are to serve as a model and their creation will be one of the major focal points of state activity. Regardless of the activity required to satisfy these needs, economic and legal conditions will be created for the development of private-rental housing. State assistance for this type of housebuilding will necessitate housing benefits for the tenants of apartments and the use of tax reductions for private developers of housing projects (designated for rental in the controlled rent sector).

The level of interest rates on loans for developers and contractors will be determined by the market. The only exception will be credits for social-rental housing organisations, and for all organisations investing in communal technical infrastructure. The loans will be granted by the National Housing Fund.

It is assumed that the system of credits for investors and contractors and the system of credits for buyers of apartments will remain separate.

Tax allowances have only been applicable to newly built dwellings, and not to 'second-hand' dwellings purchased in the market. Therefore it can be assumed that state policy over many years was based on a preference for the former type of dwellings. Table 18.9 sets out the availability of financial aid in respect of different types of house purchase.

In some instances it is possible to obtain financial aid from the state in the case of rented housing by hire contract. The forms of help anticipated in various cases are shown in Table 18.10.

DEVELOPMENTS IN HOUSING POLICY

In the near future the following changes should take place:

1 Ownership rights should be clarified and the system of management of housing stock changed.

2 Housing should become fully adapted to economic rules, together with obligatory financial assistance for housing benefits for families of the lowest financial status.

Table 18.9 Financial aid for house purchase, Poland, 1995

Type of dwelling	Possibility of financial aid in the form of:		
	Mortgage credit with indexed instalment	Income tax reduction	Bonus of guarantee
Purchasing of dwelling on the housing market for 'second-hand' stock			
One-family dwelling	None	Not vested	Vested
Dwelling in private building (part of a small apartment building)	None	Not vested	Vested
Dwelling in building belonging to a housing community	None	Not vested	Vested
Dwelling in privatised communal or factory's stock	None	Not vested	Not vested
Purchasing of newly constructed dwellings			
Being built by owner–user	Possible	Vested	Vested
Bought in co-operative	Possible	Vested	Vested
Being built by co-operative from resources of owner–user	Possible	Vested	Vested
Purchased from developer	Possible	Vested	Vested

Note: 'Vested' means 'available to purchasers'

Table 18.10 Financial aid for rented housing, Poland, 1995

Owner	Type of rent	Form of support to owners	Form of support to tenants
Gminas – communal stock	Regulated	None	Housing allowance
	Social	None	None
Factories (on the decline)	Regulated	None	Housing allowance
Social housing societies	Economic	Preferential credit	Housing allowance
Tenement co-operatives	Economic	None	Housing allowance
Private owners of rental houses			
stock included with specific mode of hiring	Regulated	None	Housing allowance
stock not included with specific mode of hiring and new constructed stock	Market	Investment reduction in income tax	None

3 Various complementary systems of financing housing construction in both public and private sectors should be introduced.
4 New financial institutions, legal regulations and organisational structures conforming to a market-oriented housing economy should be created.
5 The exclusive co-operative mode of housing development should be abandoned.
6 The conditions for competitive claims on public budget aid should be created (in respect of credits, subsidies).
7 A clear and just system of rights and duties for apartment rental should be designed.

Changes will involve the appearance of new participants in the investment process and new relations between them. New investors in housing development will be both competitive and co-operative. They will be, first of all, developers – people undertaking the construction of housing for sale, builders of social housing and infrastructure investors.

Strategic areas for reforming the system of housing needs in respect of both the existing stock and new constructed dwellings are shown in Figure 18.1.

Rent reform

It is generally accepted that rent reform and the reform of utility pricing are absolutely necessary. The main market-oriented solutions should ensure that rents will cover the full economic cost of housing, not just the maintenance of the unit, and should target subsidies directly to households in need. In the last few years rents have been raised repeatedly. An increasing number of households cannot pay because of their low income. The reform of rents will be implemented in stages because of the difficult financial situation of many households.

In the first stage, rent will be set as 3 per cent of the replacement value of the apartment per annum (i.e. the cost of building a new apartment of the same size and standard). The goal is for the rent to be at a level of 6 per cent of the replacement value. It is assumed that part of the regulatory procedure will be delegated to the level of local authorities; this will give them a tool to activate local policy in housing. The rule will apply to existing communal stock and newly constructed apartments.

The rents increase will be implemented and the right of financially disadvantaged people to obtain social assistance for housing purposes will be raised to the maximum degree. The criteria to obtain benefits are the income level of single- and multi-person households, under the general rule that housing benefits are available only to people with insufficient incomes and living in a standard apartment of a size appropriate to the number of inhabitants. Tenants of co-operative apartments are entitled to such

SYSTEMS OF SAVING FOR DWELLINGS

- building savings-loan banks (housing societies)
- rewarding with premium an annual saving payment for dwelling

NEW FORMS OF CREDITING AND SUBSIDISING

- mortage credit – mortage banks
- contractual credit – building savings-loan banks
- special funds – National Housing Fund
- co-operative banks
- investment abatements of taxes
- various forms of subsidies

Dwellings in existing stock	Newly constructed dwellings
1. Regulation of rights of property and use (private sector and public sector): – of ground – of real estate (building and premises) – infrastructure facilities 2. Reprivatisation of real estate 3. Privatisation of housing stocks – communal – factory-owned – transformation in co-operative stock 4. Reform of rents – withdrawal from subsidised, centrally regulated rents – economic rents 5. Reform of subsidy system – tax reductions – benefits and allowances – preferential credits – other forms of aid 6. New institutional solutions – managing of privatised stocks (housing communities) – dwelling agencies – estate agencies	1. New investing organisations – non-profit investing organisations – Social Housing Societies – change in the range of co-operatives' activities (construction for sale) – profitable investing organisations – developers' form of dwellings construction – pension funds, premium funds 2. Ownership transformation – privatisation, restructuring – transformation of state-owned enterprises – development of new building and design enterprises (small and medium-sized) 3. Competition between enterprises – investor's market – spreading contracts by tender – placing public orders by tender mode 4. New types of enterprises – developers' enterprises – enterprises' managing of ventures (investors' representation) – estate-consulting enterprises – design-consulting enterprises 5. New lines of fiscal and financial policy – granting credits for building and repairs – winding-up of subsidies 6. Access to grounds and town planning – elimination of regulations unfavourable for small and medium investors – market ground turnover – refused privileges in building lot granting – change of terms of dispossession

Figure 18.1 Reform of the system of housing welfare in Poland

assistance. It is financed from the *gmina*'s resources saved by the implementation of higher rents.

The system of housing benefits and tenant protection should strengthen the feeling of security in society. Legislative protection of tenants must provide a rent moratorium for those families which are temporarily in a poor financial situation (e.g. through unemployment). Intervention of the social policy will be sporadic but with swift reaction to extreme cases.

Social-rented housing

The provision of public-rental housing plays an important role in the programme of state housing policy. The lack of success of the programmes of housebuilding for the social sector which have been implemented so far, and the present condition of social housing which needs to be adapted to the requirements of the market economy are the reasons to introduce a new programme. The programme of social-rental housing presumes the creation of new investor organisations – societies of social-housing construction (TBSs), working on a non-profit basis but according to market rules. Joint companies may be created by the *gminas* (local authorities), co-operatives, economic and social organisations and individuals.

In order for rental housing to be economic, it is necessary to create a separate system of financing and credit for social-rental housing. One of the credit sources will be money from the National Housing Fund but credit from this fund cannot exceed 50 per cent of the cost of a housing project. The remaining project funds will come from the resources of the *gminas*, from the capital of societies of social-rental housing (including contributions in kind), credits and commercial loans. The societies of social-rental housing will employ contractors according to competitive bidding, and after the project is completed they will have the function of managing the stock. Rents paid by tenants will be significantly lower than market rents. Local authorities would be granted rights – to an extent equivalent to the level of their involvement in the promotion of the programme of social-rental housing – to designate the tenants for new housing according to needs and priorities. As the projects in social-rental housing progress and the ownership rights and rents are put in order in the older social stock, procedures for the integration of old and new rental stock would be implemented.

HOUSING NEEDS AND HOUSING PROVISION

The goal of state housing policy should be to improve the housing conditions of Poland's inhabitants. This could be fulfilled through the rational utilisation of old housing and through the construction of new dwellings. Under favourable economic conditions in various countries, the annual number of new housing units handed over for use does not generally exceed

2–4 per cent of the existing number of units. That is why the exlusive construction of new housing cannot satisfy housing needs within a short period of time. In the short run, considerable improvement can be achieved by matching the population to the existing mix of social units. Therefore, economic instruments should motivate the Polish population to display greater 'housing mobility' and to do away with the stereotype of 'families tied down to flats'.

Facilitating the exchange of flats in accordance with the preferences of households may considerably relieve social pressure for the construction of new housing. At the same time, 'trading in old flats' coupled with the construction of new ones can make it possible to achieve relative equilibrium in the housing market sooner than otherwise. The housing market must be treated as a whole, for this is a system of connected parts. It is necessary to preserve the inner logic of the entire system, for potential investors are not indifferent to the basic economic parameters and inter-relationships for each individual segment of the market for goods and supplies built into housing units.

A reformed housing economy should create a system for the satisfaction of housing needs which will be

- in tune with the rest of the free-market economy, but at the same time, providing housing affordable by different income groups within the population;
- pluralistic, allowing people to choose from various competitive paths, the way of acquiring a home which best suits their financial conditions, aspirations and preferences, and depends on individual effort and initiative;
- flexible, creating the opportunity for construction, rental, exchange, purchase and sale, in keeping with people's family, financial, health and occupational circumstances.

From the social point of view, what is important is the possibility for persons to acquire their first, independently owned dwelling (this applies especially to persons aged 19–29), and flats suitable for the needs of persons in the so-called 'third age'. In the initial phase of searching for a home of their own, this first group of persons generally looks for a small flat, which a new family can afford. Affluent families can afford to acquire a large dwelling in anticipation of future needs, but most families tend to acquire flats that suit their present needs. As the family grows and requires more space, these needs can be satisfied gradually, in accordance with the family's ability to pay. In other words, a system should be created to make possible, even facilitate, the exchange of flats in accordance with families' needs and their wallets.

For households, it is not important whether their first home is in an old or a newly constructed building. In each case, it is an investment outlay that generally requires previous savings. The choice of a dwelling is an independent decision dictated by economic circumstances. It may be a rented apartment or one purchased in an old or a new building. The decision should be based on economic factors – the rental or the purchase price and terms and eventual repayment of the mortgage loan. The problem of acquiring a home is presented figuratively below.

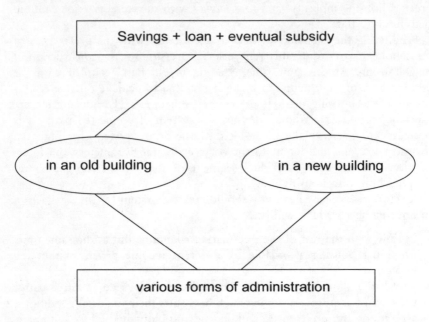

Persons of retirement age, who must find a flat suited to their actual needs, reflect adequately the importance of such a system. Many people in the so-called 'third age' are confronted with this dilemma. The model of the multi-generation family is becoming outdated. As a consequence of longer life expectancy, the period of occupational inactivity is also increasing. The result is that many people are in need of assistance which cannot be provided by the family. The demand for nursing services is increasing, as is also the willingness to exchange flats. However, the elderly have to face many barriers and difficulties when they contemplate such exchanges. A special system of nursing homes must be created to provide people with decent housing conditions.

The internal system of the housing market should be permeable, allowing people to choose from various ways of satisfying their housing require-ments. In other words, it should be possible to change dwellings, the building types, the form of use and ownership, etc. At the present time, the flexibility allowing such freedom does not exist in Poland. The market is 'petrified' by internal barriers that restrict the passage from one segment of the market to another. Ideally, all home-seekers should have the opportunity for and accessibility to the desired form of housing. They should also have the right to make their own decisions and be treated equitably.

The transformation of the housing market has been going on for five years. In order to enable housing to become a market-oriented system, it was necessary to make changes to the way existing housing was administered, as well as to build new housing units. The strategic housing areas, which should be transformed – as a cohesive system for the satisfaction of society's housing needs – are presented in Figure 18.1.

In the period 1989–95, there were three parliaments and six governments. During this time, the transformation in housing and building was regulated by acts prepared by the Minister for Ownership Changes (e.g. the Act of Enterprise Privatisation), the Minister of Finance (e.g. on the tax system), and the Minister of Spatial Economy and the Building Industry. Among the very important Acts initiated by the Minister of the Building Industry on investment in housing construction, the following can be mentioned:

- the Building Regulations Act,
- the Spatial Planning Act,
- the Land and Expropriation of Real Estate Act (several amendments).

These Acts, like those on the management of old housing, were passed in 1994. It was not until 1994 that the Diet passed bills of fundamental systematic importance for the management of existing housing resources, namely

- the Rental and Housing Allowances Act,
- the Home-Ownership Act,
- the Act of Transfer of Company-Owned Flats by Sate Enterprises,
- the Act of Co-operation pertaining to Housing Co-ops.

These Acts introduce certain essential changes, even though they are not as far-reaching as one might have expected after five years of economic reform. Generaly speaking, one can say that the reform of housing is proceeding unevenly: changes are taking place faster in the system for the regulation of new construction than in the management of old housing.

CONCLUSIONS

The period of housing-sector transformation in 1990–95 can be divided into three stages:

- the years 1990–91, when systematic macroeconomic changes occurred in financial and credit policy requiring extremely hard accommodative endeavours on the part of the housing;

- the years 1992–93 during which, in consultation with all interested parties, many programmes were formulated to provide a general framework for housing transformation;
- the years 1994–5, when legislation was introduced to reform housing economy.

Parliament often expresses its interest in improving the functioning of housing markets. But its involvement is not always useful and has possibly caused several years' delay in the implementation of measures of a constitutional nature in respect of housing. Parliament and also government, entangled with short-term considerations, are often blind to strategic problems.

The reform of the housing sector, introduced at the beginning of the 1990s, was treated as an element of general change in the economic and political system. Housing reform was not an independent issue, especially in the early period of economic transformation. It became integrated with the macroeconomic goal of financial reform and economic equilibrium.

The reforms of the housing sector in the period 1989–94 failed to solve the problems of the lack of savings banks essential for individuals eventually wishing to buy and for the provision of building credit, the lack of a mortgage credit system available to middle-income borrowers, and an absence of economic rents (initially covering running and routine repair costs and, subsequently, modernisation and depreciation costs) related to a system of housing allowances. Clearly, if a market reorientation of the housing economy is to take place, these deficiencies must be rectified.

NOTE

1. West Germany, France, the United Kingdom, Denmark and Sweden had over 400 dwellings per 1,000 inhabitants in 1986. The housing stock in Poland increased during the period 1975–92 by 12 per cent, while, for example, the housing stock in the Netherlands increased during the fifteen years 1970–85 by 30 per cent and in West Germany the corresponding increase during the same period was greater than 30 per cent. The lower increase in number of dwellings per 1,000 inhabitants in Poland was because of a lower rate of population growth.

19

SLOVENIA

Ivo Lavrač and Barbara Verlič Christensen

Transition to a market economy in all central and east European countries is proving to be more demanding and slower than was originally expected. There are still no theoretical guidelines about how to proceed with transition, and its course is a difficult political compromise between the right and the left. Our diagnosis is that on the right the new democratic parties are paradoxically in favour of a slower, more controlled, transition process and are more concerned with equity, while the democratic parties on the left, descendants from communist and related parties, are in favour of quick and less supervised transition both for reasons of efficiency and in order to secure property gain based on insider information. The development of housing policies is not very different from this overall pattern and is guided by similar motives.

With regard to the Slovenian macro-framework in which housing policy operates, independent Slovenian policy decisions have only emerged since 1991. Before that there was a common policy throughout Yugoslavia, in-spite of widely differing development levels in each of the constituent republics. Within the context of a significantly smaller macroeconomy, as in Slovenia, exchange and interest rates became sensitive to the introduction of any radical new sectoral policy. Slovenia is a fairly developed economy, its gross domestic product (GDP) per capita and wages are higher than other central and east European countries in transition and are near the level of the least developed countries of the European Union. While politics caused delays in privatisation of the economy in housing policy a compromise soon emerged and thus in Slovenia the housing sector was the first to experience radical transitional change.

TENURE

As in much of central and eastern Europe, the owner-occupied sector in Slovenia was the dominant tenure during the period of communism (Table 19.1) – in large part because of the substantial proportion of the population which still lived in rural areas even when exployed in manufacturing.

Table 19.1   Housing tenure in Slovenia, 1981–93 (percentages)

	Owner-occupation	Non-profit rented	Private-rented
1981	66	33	1
1991	68	31	1
1993	88	11	1

Source: Statistical Yearbook of Slovenia, different years

The non-profit rental sector was located mainly in urban areas. It had been based on strong tenants' rights and unclear ownership, and the social role of enterprises. It was justly and widely regarded as being incompatible with an emerging market economy. Thus the need for the privatisation of the socially-owned stock was never strongly questioned or explicitly justified. The previous system could also be criticised on both equity and efficiency grounds. Most of the non-profit housing was not distributed according to social criteria, and the system strongly hindered both labour and household mobility, and its condition was deteriorating because of poor maintenance. It permanently produced much dissatisfaction and the distribution of housing was always controversial – divorced couples, for example, were able to change residence only with great difficulty.

HOUSEBUILDING, INVESTMENT, FINANCE AND SUBSIDY

The period 1981 to 1991 was one of large-scale housebuilding, an increase in the average size of dwellings and an improvement in amenities (for example, dwellings with bathrooms) (Table 19.2). There was, however, a declining trend in output since the piecemeal introduction of the market into a command economy resulted in hyper-inflation with serious consequences for public and private saving and investment.

New dwellings were almost exclusively of two types, reflecting a dual economy. Multi-storey and multi-unit public housing in towns was financed by earmarked payroll taxation and distributed partly by local government and partly by enterprises and other employers to their employees, who as tenants of social housing were protected by very low rents. Some of

Table 19.2   Housebuilding, housing attributes and investment, Slovenia, 1981–93

	Housing completions	Average size of dwelling m^2	Dwellings with bathrooms %	Housing investment % GDP
1981	15,000	63	71	6.5
1991	6,000	69	86	3.6
1993	6,000	n/a	n/a	n/a

Source: National Housing Programme, first draft, 1995

the dwellings in these large housing complexes were sold by construction companies to private owners, who had privileged access to credit through their employers. Real interest rates for these loans were very negative and the loans were mainly financed from the profits of self-managed enterprises. This type of construction and these types of finance disappeared even before the introduction of new housing law with the emergence of the market economy.

The second type of construction was privately financed individual family houses, accounting for about 4,000 to 5,000 units a year. This was partly financed by favourable company loans, but the main sources of finance remained private savings, commercial bank loans and work by the owner and his/her relatives and friends (Kos, 1992). A major part of this output is without building permission and it places a huge demand on land and infrastructure.

Both types of supply eradicated the gross housing shortage and the 1991 Census showed for the first time an aggregate surplus of dwellings to households (Table 19.3).

Table 19.3 Housing need and supply, Slovenia, 1981–91

	Number of households	Number of dwellings	Shortage/surplus
1981	595,000	582,000	−13,000
1991	640,000	652,000	+12,000

Source: Statistical Yearbook of Slovenia, different years

But this picture is over-simplified, even if structural shortages and distribution aspects are ignored.

Official estimates contained in the National Housing Programme, 1995, which is more a statement of direction and intention than a policy statement backed by financial means, put the census housing shortage in the range of 20,000 (surplus of households over dwellings, taking account of second and vacant homes and multi-household dwellings), which is 3 per cent of the stock of dwellings or three years of housebuilding at the present low rate. Added to this is demand from an increasing number of households (average household size is getting smaller) and the demand derived from the replacement of old stock. The estimate of need in the short term is therefore almost 10,000 dwellings per annum, which is a substantial excess over the present housebuilding rate. It is considered to be a proper policy goal to translate this need into effective (cash-supported) demand, with government help. It suggests a housebuilding rate of five dwellings per thousand inhabitants, which is near the European average rate of housebuilding. We believe that this estimate is probably too high and too simplistic as it disregards the possibilities of improved use of the existing stock of dwellings.

Because of the long life of dwellings, the housing stock is large compared to the annual rate of housebuilding. A 1 per cent increase in the efficient use of stock can provide as much new supply as the total number of housing completions each year. Although it is clear that increasing efficiency has its limits and that substantial housebuilding activity is needed for structural reasons, to adjust to the composition of demand and to take account of the income elasticity of housing, the present rate of housebuilding is probably adequate for the next ten years. There are three reasons for this view.

First, the previous system of social housing with a high degree of tenant protection accumulated huge structural inefficiencies in the use of dwellings. Tenants' rights did not vary regardless of which organisation, enterprise or level of government allocated households to certain dwellings. It was practically impossible to evict tenants even for their failure to pay rent – rent which was below the cost of maintenance and consequently heavily subsidised. Tenants' rights could be inherited and dwellings could be exchanged, although this was complicated. Acquisition of these rights was often considered to be a payment in kind in addition to one's salary and it was part of an individual's calculation of benefits when changing employment. It was therefore justly regarded as a virtual title to ownership. This status was given to the socially needy and to ordinary employees, but mostly to privileged individuals, particularly in respect of the most desirable dwellings (Verlic-Dekleva, 1994). The consequences of this system were that many people remained in over-large dwellings, as they were not compelled to move even if their family circumstances or family income changed. With the privatisation of housing most of these tenants became owner-occupiers, with the possibility of subsequently selling their dwellings and moving to smaller units or subdividing and renting out part of their dwellings and thereby increasing supply.

Second, supply could increase, even if the housing stock remained relatively stable, since in the past over-investment in new housing occurred (with parallel under-investment in maintenance) because of overstated demand in both the rental and owner-occupied markets. Housing was cheap because of low, controlled rents and there were few other investment opportunities in the period of negative interest rates and within the context of the social ownership of enterprises. It can be expected that in the future people will lower their housing demand to fit higher rents and to consider other investment opportunities, for example in business.

Third, many houses remain under-utilised as they were built with children in mind, who then grew up and moved away. There was no strong initiative to subdivide and rent out part of these houses as additional dwellings. But this is changing since, for the first time, effective property taxes are being contemplated which, in effect, would penalise under-use.

DEVELOPMENTS IN HOUSING POLICY

Although there was a consensus on the need to privatise social housing, the terms of privatisation were highly controversial. Parliament decided that most of the socially-owned housing stock was to be offered for sale to the holders of tenant rights (not always tenants themselves since unregistered sub-letting was possible) at very favourable terms, amounting to 10 to 30 per cent of the market value of a dwelling (Mandic, 1992). More favourable terms were implicitly offered to tenants in good locations (as location influences market price, but not at the construction value of a dwelling – which was the basis for calculating privatisation payments). This suited members of parliament but resulted in an inequitable and inefficient outcome (Stanovnik, 1994). Many influential people in the capital city (Ljubljana) managed to secure social housing for their relatives and subsequently were able to realise windfall property gains, while in many outlying and impoverished areas people were unable to buy even at very favourable terms. Also, the provisions of privatisation law conflicted with another law passed almost simultaneously on the restitution of nationalised property to former owners. Since the present tenant and the former owner of a dwelling could not both gain out of these two provisions simultaneously and since the state was unable to compensate both of them at the same time if they were to forgo their rights of ownership, compromise compensation arrangements were found which produced further inequities between tenants as well as between former owners.

The passing of the Housing Act of 1991 was a momentous landmark which clearly marked the first step in the transition of housing policy. It has many other provisions apart from privatisation, but in this chapter only its most important aspects are considered. Its impact on the tenure structure is examined in order to define the problems of the second stage of transition, which arguably should have as its goal the normalisation of the Slovenian housing sector, making it comparable to those in developed market economies. This phase should have been clearly defined in the National Housing Programme, a policy statement to be considered by parliament in 1996. The aim of this discussion, in part, is to improve some of the initially more poorly formulated aspects of this programme.

It is necessary to start from the fact that the results of the Housing Act of 1991 have not brought Slovenia much closer to the 'normal' housing system operating in the West. Its main result so far is the increased share of owner-occupied dwellings, which is now probably the highest in Europe. The beneficial side of this is that there is very likely an increased mobility of property and households compared to the former system, and hopefully more motivation for the maintenance of the stock – which is no longer the responsibility of the state but of private owners. But the very high share of owner-occupied units brings new problems. Such stock allows less mobility

than rented units in a free-market situation, multi-owner buildings compli-
cate management and maintenance, and Slovenia is faced with a lot of new
owners who were barely able to buy their dwellings at a discount, and now
have little disposable income for maintenance, taxes and other responsibilities of ownership. On the other hand, within the social-rented sector, non-
profit tenants, who were unable to buy their dwellings, are still protected
from eviction and are paying low rents. Thus this sector retains almost all
the unsatisfactory attributes of the former system, in particular immobility,
poor maintenance and lack of equity, and still represents a burden to the
owners – companies and municipalities.

So how should Slovenia proceed to normalise its housing sector? An
obvious solution is to introduce and sharply extend a true non-profit
renting sector (private-renting is certain to increase as well but is not so
important as a policy priority). The National Housing Programme recog-
nises this need and proposes, as a solution, that the government needs to
support the development of non-profit institutions to build and buy dwellings for non-profit renting by middle-income households (Mandic, 1992).
An additional proposal is that social housing for low-income households
should be provided by municipalities. These authorities would need to be
stimulated by central government since they have been extremely reluctant
in this regard so far. Initial finance from the privatisation receipts of central
government and the municipalities is drying up, so they are expected to
provide fresh finance from their budgets (and municipalities, in addition,
should provide land and infrastructure). Some successes are apparent, but
these initiatives are falling short of the ambitious goals set in the programme. Government has also finally risked raising rents of non-profit
housing almost to cost level. In addition, government, through the National
Housing Fund, is helping mainly middle-income young families with low-
cost loans while preparing to establish a housing savings and loans institution. The programme has also adopted proposals to introduce an effective
property tax and rent subsidies, but these at the time of writing had not
been implemented.

While these developments seem to be in the right direction, some additional and different policy directions could be proposed. The programme's
emphasis on the construction of non-profit dwellings could be criticised.
More attention, arguably, should be paid to bottlenecks in the construction
of quality housing (especially in respect of land provision), and to the use of
the filtering effect in providing low-cost housing. Use should be made of
reverse mortgages to solve the problem of low-income owners. Specifically,
banks could provide owners with an additional monthly income to pay
taxes and maintenance, and in return obtain the right to participate in
inheritance procedure and to receive their share of the principal and interest
from the subsequent sale of dwellings. Government should step in to
provide guarantees in order to homogenise these securities and make

them tradable. Finally, it should be stressed vigorously that the Slovenian Land Register needs to be put in order to lower transaction costs to enable the proper functioning of the housing market, and in this way to increase property and labour mobility, which is crucial in the period of transition.

CONCLUSION

It is evident that Slovenian housing policy, at least in intention, is moving towards the social-democratic goal of governmental co-responsibility for the provision of decent housing – perhaps the only right course considering the externalities involved. It is probable that because of budgetary problems, however, this goal might only be half implemented – and the demand for decent housing will remain unsatisfied in the foreseeable future.

20

CROATIA

Zorislav Perković and Dusica Seferagić

The Republic of Croatia is a small country, covering 56,538 km^2 in 1991. It had a population of 4.7 million in that year of which 51 per cent was urban – concentrated in the capital Zagreb (with almost 1 million inhabitants) and the three regional centres of Osijek, Rijeka and Split.

Part of the Yugoslav socialist federation from 1945, the Republic of Croatia moved towards independence and socio-political change in 1990 after multi-party elections. Independence was proclaimed in 1991 after a referendum. In the second half of 1991, serious fighting broke out in many parts of the country (following the rebellion of ethnic Serbs and the intervention of the Yugoslavian army on their side). At the beginning of 1992, Croatian independence was internationally recognised and a ceasefire agreed upon, but part of Croatia remained under Serb control and a degree of UN protection. Sporadic fighting continued, as well as an influx of refugees from Bosnia. Two more elections were held, confirming the power of the ruling conservative–nationalist party.

During the period of socialism, the country went through various socio-economic phases and experiments and, accordingly, changes in housing policy. Since the political changes in 1990, a consistent housing policy has not yet emerged.

TENURE

Census publications prior to 1991 omitted tenure, only indicating the ownership of dwellings. The 1991 Census, however, provided detailed data on both ownership and tenure. It revealed that Croatia had a high proportion of owner-occupied housing, 64 per cent in 1991, with the rest of its stock consisting mainly of social-rented dwellings with only a very small private-rented sector (Table 20.1).

HOUSEBUILDING AND DEMOGRAPHIC DEMAND

The construction of dwellings had been over 30,000 units per annum for quite a long time, but after 1981 a more or less steady decline occurred, and

314

Table 20.1   Tenure, Croatia, 1991

	Number		%
Owner-occupied	932,419		64.0
Social rented:			
Protected tenancy with controlled rent	358,090	24.2 ⎫	24.7
Leased dwellings (free contract)	7,810	0.5 ⎭	
Private rented:			
Protected tenancy with controlled rent	13,092	0.9 ⎫	3.7
Leased dwellings (free contract)	41,293	2.8 ⎭	
Kinship related and other tenures	109,666		7.5
Total:	1,457,370		100.0

Source: Statistical Yearbook of Croatia, 1993 and 1994

Table 20.2   Housebuilding in Croatia, 1980–93

	Dwelling units built			
	Private	Social	Total	Private, % of total
1980	19,270	11,737	31,007	62.1
1981	18,237	12,216	30,453	59.9
1982	15,451	11,850	27,301	56.6
1983	16,601	12,323	28,924	57.4
1984	15,903	9,363	25,266	62.9
1985	13,060	9,698	22,758	57.4
1986	14,113	9,683	23,796	59.3
1987	14,762	8,006	22,768	64.8
1988	13,869	7,897	21,766	63.7
1989	13,871	6,470	20,341	68.2
1990	13,512	5,084	18,596	72.7
1991	8,470	4,153	12,623	67.1
1992	5,805	1,962	7,767	74.7
1993	7,535	808	8,343	90.3
1980–93	190,459	111,250	301,709	63.1

Source: Statistical Yearbook of Croatia, 1994

after 1991, with political changes and armed conflicts, housing completions fell to under 10,000 units a year (Table 20.2).

The output of social housing dwindled almost out of existence. Private building remained more stable, but output is less than half the rate in its best years, 1973–74.

Notwithstanding the decrease in housebuilding in the 1980s, the housing stock in Croatia increased very substantially in the two decades preceding the 1991 Census. The total number of dwellings increased by some 45 per cent and the number of dwellings for permanent residence by over 32 per cent. The increase was particularly fast in respect of holiday homes; their

Table 20.3   Increase in supply of dwellings, Croatia, 1971–91

	1971 (100) 1981	1981 (100) 1991	1971 (100) 1991
Permanent residences	116.2	114.1	132.5
Holiday homes	367.5	209.7	770.7
Total	121.4	119.8	145.4

Source: Statistical Yearbook of Croatia, 1994

number increased more than sevenfold, reflecting the comparatively high living standards for the period as well as the impossibility of investing in circulating capital. With time, however, growth has slowed down, as indicated in Table 20.3. In 1971, holiday homes made up less than 2 per cent of the total housing stock in Croatia, whereas in 1991 they were almost 10 per cent.

The greater part of the housing stock is rather new, 62 per cent of dwellings having been built since 1960 and less than a quarter being older than 50 years. With regard to the amenities provided, however, the general situation is not yet satisfactory, thought it has considerably improved in recent years (Table 20.4)

Table 20.4   Housing amenities, Croatia, 1981–91

	1981 % dwellings	1991 % dwellings
Electricity supply	96.8	98.6
Piped water	60.5	86.2
Central heating	13.4	24.7
Inside WC	63.6	80.4
Bath/shower	57.1	75.7

Source: Statistical Yearbook of Croatia, 1994

The size of the average dwelling has been increasing constantly (2.29 rooms in 1981, 2.66 rooms in 1991), but it remains unsatisfactory compared to the number of persons occupying the average dwelling unit (3.3 in 1981, 3.2 in 1991).

On the demand side, there was a gradual increase in the population of Croatia over 1971–91, and also an increase in the number of households. From 1971 to 1991 the resident population increased from 4,426,000 to 4,784,000, an increase of 8.1 per cent.

The principal demographic components have been a constantly falling birth-rate and considerable migration (from Croatia to western countries, from Bosnia and Herzegovina to Croatia, and from rural to urban and

Table 20.5 Demographic change, Croatia, 1991–93

	Live births	Deaths	Natural change
1991	51,829	54,832	−3,003
1992	46,970	51,800	−4,830
1993	48,535	50,846	−2,311

Source: Statistical Yearbook of Croatia, 1994

coastal areas). After 1991, only preliminary estimates based on the 1991 Census and on natural change are available. They do not include migrations (as a result of the war there are no reliable data). The population has been decreasing through an excess of deaths over births (Table 20.5), the preliminary estimate of residential population falling to 4,779,000 in 1993.

The number of households, however, had increased ahead of population (by 19.8 per cent over 1971–91), thanks to constantly diminishing household size (Table 20.6).

Table 20.6 Household size, Croatia, 1971–91

Census	Households	Average size
1971	1,289,000	3.43
1981	1,424,000	3.23
1991	1,544,000	3.10

Source: Statistical Yearbook of Croatia, 1994

By 1995, with a stagnant or declining population, the number of households probably also diminished (there are no current data on households), although the change has possibly not been statistically important. Of course, the dislocations of war could be very significant but as yet it is extremely difficult to assess their demographic consequences.

DEVELOPMENTS IN HOUSING POLICY

To describe, explain and understand recent changes in housing policy, it is necessary initially to consider the socialist heritage of almost half a century which has left serious marks on the present housing situation.

The Republic of Croatia was part of the Yugoslav federation, sharing its main characteristics in all political, economic and social spheres including housing. Yugoslavia was similar to other socialist countries in respect of its political system (being governed exclusively by the communist party), its regulatory mechanism (state planning), its ideology (rule by the working class), and its reality (a generally inefficient economy, persistence of social inequalities, and frequent but not fundamental changes). Territorially, the main strategy was industrialisation, a reduced emphasis on agriculture, and urbanisation – resulting in a pattern of 'under-urbanisation' marked by a huge influx of the agricultural population into cities without an adequate urban (technical and social) infrastructure and with an inadequate supply of housing.

Yugoslavia, however, was different from the other socialist countries. It was open to the western world (there was no Iron Curtain), and it was well known for devolved politics and self-management in economic and territorial organisation. It was also known for its non-aligned foreign policy and for Tito's personal popularity. According to wide-ranging socio-political and economic changes, housing policy changed approximately every ten years.

But throughout the period of socialism some of the main features resisted change, notably planning, the distribution of resources, and the orientation towards social housing. Even though Croatia inherited an almost completely private housing stock in 1945 and even though throughout the socialist period of government private-sector housing constituted most of the housing stock (up to 100 per cent in villages and about 50 per cent in towns), housing policy all the time focused on social housing. Simultaneously, the federal state, the local state, and other public organisations were concerned with housing production, allocation, maintenance and the sale of dwellings to social organisations and individuals. This was a specifically urban feature. Big public firms would build housing according to state-defined household needs (first for working-class families and then for the middle and upper classes and politically important employees). Public or state organisations would distribute dwellings to their employees according to social and political criteria.

Tenants were a protected group of households as they occupied subsidised low-rent housing, the accommodation could be inherited, and it was easy to exchange dwellings. At the same time, private owners were under pressure from a housing policy hostile to their interests (it was difficult and time-consuming to get building permission, there were heavy taxes on exchange, there were problems of affordability, and there were high taxes on inheritance). Yet the quantity of private housing eased the problem of supply throughout the period of socialism even in the towns and big cities. The state even tolerated illegal building because it was needed.

With all the mistakes of 'socially directed housing', housing policy succeeded in decreasing the housing shortage. It improved the quality of housing and even produced a crude surplus. All the time, private housebuilding, swapping and buying and selling under normal market laws existed in parallel with the provision of social housing.

In the last decade of socialism, housing policy changed in accordance with other changes associated with the development of a market economy and the liberalisation of economic activity. Transitional Croatia, however, inherited problems such as large inequalities between villages and cities, a

318

lack of local property taxation, and soaring house prices in the private sector. But apart from privately built owner-occupied and privately rented dwellings, the housing stock still comprised a sizeable proportion of publicly built dwellings for rent, previously nationalised dwellings with tenants, and some state property for special households (state bureaucrats, army leaders and veterans).

It is important to bear in mind that substantial changes in housing (as in other matters), started in Croatia prior to independence. Changes were associated with the last economic reform to occur in Yugoslavia before its break-up. Suggested by the Federal Executive Council and Prime Minister A. Markovic, the reform introduced more pluralism and democracy in the political sphere as well as a market-oriented economy. After the falling apart of Yugoslavia, some things in Croatia continued as before, while other things changed. On the one hand, the state even strengthened its role while, on the other hand, privatisation and the development of the market became more open.

Privatisation

It is not appropriate to talk about a 'new housing policy'. Instead there is a patchwork of different elements. In general they are compatible with the transitional changes that are occurring in other parts of the Croatian economy as well as in other ex-socialist countries. The so-called 'Integral Concept of Housing Reform' voted for in the Croatian parliament as recently as 1995 shows all the weaknesses of this approach.

First, it is not integral at all, because it concerns mainly the privatisation of social housing and is not about the denationalisation of houses and urban land. The main concern of this approach is the privatisation of social housing through sales to sitting tenants. Under quite good conditions, tenants can buy dwellings for cash or foreign currency or by means of long (30-year) instalments. Each form of payment is at a discount. Buyers are eligible for discounts on social and employment criteria. Most buyers choose instalment payments since they do not have enough capital to buy outright. The general poverty of most of the population makes privatisation impracticable except for better-off households who actually profit from it.

Second, the deregulation of the housing market is not compatible with an integral approach to reform. Housing is no longer rented at 'use-value', nor is it planned, produced or allocated by the state, but is a commodity submitted to market rules. Out of 400,000 social dwellings, 145,000 cannot be sold to their tenants (since they are nationalised, with special status, or in occupied zones and/or severely damaged). The rest, however, are in the process of being sold off under ever-changing rules.

A massive sale of purpose-built social dwellings to sitting tenants began in 1991. Between 19 June 1991 and 31 December 1993, 162,781 dwellings were sold off at terms favourable to the tenants (notably large discounts and instalment payments). Of the originally socially-owned housing stock, 46 per cent was privatised, increasing owner-occupation to some 75 per cent, while the proportion of social-rented dwellings fell to only 13 per cent. This process continues, albeit at a much slower rate, since many of the remaining tenants lack capital for even a discounted purchase. Also, a significant part of the socially-owned stock consists of previously national-ised or military property, and the legal status as well as the future of both have not yet been decided. This remains on the legislative agenda, with conflicting claims and intentions (among tenants, former owners, parties with political resentment, etc.). Yet, in comparison with the 1,457,370 dwellings for permanent residence (1991 Census) the sale is not impressive. It is, moreover, only a change of ownership, not an improvement of the housing situation either for individuals or for society. Previous goals such as more public money for housing, better maintenance, improvement of housing conditions and the 'restoration of social justice' were not realised by privatisation.

CONCLUSION

Several decades of considerable housebuilding effort and investment (both social and private) have resulted in a very substantial enlargement and improvement of the Croatian housing stock. The decrease in the rate of population growth by itself would have made it easier for housebuilding to catch up with demand, but the trend towards smaller households produced a greater demand for dwellings. Nevertheless there was a small crude surplus of dwellings for permanent residence in 1991 and the number of vacant dwellings more than doubled in ten years. The housing problem was alleviated but not completely solved, especially in respect of quality and adequate floorspace. The problem remained more serious in urban areas where in some localities there was still a crude deficit.

The ageing of the population resulted in specific problems, such as under-occupancy and difficulties in the maintenance and repair of part of the housing stock. This is evident especially in old city centres but also occurs in some under-populated rural areas (notably islands in the Adriatic). Emigration created an inflow of money (from the earnings of emigrant workers) which was used to a considerable degree for housebuilding, especially in their home villages (to which they usually did not return). Therefore a proportion of these houses was under-used and provided secondary residences or – on the coast – temporary rented accommodation for tourists.

The social structure of migration to the larger cities also had an influence on how housing demand was met. Being predominantly rural, migrants were more likely (and more able) to build their own houses, often illegally and with the help of neighbours or relatives – thus creating semi-rural and substandard settlements on the outskirts of cities. Pressure on publicly organised building was consequently less acute.

War damage seriously worsened the general housing situation. Many dwellings were destroyed or made otherwise unfit for habitation. Large numbers of refugees or displaced persons created pressure for temporary housing in safer parts of Croatia. This particular problem changed the whole concept of housing reform. As long as people do not occupy their own homes, the state cannot withdraw from the field; on the contrary, as far as possible it must find ways to help these people – host families cannot provide the only source of temporary accommodation. Another serious problem affecting housing is the unequal urbanisation of the whole country. The quality of the housing stock does not just depend upon the condition of individual dwellings but upon the whole infrastructure – technical as well as social. Remedying this must be the main goal of housing reform.

It is hard, under these circumstances, to predict the future. Until now it has been evident that Croatia has chosen the so-called 'dualistic model' (Kemeny, 1995) akin to that of countries where free markets regulate demand and supply, rather than the unitary model typical of social-democratic countries where competition and collaboration of the different housing sectors give more variety and a better supply for all. Within Croatia, this means that better-off households will be able to arrange their housing needs or build and rent to others easily but it also means that most of the population will have serious housing problems while the state will take care of only the very poor. One can only predict that the worst housing built under socialism (in large poor-quality housing estates) will become social ghettos for the poor, the young and the displaced who cannot return to their former homes.

21

CONCLUSIONS

Paul Balchin

The early years of the twenty-first century might witness an increase in the spatial extent of the European Union (EU) with the inclusion of a number of central European countries among its members. Although this will have a substantial effect on the European economy, particularly with regard to the mobility of capital and labour and the flow of goods and services, it, or any other general development, is unlikely to have any direct impact on housing policy. The Treaty of Rome of 1958, the Single European Act of 1985 and the Treaty of Maastricht of 1992 each failed to include housing within the areas of responsibility of the European Commission. Although the European Parliament has passed resolutions demanding action on housing policy by the Council of Ministers and the Commission, within the arena of housing the powers of the Commission are restricted to issuing directives on the procurement of building contracts, the construction industry and on the environment, and on the management of the EU's regional and social funds – resources which have only an indirect relevance to housing (McCrone and Stephens, 1995). Clearly, it has been deemed that housing policy can only be effectively formulated and applied within the boundaries of a national or local state, rather than at a supra-national level.

Certain policy developments within Europe, however, are of considerable relevance to housing supply and housing need. Since the early 1970s, the European Commission has aimed to create a single market in mortgage finance – the Cecchini Report (CEC, 1988) arguing that cross-border competition between financial institutions would be desirable if economies of scale and associated efficiency (in terms of low-cost mortgages) were maximised at a European, rather than at a national or sub-national level. There is little evidence, however, that this would result from the establishment of a single market (see McCrone and Stephens, 1995). In respect of many mortgage institutions, full economies of scale might be achieved within comparatively small markets, while the larger institutions might exploit opportunities to increase their monopoly power, leading possibly to cheaper finance in some countries but undoubtedly to more expensive mortgages elsewhere. Contrary to the central assumption of the Cecchini

323

Report, greater efficiency in the provision of finance would not necessarily lead to lower mortgage prices. Even if it could be demonstrated, however, that a single mortgage market in Europe was of greater benefit to borrowers than national markets, it needs to be questioned whether or not it is every-where desirable to extend home-ownership beyond present levels – parti-cularly in a period when government expenditure and subsidies are (with the exception of housing allowances) being reduced within the field of social housing, and at a time when it is very apparent that the market for owner-occupied houses can fluctuate substantially – alternating between booms (which arguably divert an increasing share of financial resources away from other areas of the economy) and slumps (which lead to negative equity, mortgage foreclosure and dispossession).

The mobility of labour is another issue of direct concern to the European Commission, with considerable relevance to housing. Under the Single Euro-pean Act, qualifications are scheduled to become interchangeable, easing the flow of skilled labour between member countries, and the Treaty of Maas-tricht subsequently provided for an unrestricted movement of EU citizens where those involved have secured employment in another member coun-try, or are students or retired (McCrone and Stephens, 1995). Apart from an increased movement of population across the boundaries of member states, there is also the continuing likelihood of large-scale immigration into the EU from countries elsewhere – for both economic and political reasons and on a temporary or permanent-stay basis. In 1991, for example, whereas nearly 5 million European Community (EC) citizens migrated from one member country to another, nearly twice that number of people emigrated to the countries of the EC from outside (the EC being the predecessor of the EU) (Table 21.1). Clearly, post-Maastricht, the amount of migration by EU citizens within the EU will predictably increase, and will increase yet again when the membership of the EU expands in the twenty-first century.

Whether migration is associated with short- or long-term residence, in each member country the increased demand for housing in 1991 (in gross terms) ranged from 1.1 to 29.8 per cent above that emanating from the national population, or from 88,000 in Ireland to 5,343,000 dwellings in Germany. In aggregate, nearly 15 million homes had to be found for immigrants in the EC in 1991 (Table 21.1). Clearly any marked increase in owner-occupation and decrease in the supply of rented housing within the host countries would be less likely to satisfy the temporary needs of a high proportion of immigrants than access to a readily available supply of rented dwellings in both the private and social sectors, but, as yet, the European Commission has failed to take a lead in the development of housing policy in relation to labour mobility.

As the chapters in this book reveal, across Europe economic pressures in recent years (such as high rates of inflation and large budget deficits) have persuaded governments to reduce the level of housing investment

324

CONCLUSIONS

Table 21.1 Migration within and into the European Community, 1991

	EC immigrants (000s)	Non-EC immigrants (000s)	Total (000s)	% total population
Germany	1,439	3,904	5,343	6.7
France	1,312	2,285	3,597	6.3
UK	781	1,647	2,428	4.2
Netherlands	168	524	1,092	4.6
Belgium	551	353	904	8.8
Italy	149	632	781	1.4
Spain	273	211	484	1.1
Greece	54	175	229	2.2
Denmark	28	133	161	3.0
Luxembourg	103	13	116	29.8
Portugal	29	79	108	1.1
Ireland	69	19	88	2.5
Total	4,956	9,975	14,931	4.3

Source: Eurostat, *Demographic Statistics, 1993*

specifically in the social sector and, in some countries, to shift resources from new housebuilding to rehabilitation – deemed to be more cost-effective, at least in the short term. Because of the same pressures, there has also been a shift of emphasis from 'bricks and mortar' or object subsidies to more targeted demand-side or subject subsidies particularly in the rented sectors – although in some countries regressive forms of tax relief and exemptions in the owner-occupied sector remain, for reasons possibly more ideological than economic. To help curb public expenditure, to raise revenue and to respond to revised political nostrums, governments throughout much of western and central Europe have also undertaken (or are about to undertake) large-scale programmes of housing privatisation.

Whereas the introductory chapter of this book suggests that in social policy there were three distinct regimes in western Europe: social-democratic states, corporatist states and liberal-welfare states – regimes very broadly compatible with the 'left of centre', 'middle of the road' and 'right of centre' traditions, in subsequent chapters an examination of housing policy in western Europe in the 1980s and 1990s indicates that there has been a lurch to the right, with the state gradually withdrawing from the housing market – most notably in respect of government expenditure and taxation and in its overarching responsibility for social housing. Policies promoting tenure-neutrality in social-democratic regimes are giving way to more discriminatory measures in favour of owner-occupation and private renting; policies favouring both private- and social-rental housing in corporatist regimes are being superseded by initiatives to expand owner-occupation; and in liberal welfare regimes privatisation and the general development of the 'free market' have been taking countries further to the

325

right, or more correctly back to a period of mid-nineteenth-century *laissez-faire*.

It was understandable that in central Europe in the early 1990s, at a time when it seemed imperative to shake off the constraints of a command economy formerly imposed under communism, that the liberal welfare approach to housing (and to other aspects of the economy) was selected in preference to the corporatist pathway or social democracy. There was a perception – engendered by the International Monetary Fund and western governments – that the period of transition to an alternative economy and political system would be shorter if the state withdrew from as much of the economy as possible and (as in a liberal-welfare regime) allow market forces to dominate the allocation of resources and the production of goods and services – including the provision of housing. Since corporatist regimes require a balance between their many constituent parts and policies which very carefully reflect the interests of all or most of these parts, a comparatively long period of evolution – often under coalition governments – might be needed before economically strong corporatist states can emerge. A social-democratic approach was similarly eschewed since it could involve a degree of state intervention into markets considered unacceptable to countries wishing to minimise state involvement in economic and social issues – even though the aims and objectives of social-democratic policy would be very different from the policies of communist government. Nevertheless, the adoption of a liberal-welfare approach has been fraught with problems including a severe reduction in the level of housebuilding, a substantial rise in the level of unemployment and homelessness, and the loss (through privatisation) of an increasing proportion of social housing – reducing household choice of tenure and with predictably harmful effects on the maintenance and repair of the privatised stock, and on its eventual condition.

Thus, at the end of the twentieth century, in western Europe 'dualist' rental systems are supplanting 'unitary' markets, while in central Europe they are replacing communist rental systems. The social-rental sector (except possibly in Sweden) is under threat throughout the continent. 'Bricks and mortar' subsidies are being dramatically reduced, housebuilding is being cut back, and the stock is being increasingly privatised. While private institutional finance might cushion some of the impact of reduced state assistance, at best the sector might survive in the form of welfare housing for the marginalised – as in the United States. As such, social housing will be prevented from competing with private-rental housing to the detriment of the rental sector overall in terms of its condition and appeal – leaving owner-occupation as an increasingly dominant tenure and denying households a choice of renting or owning. Consumer choice, central to the efficient functioning of any advanced capitalist country, is thus being sacrificed on the altar of liberal welfarism.

By the end of the first decade of the twenty-first century, major changes could occur in the political philosophy of the major political parties in Europe (there could, for example, be a lurch back to the left or in some countries a reversion to the centre), or alternatively Europe could enjoy a period of substantial inflation-free growth, or both could happen simultaneously. In these scenarios, Keynesian demand-management strategies could reverse cuts in government expenditure and increase public investment in social housing, both to satisfy need and reduce unemployment. The privatisation of the social-rented stock might only proceed if the receipts therefrom were used to replace the lost stock in areas of housing shortage – through either new-build or acquisitions in the open market. The shift of emphasis, in some countries, away from new housebuilding to rehabilitation might similarly be reversed – again to satisfy growing demand more directly than hitherto. Subsidies will, in general, need to be maintained, or preferably increased. It has been acknowledged for some time that:

> they are needed to ensure that all households can obtain housing of some minimum standard. Without financial assistance to reduce housing costs, a significant proportion of households would be unable to pay for decent housing. Left to itself, the private market would produce both insufficient homes, often of inadequate standard, and produce a very unequal distribution of housing resources.
>
> (Lansley, 1982)

Social renting, in particular, could be selectively subsidised to ensure that at the minimum it survives, but at the maximum it is able to compete on equal terms with private renting and owner-occupation through the process of maturation. Social-rented housing should no longer be systematically disadvantaged by the state, least of all through indiscriminate privatisation, and instead be de-politicised to enable it to offer households an accessible and attractive alternative tenure to private sector renting or to owner-occupation (Kemeny, 1995). Clearly, tenure-neutral subsidies and the appropriate taxation of capital gain would go some way to creating a balanced housing market, but, to enable social renting to respond more effectively to housing need, and to shift away from a dualist to a unitary rental market, the transfer of the public-rented component to 'arms-length companies' might ensure that the provision and management of the stock becomes – as in Sweden – comparatively isolated from the vagaries of governmental economic policy and cosmetic changes in ideology (see Raynsford, 1992). Alternatively, tranches of public-rental housing – with a debt profile matching the debt profile of the total municipal stock – could be transferred to non-profit housing organisations at cost (Kemeny, 1995).

It must be recognised that private renting, at any time in history or in any country, has not been able to satisfy the demand for rental housing, and when governments reduce the scale and ability of the social-rented sector to compete in the rental market there is even less possibility of rental housing providing an alternative to owner-occupation. The private-rental stock with its near-market rents and comparative insecurity of tenure is ill-suited to compete with owner-occupation with its prospect of long-term capital appreciation and its security of tenure. Nevertheless, for reasons of both equity and minority consumer needs, the private-rental sector deserves to survive, be able to compete with social renting, and in some countries notably expand – conditions that are met in a number of EU countries through a combination of rent controls and tax breaks such as depreciation allowances.

In a small number of countries in western Europe, owner-occupation has undoubtedly expanded to a scale which it probably would not have reached under comparatively free-market conditions. In some central European countries, the increase in the size of the sector in recent years has similarly not been an outcome of the free market. While the availability of finance and tax concessions have undoubtedly inflated demand for home-ownership, so too has the unsatisfactory supply of and access to rented housing in both the private and social sectors. Through these factors, and because of the very distorting effects of tax relief and exemption on both the housing market and the wider economy (see Balchin, 1995), eligibility for mortgage-interest tax relief (often regressively distributed) and for tax exemptions where they occur cannot be justified in full, if at all. Overall, however, and regardless of tenure, personal housing allowances should be available to all households depending on income and family responsibilities, and be graded in such a way as to reduce the poverty trap.

Clearly, if significant political changes do not occur in Europe in the near future, at either a national or the EU level, or if macroeconomic policies are not introduced across the continent to stimulate growth, few if any of the above housing policies will be introduced. Housebuilding will remain at a low level and the condition of a high proportion of the stock will deteriorate because of unaffordable maintenance, the supply of rented housing will diminish, and the population of Europe will become increasingly polarised by being housed either in the owner-occupied sector with a variable degree of consumer satisfaction or in the 'safety net' of a dwindling supply of welfare housing (supported by means-tested personal allowances), but without the availability of attractive, financially viable and widely accessible rental sectors – not a situation compatible with social democracy or corporatism.

REFERENCES

1 INTRODUCTION

Barlow, J. and Duncan, S. (1994) *Success and Failure in Housing Provision: European Systems Compared*, Pergamon, Oxford

Boelhouwer, P. (1991) 'Convergence or divergence in the general housing policy in seven European countries?', Paper presented at the conference on Housing Policy as a Strategy for Change, European Network for Housing Research, Oslo, 24–27 June 1991

Esping-Andersen, G. (1990) *The Three Worlds of Welfare Capitalism*, Polity Press, Cambridge

Gilbert, M. and Associates (1958) *Comparative National Products and Price Levels*, Organisation for European Economic Co-operation, Paris

Kemeny, J. (1995) *From Public Housing to the Social Market: Rental Policy Strategies in Comparative Perspective*, Routledge, London

McCrone, G. and Stephens, M. (1995) *Housing Policy in Britain and Europe*, UCL Press, London

Oxley, M. and Smith, J. (1993) 'Housing investment in the UK: A European comparison', European Housing Research: Working Paper Series, No. 1, School of the Built Environment, De Montfort University, Milton Keynes

Saunders, P. (1990) *A Nation of Home Owners*, Unwin Hyman, London

2 INTRODUCTION TO PRIVATE RENTED HOUSING

Duvigneau, H. J. and Schönefeldt, L. (1989) *Social Housing Policy: Federal Republic of Germany*, Confederation of European Community Family Organisations (CO-FACE), Brussels

Emms, P. (1990) *Social Housing: a European dilemma?*, School for Advanced Urban Studies, Bristol

Gurtner, P. (1988) 'Switzerland', in H. Kroes, F. Ymkers, and A. Mulder (eds) (1988) *Between Owner Occupation and the Rented Sector: Housing in Ten European Countries*, the Netherlands Christian Institute for Social Housing (NCIV), De Bilt

Jaedicke, W. and Wollman, H. (1990) 'Federal Republic of Germany', in W. van Vliet (ed.), *International Handbook of Housing Policies and Practices*, Greenwood Press, New York

Kemeny, J. (1995) *From Public Housing to the Social Market. Rental Policy Strategies in Comparative Perspective*, Routledge, London

Leutner, B. and Jensen, D. (1988) 'German Federal Republic', in H. Kroes, F. Ymkers, and A. Mulder (eds) (1988) *Between Owner Occupation and the Rented*

329

Sector: Housing in Ten European Countries, the Netherlands Christian Institute of Social Housing (NCIV), De Bilt

McCrone, G. and Stephens, M. (1995) *Housing Policy in Britain and Europe*, UCL Press, London

OECD (1994) *OECD Economic Surveys 1993–1994: Switzerland*, OECD, Paris

Schips, B. and Müller, E. (1993) 'Mietzinsniveau bei Marktmieten', in *Office Fédéral du Logement*, Materialien zum Bereicht der Studienkommission Markmiete, Rapport de travail sur le logement, Vol. 29, Berne

Schulz, H. R. *et al.* (1993) 'Wohneigentumsforderung durch den Bund', *Bulletin du logement*, Vol. 55, Berne

Tomann, H. (1992) 'Towards a housing market in eastern Germany', Unpublished paper, Freie Universität, Berlin

3 SWITZERLAND

Bassand, M., Chevalier, Ch. and Zimmermann, E. (1984) *Politique et logement*, Presses polytechniques romandes, Lausanne

Biélier, P., Ghelfi, J.-P., Lachat, D. and Moutinot, L. (1993) *Faut-il libéraliser les loyers?* Les éditions de l'Association suisse des locataires, Lausanne

Lawrence, R. (1986) *Le Seuil franchi: Logement populaire et vie quotidienne en Suisse romande, 1860–1960*, Georg Editeur, Geneva

Lawrence, R. (1989a) 'Constancy and change in household demography, dwelling designs and home life – the case of Geneva', *Housing Studies*, 4(1): 36–43

Lawrence, R. (1989b) 'Habitat et habitants à Genève: une perspective historique du logement populaire', in N. Haumont and M. Segaud (eds) *Familles, modes de vie et habitat*, Editions l'Harmattan, Paris

OECD (1994) *OECD Economic Surveys 1993–1994: Switzerland*, OECD, Paris

Office fédéral du logement (Federal Housing Office) (1986) *Evaluation de logments: Système d'évaluation de logements (SEL), Edition 1986*, Berne, 2nd edition

Office fédéral du logement (Federal Housing Office) (1990) *Groupes défavorisés sur le marché du logement: Problèmes et mesures*, Berne

Office fédéral du logement (Federal Housing Office) (1993) *Le logement en Suisse, 1992*. Paper prepared for the 54th session of the Committee on Housing, Building and Planning of the Economic Commission for Europe, held in Geneva from 21 to 23 September 1993

4 GERMANY

Bartholmai, B. *et al.* (1994) Zeitreihen für das Bauvolumen in der Bundesrepublik Deutschland, Deutsches Institut für Wirtschaftsforschung, *Beiträge zur Strukturforschung* 154, Berlin

Hills, J. (1990) 'Shifting subsidies from bricks and mortar to people', *Housing Studies* 5(3)

Tomann, H. (1992) 'Towards a housing market in Eastern Germany', *Housing Finance International*, December

Wohnungspolitik auf dem Prüfstand, im Auftrag der Bundesregierung verfasst von der Expertenkommission Wohnungspolitik, Tübingen 1995

Wohnungspolitik für die neuen Länder, im Auftrag der Bundesregierung verfasst von der Expertenkommission Wohnungspolitik, Tübingen 1995

5 INTRODUCTION TO SOCIAL HOUSING

Boelhouwer, P. and van der Heijden, H. (1992) 'On shaky grounds: the case for the privatisation of the public sector in the Netherlands', *The Netherlands Journal of Housing and the Built Environment*, 3, 319–333

Bull, G. (1996) 'Implications of the changing social–private housing mix on housing provision, affordability and social exclusion'. Paper presented at the conference on Housing and European Integration, European Network for Housing Research, Helsingor, 26–31 August 1996

Duclaud-Williams, R. H. (1978) *The Politics of Housing in Britain and France*, Heinemann, London

Emms, P. (1990) *Social Housing: A European dilemma?* School for Advanced Urban Studies, Bristol

Geindre, F. (1993) *Le logement: une priorité pour le XIème plan* [report to the Prime Minister], Documentation Française, Paris

Ghékiere, L. (1991) *Marchés et politiques du logement dans la CEE*, Documentation Française, Paris

Kemeny, J. (1995) *From Public Housing to Social Market. Rental Policy Strategies in Comparative Perspective*, Routledge, London

Lindecrona, T. (1991) 'Non-profit housing: a fifty-year full-scale experiment in Sweden'. Unpublished paper, Swedish Association of Municipal Housing Companies (SABO)

Lundqvist, L. (1988) *Housing Policy and Tenures in Sweden*, Avebury, Aldershot

McCrone, G. and Stephens, M. (1995) *Housing Policy in Britain and Europe*, UCL Press, London

Matznetter, W. (1992) 'Organisational networks in a corporatist housing system: non-profit housing associations and housing politics in Vienna, Austria', in *Scandanavian Housing and Planning Research*, supplement 2

Ministerie van VROM (Ministerie van Volkshuisvesting Ruimtelijke Ordening en Milieubeheer) (1989) White Paper, *Volkshuisvesting in de jaren negentig* (Policy Document on Housing in the 1990s) – the Heerma Memorandum

Papa, O. (1992) *Housing Systems in Europe, Part 2: A Comparative Study of Housing Finance*, Delft University Press, Delft

Petersson, A. (1993) 'The Swedish housing allowance system: effects and effectiveness': Unpublished paper, BOVERKET

van Weesup, J. (1986) 'Dutch housing, recent developments and policy issues', *Housing Studies*, 1(1), 61–66

VROM (1993) *Volkshuisvesting in Cijfers 1992*, VROM, The Hague

Ymkers, F. and Kroes, H. (1988) 'The Netherlands' in H. Kroes, A. Ymker and A. Mulder (eds) (1988) *Between Owner Occupation and the Rented Sector*, The Netherlands Christian Institute of Social Housing (NCIV), De Bilt

6 THE NETHERLANDS

Adriaansens, C. A. and Priemus, H. (1986) *Marges van volkshuisvestingsbeleid, naar een flexibeler juridische vormgeving van een marktgevoelige beleidssektor* (Margins of housing policy: towards a more flexible legal shaping of a policy sector sensitive to market changes), Staatsuitgeverij, The Hague

Boelhouwer, P. and van der Heijden, H. (1992) *Housing Systems in Europe: Part I: A Comparative Study of Housing Policy*, Delft University Press, Delft

Brouwer, J. (1988) 'Rent policy in the Netherlands (1975–1985)', *The Netherlands Journal of Housing and Environmental Research*, 3(4), 295–307

Haffner, M. E. A. (1992) *Eigen woning in de EG: fiscale en overige financiële instrumenten* (Owner-occupied dwellings in the EC: fiscal and other financial instruments), Delft University Press, Delft

Heerma, E. (1992) *Beleid voor stadsvernieuwing in de toekomst* (Policy for urban renewal in the future), Tweede Kamer 1991–1992, 22396, 3, Staatsuitgeverij, The Hague

Klunder, R. (1988) 'Private financing in the Dutch public housing sector: the rise and fall of the accumulative loan', *The Netherlands Journal of Housing and Environmental Research*, 3(4), 309–318

Ministerie van VROM (Ministerie van Volkshuisvesting Ruimtelijke Ordening en Milieubeheer) (1989) *Volkshuisvesting in de jaren negentig* (Housing in the Nineties), Tweede Kamer 1988–1989, 20691, 3, Staatsuitgeverij, The Hague

Priemus, H. (1995) 'How to abolish social housing? The Dutch case', *International Journal of Urban and Regional Research*, 19(1), 145–155

Priemus, H. and Metselaar, G. (1992) *Urban Renewal Policy in a European Perspective: An International Comparative Analysis*, Delft University Press, Delft

van der Schaar, J. (1987) *Groei en bloei van het Nederlandse volkshuisvestingsbeleid* (Growth and flourishing of the Dutch housing market), Delft University Press, Delft

de Vreeze, N. (1993) *Woningbouw, inspiratie en ambities; Kwalitatieve grondslagen van de sociale woningbouw* (Housebuilding, inspiration and ambitions; qualitative bases of social housing construction), Nationale Woningraad, Almere

7 SWEDEN

Boverket (1994) 'Bostadsmarknaden och 90-talets förändringar' (Housing market adjustment in the nineties), E. Hedman (ed.), Boverket, Rapport 1994:1

Englund, P., Hendershott, P. H. and Turner, B. (1995) 'The tax reform and the housing market', Tax Reform Evaluation Report no. 20, (November), National Institute of Economic Research, Economic Council, Stockholm

Hansson, I. (1977) 'Bostandsfinansiering och bostadsbeskattning under inflation' (Housing finance and housing taxation under inflation), Byggforskningsrådet, Rapport T26

Hendershott, P. H., Turner, B. and Waller, T. (1993), 'Computing expected housing finance subsidy costs: An application to the current and proposed Swedish housing finance systems', *Scandinavian Housing and Planning Resarch*, 10: 105–114

Jacobsson, J. (1995) 'Housing finance and interest subsidies in Sweden: A computer model', Research Report, Institute for Housing Research, Uppsala University

Kearl, J. R. (1979 'Inflation, mortgages and housing', *Journal of Political Economy* 87, 1115–1138

Turner, B. (1988) 'Economic and political aspects of negotiated rents in the Swedish housing market', *Journal of Real Estate Finance and Economics*, 1: 257–276

Turner, B. (1990) 'Housing finance in the Nordic countries', *Housing Studies* 5: 168–183

8 AUSTRIA

Donner, C. (1990) *Wohnen und was es kostet*, Christian Donner, Vienna

Köppl, F. (1994) 'Basic principles of the Austrian housing policy', in WBSF/UN-ECE (ed.) *Vienna Paper on Urban Renewal*, WBSF, Vienna

332

REFERENCES

Österreichisches Statistisches Zentralamt (1993) 'Volkszählung 1991 – endgültige Ergebnisse: Wohnbevölkerung', in *Statistische Nachrichten 1/1993*, Österreichische Staatsdruckerei, Vienna
Statistisches Amt der Stadt Wien (1994) *Statistisches Jahrbuch der Stadt Wien 1993*, Magistrat der Stadt Wien, Vienna

9   FRANCE

Barre, R. (1976) *La Réforme du financement du logement*, Documentation française, Paris
Blanc, M. (1992) 'From substandard housing to devalorized social housing: ethnic minorities in France, Germany and the UK', *European Journal of Intercultural Studies* 3(1)
Blanc, M. (1993) 'Housing segregation and the poor: new trends in French social rented housing', *Housing Studies* 8(3), 207–213
Blanc, M. and Stébé, J. M. (forthcoming) 'High-rise housing estates in France', in R. van Kempen, R. Turkington and F. Wassenberg (eds) *Future of High-Rise Housing Estates: A European Perspective*
Cour des comptes (1994) *Enquête sur le logement (I): les aides au logement dans le budget de l'Etat 1980–1993*, Rapport au Président de la République, Paris
Daly, M. (1994) *Le Droit à un logement, le droit à un avenir*, Troisième rapport de l'Observatoire européen des Sans-Abri, FEANTSA, Brussels
Lacroix, T. (1992) 'Quels besoins en logement?', Proceedings of the 29th Conference of the Fédération nationale des promoteurs–constructeurs, FNPC–INSEE, Paris
Raillard, A. (1995) 'La loi Besson à l'épreuve de l'usage et du temps', *Fondations*, no. 2

10   INTRODUCTION TO OWNER-OCCUPATION

Banco Hipotecario *Nota* (various issues), Madrid
Blackwell, J. (1988) *A Review of Housing Policy*, No. 87, National Economic and Social Council, Dublin
Department of the Environment (1987) *Housing: The Government's Proposals*, Cm 214, HMSO, London
Department of the Environment (1991) *A Plan for Social Housing*, Dublin
Grieve, Sir Robert (chairman) (1986) *Inquiry into Glasgow Housing*, Glasgow District Council, Glasgow
Holmans, A. E. (1987) *Housing Policy in Britain*, Croom Helm, London
Kemeny, J. (1995) *From Public Housing to the Social Market. Rental Policy Strategies in Comparative Perspective*, Routledge, London
McCrone, G. and Stephens, M. (1995) *Housing Policy in Britain and Europe*, UCL Press, London
Power, A. (1993) *Hovels to High Rise: State Housing in Europe since 1850*, Routledge, London
Scottish Development Department (1987) *Housing: The Government's Proposals for Scotland*, Cm 242, HMSO, Edinburgh

11   IRELAND

Blackwell, J. (1988) *A Review of Housing Policy*, The National Economic and Social Research Council, Dublin

REFERENCES

Blackwell, J. (1990) 'Housing finance and subsidies in Ireland', in D. Maclennan and R. Williams (eds), *Affordable Housing in Europe*, Joseph Rowntree Foundation, York

Department of the Annual Housing Statistics Bulletins (various), Stationery Office, Environment, Dublin

Department of the Environment (1991) *A Plan For Social Housing*

Murphy, L. (1995) 'Mortgage finance and housing provision in Ireland 1970–90', *Urban Studies* 32(1), 135–154

Power, A. (1993) *Hovels to High Rise*, Routledge, London

Quinlan, G. (1995) *Aspects of housing policy in Ireland*, unpublished BSc (Hons) dissertation, University of Greenwich

12 SPAIN

Instituto Nacional de Estadística (1991) *Housing Census*

Ministerio de Obras Públicas (1989) *Encuesta sobre vivienda en alquiler* (Survey on rent housing)

Naredo, J. M. (1993) 'La riqueza inmobiliaria española en el Balance Nacional', *Revista Española de Financiación a la Vivienda*, no. 24/25, December

San Martin, I. (1993) 'Housing supply and demand', *Revista Española de Financiación a la Vivienda*, no. 24/25, December

13 ITALY

ANIACAP (Associazione Nazionale IACP) (1993) 'Nuovi modelli procedurali per migliorare strutture e procedure dell'intervento pubblico nell'edilizia abitativa', E Piroddi (ed), ANIACAP, Rome

ANIACAP (1994) La vendita del patrimonio di edilizia pubblica, prime valutazioni, ANIACAP, Rome

Bagnasco, A. (1977) *Tre Italie: La problematica territoriale dello sviluppo italiano*, Il Mulino, Bologna

Boelhouwer, P. and van der Heijden, H. (1992) *Housing Systems in Europe, Part 1: A Comparative Study of Housing Policy*, Delft University Press, Delft

CENSIS (1985) *19° rapporto sulla situazione sociale del paese*, Franco Angeli, Rome

Coppo, M. (1994a) 'Mercato, politiche ed evoluzione del sistema abitativo', *Urbanistica* 102: 10–15

Coppo, M. (1994b) 'L'alienazione del patrimonio abitativo pubblico', *Urbanistica Informazioni* 134: 38–43

Cremaschi, M. (1994) 'La denazionalizzazioinme del problema abitativo', *Urbanistica* 102: 23–28

De Rita, G. (1994) (Paper presented at the) Conferenza nazionale programmatica sulle politiche abitative, Roma 14–16 febbraio 1994, Ministero dei Lavori Pubblici

Mortara, C. A. (1975) 'Venti anni di edilizia pubblica in Italia', *Economia pubblica* 2–3: 34–51

Padovani, L. (1984) 'Italy', in M. Winn (ed.) *Housing in Europe*, Croom Helm, London

Padovani, L. (1990) 'L'esperienza "pilota" dei programmi integrati di recupero in Lombardia', *Recuperare*, 49: 472–476, or 'New relationships between public and private sectors in Italian urban renewal and housing policy', Research Conference, Housing Debates – Urban Challenges (CILOG), Paris, 3–6 July

Padovani, L. (1991) 'Qualità dell'abitare nella città di fine secolo', in *La costruzione della città europea negli anni '80*, Vol. I, Credito Fondiario, Saggi, Rome
Secchi, A. (1993) 'I programmi integrati di recupero in Lombardia', *Urbanistica Informazioni* 129–130: 37–39
Tosi, A. (1990) 'Italy' in W. van Vliet (ed.), *International Handbook of Housing Policies and Practices*, Greenwood Press, London
Tosi, A. (1994) *Abitanti: Le nuove strategie dell'azione abitativa*, Il Mulino, Bologna

14   UNITED KINGDOM

Association of Metropolitan Authorities (1983) *Defects in Housing Part 1*, AMA, London
Association of Metropolitan Authorities (1984) *Defects in Housing Part 2*, AMA, London
Association of Metropolitan Authorities (1985) *Defects in Housing Part 3*, AMA, London
Aughton, H. and Malpass, P. (1991) *Housing Finance: a basic guide*, Shelter Publications, London
Beveridge Report (1942) *Social Insurance and Allied Services*, Cmnd 6404, HMSO, London
Carvel, J. (1985) '£19 bn needed to right council house defects', *Guardian*, 5 March
Crook, A., Hughes, J. and Kemp, P. (1995) 'The supply of privately rented housing', *Housing Research 139*, Joseph Rowntree Foundation
Daniel, T. (1987) letter to *Guardian*, 3 October
Department of the Environment (1977) *Housing Policy: A Consultative Document*, HMSO, London
Department of the Environment (1983) *The English House Condition Survey 1981*, HMSO, London
Department of the Environment (1987) *Housing: The Government's Proposals*, Cmnd 214, HMSO, London
Department of the Environment (1995a) *Annual Report*, HMSO, London
Department of the Environment (1995b) *Our Future Homes*, HMSO, London
Hills, J. (1991) *Unravelling Housing Finance: Subsidies, Benefits and Taxation*, Clarendon Press, Oxford
Holmans, A. (1995) 'Housing demand and need in England 1991–2011', *Housing Research 157*, Joseph Rowntree Foundation
House of Commons (1982) *The First Report of the Environmental Committee*, Session 1981/82, HMSO, London
Joseph Rowntree Foundation (1991) *Inquiry into British Housing: Second Report*, JRF, York
Kemeny, J. (1995) *From Public Housing to the Social Market. Rental Policy Strategies in Comparative Perspective*, Routledge, London
Leather, P. and Mackintosh, S. (1993) 'Housing renewal in an era of mass home ownership', in P. Malpass and R. Means (eds) *Implementing Housing Policy*, Open University Press, Buckingham
Leather, P., Mackintosh, S. and Rolfe, S. (1994) *Papering over the Cracks: Housing Conditions and the Nation's Health*, National Housing Forum, London
McKechnie, S. (1987) letter to *Guardian*, 17 August
National Federation of Housing Associations (1985) *Inquiry into British Housing: Report*, NFHA, London
Niner, P. (1988) Unpublished Department of the Environment Report on homelessness

Roof (1994) 'Housing's new hit list', March–April
Shelter (1982) *Housing and the Economy – A Priority for Reform*, Shelter, London
Whitehead, C. (1995) 'Housing associations, private finance and market rents', *Housing Research 154*, Joseph Rowntree Foundation

15 INTRODUCTION TO HOUSING IN TRANSITION

Baross, P. and Struyk, R. (1993) 'Housing transition in Eastern Europe', *Cities*, August
Bogdanowicz, W. (1993) 'Research Report', Housing Renewal in the Context of Privatisation, Manuscript, School of Land and Construction Management, University of Greenwich
Clapham, D. (1993) 'Privatisation and the East European housing model'. Paper presented at the conference on Housing Policy in Europe in the 1990s, European Network for Housing Research, Budapest, September 1993
Csomos, J. (1993) 'Principles of the Hungarian housing reform'. Paper presented at the conference on Housing Policy in Europe in the 1990s, Budapest, September 1993, European Network of Housing Research.
Czerny, M. (1992) 'The current crisis condition in the building industry (CEE and CIS countries)'. Paper presented at the Expert Seminar on the Building Industry in Transition Economies, Vienna, 23–25 November 1992
Elter, I. and Baross, P. (1993) 'Budapest', *Cities*, August
Gowan, P. (1995) 'Neo-liberal theory and practice for Eastern Europe', *New Left Review*, 213, September/October
Hajduk, H. (1992) 'Strategy for restructuring and privatisation: Country Report: Poland'. Paper presented at the Expert Seminar on the Building Industry in Transition Economies, Vienna, 23–25 November 1992
Hegedüs, J., Mark, K. and Tosics, I. (1993) 'Restructuring the housing subsidy system in Hungary'. Paper presented at the Conference on Housing Policy in Europe in the 1990s, European Network for Housing Research, Budapest, September 1993
Hegedüs, J., Mark, K., Struyk, R. and Tosics, I. (1992) *The Privatisation Dilemma in Budapest's Public Rental Sector*, Metropolitan Research Institute, Budapest
Kahout, J. and Zajicova, P. (1992) 'Main problems of social housing in Czechoslovakia'. Paper presented at the housing workshop on Housing in Transition, Hungarian Ministry of Welfare and the United Nations Economic Commission for Europe, May 1992
Kemeny, J. (1995) *From Public Housing to the Social Market: Rental Policy Strategies in Comparative Perspectives*, Routledge, London
Kingsley, G. T., Tajcman, P. and Wines, S. W. (1993) 'Housing reform in Czechoslovakia', *Cities*, August
Kozlowski, E. (1992) 'The housing system in Poland: changes and direction' in B. Turner, J. Hegedüs, J. and I. Tosics (eds), *The Reform of Housing in Eastern Europe and the Soviet Union*, Routledge, London.
Kulesza, H. (1990) 'Zle sytuowani' (The badly off), *Polytika*, vol. 49, 8 December 1989
Long, A. (1993) 'The changing economy of housing in central Europe with special reference to Poland', unpublished MSc Dissertation, School of Land and Construction Management, University of Greenwich
Mandic, S. (1992) 'Reformation in Yugoslavia: introductory remarks', in B. Turner, J. Hegedüs, and I. Tosics (eds) *The Reform of Housing in Eastern Europe and the Soviet Union*, Routledge, London

REFERENCES

Millard, F. (1992) 'Social policy in Poland', in B. Deacon *et al.*, *The New Eastern Europe*, Sage, London

Nord, L. (1992) 'Yugoslavia: an example reconsidered', in B. Turner, J. Hegedüs, and I. Tosics (eds) *The Reform of Housing in Eastern Europe and the Soviet Union*, Routledge, London

Seferagić, D. (1985) *Problemi kvalitete zivota u novim stambenim naseljima*, IDIS, Zagreb

Turner, B. (1992) 'Housing reforms in Eastern Europe' in B. Turner, J. Hegedüs, and I. Tosics (eds), *The Reform of Housing in Eastern Europe and the Soviet Union*, Routledge, London

UNICEF (1995) 'Poverty, children and policy: Responses for a brighter future', *Central and Eastern Europe in Transition*, United Nations Children's Fund, Florence

16 HUNGARY

BRHPS (1992) Budapest Rental Housing Panel Survey, MRI

Hegedüs, J. (1987) 'Reconsidering the roles of the state and the market in socialist housing systems', *International Journal of Urban and Regional Research*, 11(1), 79–97

Hegedüs, J. (1988) 'Inequalities in east European cities: A reply to Iván Szelényi', *International Journal of Urban and Regional Research*, 12(2)

Hegedüs, J. and Kovács, R. (1993) 'Housing situation in 1992', Metropolitan Research Institute

Hegedüs, J. and Tosics, I. (1992) 'Conclusion: Past tendencies and recent problems of the east European housing model', in B. Turner, J. Hegedüs and I. Tosics (eds) *The Reform of Housing in Eastern Europe and the Soviet Union*, Routledge

Hegedüs, J., Mark, K., Struyk, R. and Tosics, I. (1993) 'Local options for transforming the public rental sector. Empirical results from two cities in Hungary', *Cities*, August 1993

Hegedüs, J., Mark, K. and Tosics, I. (1995) 'Uncharted territory: Hungarian housing in transition', in R. Struyk (ed) *Economic Restructuring in the Former Soviet Bloc: Evidence from the Housing Sector*, The Urban Institute Press, Washington DC

Hegedüs, J., Mark, K., Sárkány, Cs. and Tosics, I. (1996) 'Hungary: the dilemmas of "give-away" privatisation', in D. Clapham, J. Hegedüs, K. Kintrea and I. Tosics (eds) *Chasing the market: Housing Privatisation in Eastern Europe*, Greenwood

Household Survey (1983) Survey carried out by the Institute of Sociology, Hungarian Academy of Sciences in 1982

Tosics, I. (1987) 'Privatisation in housing policy: The case of the western countries and that of Hungary', *International Journal of Urban and Regional Research*, 11(1)

Tosics, I. (1988) 'Inequalities in East European cities: Can redistribution ever be equalising, and if so, why should we avoid it? A reply to Iván Szelényi', *International Journal of Urban and Regional Research*, 12(2)

17 THE CZECH REPUBLIC

ABF (1992) *Architektura, bydlení, finance 1993 (Architecture, housing, finance 1993)*, ABF nadace, Prague.

Anderle, A. (1991) 'Ke koncepčnímu řešení rozvoje bydlení' (Towards a conceptual solution of housing development), Územní plánování a urbanismus 18(2): 57–66.

Clapham, D. (1995) 'Privatisation and the east European housing model', *Urban Studies* 32(4–5): 679–694.

Daněk, M. (1994) 'Současná česká bytová politika, zkušenosti s její aplikací v Českých Budějovicích a restituce domovního fondu' (Contemporary Czech housing policy, its application in České Budějovice and restitution of housing stock), BA thesis, Department of Social Geography and Regional Development, Charles University of Prague.

ESHOU (1995) 'Social housing in Europe, a permanent building site', Special issue, June 1995, European Social Housing Observation Unit.

Eskinasi, M. (1994) 'Transforming Prague: A step over the threshold (A study of privatisation, housing and urban renewal in a city in transition)', MA thesis, Universiteit van Amsterdam, Planologisch en Demografisch Instituut.

Eskinasi, M. (1995) 'Changing housing policy and its consequences: The Prague case', *Housing Studies* 10(3): 383–398.

Kingsley, G. T., Tajčman, P. and Wines, S. W. (1993) 'Housing reform in Czechoslovakia: Promise not yet fulfilled', *Cities* 10(3): 224–236.

Michalovic, P. (1992) 'Housing in Czechoslovakia: Past and present problems', in B. Turner, J. Hegedüs and I. Tosics (eds) *The Reform of Housing in Eastern Europe and the Soviet Union*, Routledge, London

Musil, J. (1987) 'Housing policy and the sociospatial structure of cities in a socialist country: The example of Prague', *International Journal of Urban and Regional Research* 11 (1): 27–36.

Musil, J. (1992) 'Recent changes in the housing system and policy in Czechoslovakia: An institutional approach', in B. Turner, J. Hegedüs and I. Tosics (eds) *The Reform of Housing in Eastern Europe and the Soviet Union*, Routledge, London

Reynolds, D. (1995) 'Lazard Freres puts up housing infrastructure loan', *Estate News*, no. 5 (45), May

Short, D. (1990) 'Housing policy in Czechoslovakia' in J. A. A. Sillince (ed.) *Housing Policies in Eastern Europe and the Soviet Union*, Routledge, London.

Sýkora, L. and Šimoníčková, I. (1994) 'From totalitarian urban managerialism to a liberalized real estate market: Prague's transformations in the early 1990s', in M. Barlow, P. Dostál and M. Hampl (eds) *Development and Administration of Prague*, Universiteit van Amsterdam, Instituut voor Sociale Geografie.

Telgarsky, J. P. and Struyk, R. (1991) *Toward a Market-Oriented Housing Sector in Eastern Europe: Developments in Bulgaria, Czechoslovakia, Hungary, Poland, Romania, and Yugoslavia*, Urban Institute Report 90–10, The Urban Institute Press, Washington DC.

TERPLAN (1993) *Obyvatelstvo, bydlení a bytový fond v územích České Republiky 1961–1991* (Population and housing in territories of the Czech Republic 1961–1991), TERPLAN, Prague.

19 SLOVENIA

Kos, D. (1992) 'Informal activities in the formal housing system', in B. Turner, J. Hegedüs and I. Tosics (eds) *The Reform of Housing in Eastern Europe and the Soviet Union*, Routledge, London

Mandic, S. (1992) 'Restructuring the housing policy system in Slovenia', in I. Svetlik (ed.), *Social Policy in Slovenia*, Avebury, Aldershot

Stanovnik, T. (1994) 'The sale of the social housing stock in Slovenia: What happened and why', *Urban Studies*, 9

REFERENCES

Verlic Dekleva, B. (1994) 'Never-ending transition of housing policy in Slovenia', in T. Tanninen, I. Ambros and O. Siksio (eds); *Transitional Housing Systems*, Bauhaus Dessau, Dessau

20 CROATIA

Kemeny, J. (1995) *From Public Housing to the Social Market. Rental Policy Strategies in Comparative Perspective*, Routledge, London

21 CONCLUSIONS

Balchin, P. (1995) *Housing Policy: An Introduction*, Routledge, London
CEC (1988) 'The economics of 1992' (Cecchini Report), *European Economy*, vol. 35
Kemeny, J. (1995) *From Public Housing to the Social Market. Rental Policy Strategies in Comparative Perspective*, Routledge, London
Lansley, S. (1982) *Housing Finance: A Policy for Labour*, Labour Housing Group, London
McCrone, G. and Stephens, M. (1995) *Housing Policy in Britain and Europe*, UCL Press, London
Raynsford, N. (1992) 'Arm's length companies: an option for local authority housing', *Housing Review*, 41(2), 26–28

INDEX

Learning Resources
Centre

2741261